UNRAVELLING COMPLEXITY

The Life and Work of Gregory Chaitin

UNRAVELLING COMPLEXITY

The Life and Work of Gregory Chaitin

Editors

Shyam Wuppuluri
R N Podar Institute, India

Francisco Antonio Doria
Universidade Federal do Rio de Janeiro, Brazil

World Scientific

NEW JERSEY • LONDON • SINGAPORE • BEIJING • SHANGHAI • HONG KONG • TAIPEI • CHENNAI • TOKYO

Published by

World Scientific Publishing Co. Pte. Ltd.
5 Toh Tuck Link, Singapore 596224
USA office: 27 Warren Street, Suite 401-402, Hackensack, NJ 07601
UK office: 57 Shelton Street, Covent Garden, London WC2H 9HE

British Library Cataloguing-in-Publication Data
A catalogue record for this book is available from the British Library.

UNRAVELLING COMPLEXITY
The Life and Work of Gregory Chaitin

ISBN 978-981-120-006-9

For any available supplementary material, please visit
https://www.worldscientific.com/worldscibooks/10.1142/11270#t=suppl

Desk Editors: V. Vishnu Mohan/Kwong Lai Fun

Typeset by Stallion Press
Email: enquiries@stallionpress.com

To João Bernardo and Maria Clara,
Virginia and Greg's children

Foreword

One day in August 2007 I attended a meeting *Gödel: Logic and Time* at the Brazilian Center for Research on Physics, organized by Francisco Doria and Mario Novello. I was then a doctoral student at the Federal University of Rio de Janeiro, working on epistemology, different forms of rationality, and conceptual migration in interdisciplinary research. I had already heard of Gregory Chaitin from a close friend of his, the late Roberto Lins de Carvalho, a logician, whom I befriended during the period we were both teaching at a post-graduate course on giftedness at the State University of Rio de Janeiro. I had also bumped into Greg's work writing a paper on complexity for non-mathematicians and again, a couple of years later, preparing my doctoral research proposal which was initially connected to complexity. I remember I did not attend the Gödel meeting expecting I would meet Gregory Chaitin, but my friend Roberto was thrilled to see Greg again and wanted to introduce us. I thought Greg was in his late seventies or early eighties.

So, there I was, sitting in one of the last rows of the auditorium. When Greg Chaitin was announced, to my surprise I saw a teenager jumping onto the podium, full of energy and waving his arms, overflowing happiness and enthusiasm for life and for the ideas he went on presenting as if he were having an informal conversation during a break between classes. Yes, between ordinary classes, not at an international event. I felt him talking to us as if we were all known to each other, fellow travelers in the search for understanding, and he was wholeheartedly sharing some of that with us; an understanding which was to be taken lightly and joyfully, despite its staggering depth and clarity. In the midst of Greg's talk, an idea that sounded like music to my ears: Gödel was actually a philosopher who expressed his philosophical ideas with the sharpness of mathematical language. So, I concluded, Gregory Chaitin's cognitive style intuitively captures and relates concepts from different domains and then binds them mathematically into theorems, and this is a beautiful example of interdisciplinary creative thinking. He was a live example of my doctoral research topic.

The unfolding of this personal connection and intellectual interaction provided me with a particularly close perception of how Greg relates to life and, within life, to his work. There is no recipe for a new idea or for a new theorem, but the sense of freedom from rules and regulations, and the playful fulfillment of the joys of life with a *soupçon* of intelligent intellectual exchange are definitely part of this endeavor. His is a very obstinate mind, which can be seen at work concentrating on going up a mountain, writing down notes in bed when he has an idea, ordering from an interesting menu, or thinking non-stop how to put ideas together precisely so that a theorem can be proved, eventually in more than one form. But this obstinate abstract mathematical thinking is, however, intimately connected with strong emotions and an ever present aesthetic sense of beauty and sensual pleasure. Intensity and exaggeration are always backstage, at any moment ready to appear. For all this I feel Greg is a Dionysian Apollo, combining unpredictability, disorder, lack of structure or extreme randomness with precision, minimalist elegance and simplicity.

Collaborating with Greg, I have also had the chance to see his liberal attitude towards others developing his ideas in different directions from those he had established. Metabiology, as Greg says in the preface of *Proving Darwin: Making Biology Mathematical* (2012), "is very much a joint effort; our [Greg's and mine] three-year-old child as it were." Metabiology is about evolving single software organisms with no bodies and no population. My participation in this joint effort has been to epistemically appraise metabiology and in so doing provide broader understanding of the metabiological proposal as a mathematical metatheory for biological evolution. However, my concept of the "metabiological body" is considerably foreign to this original setting, and Greg made me feel entirely free to extend metabiology to a non-mathematical domain. The same can be said about his former doctoral student Felipe Abrahão's work on metabiology without oracles or with populations of metabiological organisms. Although Greg does want his ideas to be understood well, he greatly appreciated and supported our intellectual independence.

In developing metabiology, at first Greg felt uneasy about using oracles. I pointed out to him that "oracle" in his mathematical model is the conceptual migration of "environment" in the Darwinian theory of evolution. Later, for aesthetic reasons, Greg introduced very general algorithmic mutations — which seemed to him very unbiological — that greatly simplified the proofs. I felt this raised an important issue: Could we migrate algorithmic mutations back to biology? In fact, it may turn out that algorithmic mutations are the greatest contribution of metabiology to biology. For instance, they could be useful for explaining the Cambrian explosion, the major transitions in evolution, and punctuated equilibrium. This intuition has recently been supported by the highly suggestive computer experiments undertaken by Hector Zenil and his collaborators (http://dx.doi.org/10.1098/rsos.180399).

We all know how dear Plato is to mathematicians, Greg included. Plato's ontology is universal, perfect, static and eternal. This was supposed to be the ontology for mathematics and this ontology supported Hilbert's dream. However, Georg Cantor, Kurt Gödel, Alan Turing and Greg Chaitin have revealed a different, troubling, open-ended, in *statu nascendi*, creative ontology for mathematics! As Vladimir Tasić points out in his book *Mathematics and the Roots of Post-Modern Thought*, post-modernism did not, but certainly could have emerged from this mathematical work. "Toto," Dorothy says to her dog in *The Wizard of Oz*, "I think we are not in Kansas anymore."

So, this book celebrates Greg's deterritorializing journey from platonism to post-modernity, from the United States to Brazil, from suburbia to Rio, from being single to being at seventy a first-time father of our little boy João Bernardo — an incommensurable joy in our lives. All of this, a not at all obvious, rhizomatic deleuzian "line of flight" from that 2007 meeting on Gödel. And it solves, at least temporarily, Greg and Virginia's personal incompleteness problem. Enjoy the book!

Virginia Maria Fontes Gonçalves Chaitin

Rio de Janeiro
29 June 2018

Preface

We are but dwarfs mounted on the shoulders of giants, so that we can see more and further than they; yet not by virtue of the keenness of our eyesight, nor through the tallness of our stature, but because we are raised and borne aloft upon that giant mass.

— Bernard of Chartres

No talk about the foundations of mathematics and about Gödel is complete without referring to Chaitin. As a teenager at the Bronx High School of Science, he puts forth in his second paper a seed of an idea on complexity, which later permeated into so many fields — metamathematics, theoretical physics, economics, biology, theory of computation, to name a few — so much so that the probability of randomly picking up a text in computer science or mathematics (or perhaps biology in the future) and not finding the name of Gregory Chaitin in it is quickly tending towards zero. Such is the interdisciplinary nature of his work. He is unconventional in every sense of the word. He never had a degree beyond his high school and has published groundbreaking results when he was a teenager. He is a *rebel.* His personality justifies the title of the documentary done on Chaitin's life and work by Karol Jałochowski, *Against Method.* This word has to be understood carefully. There are two ways in which one can be a rebel. One is when one truly goes against the system with an intent to do so. The other is pareidolic in nature. When one is an outstanding genius, transcending all the humdrum conventions, there is literally no definition that one can impose on him. He — in the latter sense — does look like an outlier or a rebel. He is a mathematician with the soul of a poet.

I don't think I can describe his work as beautifully as others did elsewhere, but I would like to comment on the nature of this volume. Let's start by asking the question: is there any definition/article/volume that can describe "Greg Chaitin" in totality? This is similar to Chaitin's own motivation of finding the smallest computer program (best theory) that can generate the experimental data. The answer, perhaps due to incompleteness (?), is unfortunately no. Neither this volume nor anything else can describe him accurately. This *festschrift*, unconventional as it is, like the man himself, just showcases a slice of his personality and furnishes us with some views. Hopefully, assembling them all gives some understanding of the "Chaitin" phenomenon. Articles within the volume are both biographical and scientific in nature. They are written by leading scholars and dear friends of Chaitin, who greatly admire his work. The contributions within the volume primarily fall into one of these three categories: Technical contributions, which include articles stating the importance and relevance of Chaitin's work to many domains, including theoretical physics, mathematics, computation theory, (Meta) biology algorithmic information theory and economics; philosophical essays, epistemological and ontological issues pertaining to humanistic thought, knowledge and their limitations; Biographical memoirs, which include Chaitin's autobiography and other expository essays including Greg's own expository pieces and lectures to further the reader's understanding on the topic at hand. The volume also furnishes a gallery of images of Chaitin and his family arranged in chronological order.

This book is the brainchild of Francisco Antonio Doria (Chico), who is primarily responsible for its conception. A wonderful human being and a dear friend of Chaitin, he has supported me in all possible ways in finishing this book and in life in general. I would like to thank Virginia Chaitin for her support and also for providing the foreword to this book. She is the context of Greg

Chaitin and reminds me of Yashodhara, who always cared for the Prince Siddhartha, pre and post enlightenment, supported him and nourished him with love and care. I would of course like to thank all the authors for taking out time and writing these wonderful articles. I also thank Nisha Rahul, V. Vishnu Mohan and other editors at World Scientific for their editorial support. During the span of the past two years, I gained a deeper association with Greg Chaitin and was able to witness his human side, and I was deeply touched by his gentle nature. Chaitin's work has been a constant source of inspiration not just for Chico but also for many of us. I couldn't have asked the universe for anything more than these vivid memories that I will eternally cherish. These memories and friendships are the only things that matter. They transcend destiny. They even give the lie to death.

Shyam Wuppuluri

About the Editors

Shyam Wuppuluri is an independent researcher working in the domain of foundations of science. He is currently working as a research associate at R. N. Podar institute. His research interests range from philosophy to theoretical physics, mathematics and cognitive science. He is the recipient of Honra ao Merito from the Brazilian Academy of Philosophy (*Academia Brasileira de Filosofia*) and is also a corresponding member of the academy since 2018. He is also the recipient of prestigious Albert Einstein Fellowship 2020 from the Einstein Forum at Potsdam & Caputh and is a fellow of Royal Society of Arts. As a primary editor, he has edited various highly interdisciplinary volumes to disseminate and address the issues underlying the foundations of sciences by gathering world's leading scientists and intellectual leaders including Noam Chomsky, Sir Roger Penrose, Sir Martin Rees, Daniel Dennett, A. C. Grayling, Nicholas Humphrey, Gregory Chaitin, Gerard 't Hooft, Ian Stewart, Barry Mazur, Stephen Wolfram, Paul Davies, Tyler Burge, Doron Swade, Julian Barbour, Newton Da Costa, Francisco Antonio Doria, Reuben Hersh, Nicholas Maxwell and many others.

Francisco Antonio Doria is a Brazilian physicist. Doria is a Professor Emeritus at the Federal University of Rio de Janeiro, where he currently teaches on the foundations of economic theory at the graduate School of Engineering (UFRJ COPPE). Doria has a B.Sc. in chemical engineering and a Ph.D. in mathematical physics (under the guidance of Leopoldo Nachbin). He has made contributions to the gauge field copy problem in quantum field theory and proved with Newton da Costa several incompleteness theorems in mathematics, physics and mathematical economics, including the undecidability of chaos theory. Doria is a member of the Brazilian Academy of Philosophy, was a Senior Fulbright Scholar at Stanford University, 1989–1990, and a visiting researcher at the mathematics department, University of Rochester, 1979–1981. He thinks of himself as a philosopher and literary scholar with a humanist education, and has had as students Marcelo Gleiser and José Acacio de Barros, among other noted researchers. He likes to trace his interdisciplinary interests to a 17th century relative, the noted Portuguese writer Father Antonio Vieira (1608–1697).

About Dr. Gregory Chaitin

1. Personal Information

Gregory Chaitin is an Argentine-American mathematician living in Rio de Janeiro.

In 2007, he was the recipient of the *Leibniz Medal* awarded by Wolfram Research — whose obverse commemorates Leibniz's discovery of binary arithmetic and whose reverse commemorates Chaitin's discovery of the halting probability Omega — shown here:

- https://vimeo.com/273200092.

On the ephemerality of accomplishment, Jacob T. Schwartz once admonished him thus: "You've put a scratch on the rock of eternity." And Stephen Wolfram himself points out that Leibniz is now a cookie!

Gregory Chaitin's father was the playwright Norman Chaitin, whose 1962 feature film *The Small Hours* is preserved in the Film Library of the Museum of Modern Art in New York and was screened there in April 2016 with the presence of its writer/director, Norman Chaitin, who died later that year at the age of 92.

A documentary by Karol Jałochowski, *Against Method*, in which Gregory and Virginia Chaitin discuss creativity at their philosophical retreat on the tropical island of Paquetá was broadcast on Polish TV in 2015. (It is now on Amazon Prime Video.)

And in 2017 their son João Bernardo Fontes Chaitin was born in Rio de Janeiro. (A daughter is on the way!)

2017 also marked Chaitin's 70th birthday and the 30th anniversary of the Cambridge University Press publication of his first book, *Algorithmic Information Theory*, coincidentally the first volume in the *Cambridge Tracts in Theoretical Computer Science.*

Two unconventional reactions to his work are: *Pythagoras' Revenge*, a mathematical mystery novel written by Arturo Sangalli, published by Princeton University Press and translated into many languages; and *Approximating Omega*, a CD issued by Wandelweiser Records containing two musical tributes to Gregory Chaitin by the composer Michael Winter.

Videos of Chaitin interviews are included in the *Why Are We Here?* and *Closer to Truth* websites, both associated with television series:

- https://www.whyarewehere.tv/people/gregory-chaitin/.
- https://www.closertotruth.com/contributor/gregory-chaitin/profile.

Chaitin also appears in David Malone's superb documentary *Dangerous Knowledge*, "about four of the most brilliant mathematicians of all time, Georg Cantor, Ludwig Boltzmann, Kurt Gödel and Alan Turing, their genius, their tragic madness and their ultimate suicides." However his attitude to math is best captured by the musical comedy *Fermat's Last Tango* at

- https://www.youtube.com/watch?v=RNqQPcMYcG8.

A number of his lectures are available for viewing free of charge through YouTube. His personal favorite: the 2016 Nanyang Technological University talk at

- https://www.youtube.com/watch?v=naZHlXgeXkc.

with the unassuming title "Leibniz on complexity," perhaps the best introduction to Chaitin's approach to mathematics and to philosophy.

Last but not least, part of Gregory and Virginia Chaitin's art collection is visible in their former New York residence in these photos taken with a fisheye camera by Veikko Keränen:

- http://imageevent.com/veikkojkeranen/miscellaneousfolder/miscellaneous2.

2. Favorite Reviews

2.1. *An unsuspected artist*

Buenos Aires book review of the Spanish edition of Meta Math! (*translated from the Spanish*)

"In *Conversations with Contemporary Artists*, by Hans-Ulrich Obrist, a publication of the prolific Diego Portales University, appears Marina Abramović, the queen of the performance, who in these days stirs Buenos Aires with her presence and her workshops. In the first interview of the book, Abramović meets Gregory Chaitin in Japan and surprises him saying: 'I discover that you are very poetic and have enormous erotic energy.' One could imagine Chaitin painting half-naked like Nick Nolte in that (bad) Scorsese movie, but Chaitin is a mathematician. However, his mating with Abramović is not capricious, because both are at the limit of their disciplines."

"Although the book does not mention it, Chaitin lived as a boy in Argentina (he is a kind of Viggo Mortensen of science) and was here a few more times. In fact, on his website [his defunct UMaine website] he can be seen in a photo with Hernán Lombardi, the minister of culture macrista fallen in disgrace, in a recent festival of mathematics. I knew him a long time ago, when I was studying in Exact Sciences and one day a short guy appeared who spoke a strange Spanglish and who demonstrated Gödel's theorem using computer tools. In the following class, Chaitin gave two further demonstrations of Gödel's theorem; in the next, another four, and I do not know how many classes there were. Like many people, I was always drawn to Gödel's famous theorem on the incompleteness of arithmetic and the existence of true but unprovable propositions, although I only understood it many years later in a seminar with Carlos Lungarzo at the Faculty of Philosophy whose other student was Horacio Arló, who died recently in the United States."

"Chaitin tells his own story with the demonstration of Gödel (which he considers poor in ideas), in a book that just appeared under the title *The Number Omega: Limits and Enigmas of Mathematics*, and which is the best that mathematical popularization can offer, in the antipodes of a deceptive account like Simon Singh's on Fermat's theorem. *The Number Omega* is an extraordinary book: it gives an idea of Chaitin's work as a pioneer of the Algorithmic Theory of Information, of random numbers, of Turing's halting problem, of the LISP programming language, of the mathematization of the theory of Darwin and many things outside of the technical, such as his brilliant defense of Leibniz against Newton (and against Voltaire) or a conception of mathematics that privileges computer search rather than routine demonstration and paper."

"Chaitin has always earned his living working as a researcher for IBM, allowing him a remarkable degree of freedom to do math and talk about it, a little outside the world of his colleagues, among whom he arouses different degrees of rejection. As in the case of Abramović, Chaitin wondered if his discipline could be something other than what was supposed. But while the visual arts admit dissent as an essential part of their evolution (as is also the case in physics and the natural sciences), it is rare for anyone to suppose that mathematics could be otherwise, as suggested in this very singular passionate and personal book."

— Quintín in PERFIL, 2015

2.2. *Review of the limits of mathematics*

"Chaitin is less a mathematician or a programmer or a scientist, but a philosopher who happens to do mathematics, programming and science. How many philosophers would write and publish a book with a bundled copy of their proof coded in LISP? (see *The Limits of Mathematics*)... The reason Chaitin is a philosopher, is that he aims for a type of metaphysical commitment one rarely sees in the field of pure math. That's the biggest reason why mathematicians have never liked what he is doing and why his work is controversial; because he exposes something that can never fit into the long term projects of the field. A field which fully believes that mathematical theorems of reason are achievable by finding the stable axioms and deducible theorems."

— Robert Jackson, 2011 Blog

2.3. *Review of algorithmic information theory*

"Gregory Chaitin... has proved the ultimate in undecidability theorems... that the logical structure of arithmetic can be random... The assumption that the formal structure of arithmetic is precise and regular turns out to have been a time-bomb, and Chaitin has just pushed the detonator."

— Ian Stewart in NATURE, 1988

3. Previous Attempts at Autobiography

- First/last chapter of Chaitin, *Information-Theoretic Incompleteness*, World Scientific, 1992.
- "Recollections" in Calude, *Randomness and Complexity, From Leibniz to Chaitin*, World Scientific, 2007.

4. Favorite Publications

1. "Consciousness and information," critique of Chalmers, *The Conscious Mind*, 1996. Video at

   https://www.youtube.com/watch?v=RlotKFaHECU.

   Paper in Pinguelli Rosa *et al.*, *A Transdisciplinaridade da Consciência*, Edite, 2018, pp. 43–50.

2. With Virginia Chaitin, "A philosophical perspective on a metatheory of biological evolution," in Wuppuluri, Dória, *The Map and the Territory*, Springer, 2018, pp. 513–532.
3. "Building the world out of information and computation," in Wuppuluri, Dória, *The Map and the Territory*, Springer, 2018, pp. 431–438.
4. "Conceptual complexity and algorithmic information," *La Nuova Critica*, No. 61–62 (2015), pp. 9–28.
5. "The perfect language," published 2015 at

   http://inference-review.com/article/the-perfect-language.

   Reprinted in Dória, *The Limits of Mathematical Modeling in the Social Sciences*, World Scientific, 2017, pp. 91–109.
6. With Virginia Chaitin and Felipe Abrahão, "Metabiología: los orígenes de la creatividad biológica," *Investigación y Ciencia*, January 2014, pp. 74–80.
7. "A mathematical theory of evolution and biological creativity," included as a chapter in Chaitin, *Proving Darwin*, Pantheon, 2012, pp. 52–70.
8. *Proving Darwin: Making Biology Mathematical*, Pantheon, 2012. Also published in Spanish, Italian, Japanese and Chinese.
9. *Thinking about Gödel and Turing: Essays on Complexity, 1970–2007*, World Scientific, 2007.
10. *Meta Maths: The Quest for Omega*, UK edition of *Meta Math!*, Atlantic Books, 2006.
11. *Meta Math!: The Quest for Omega,* Pantheon, 2005. Also published in Spanish, Portuguese, Italian, Greek, Japanese and French.
12. *Conversations with a Mathematician: Math, Art, Science and the Limits of Reason,* Springer, 2002. Also published in Portuguese and Japanese.
13. *The Limits of Mathematics: A Course on Information Theory and the Limits of Formal Reasoning,* Springer, 1998. Also published in Japanese.
14. *Algorithmic Information Theory*, Cambridge University Press, 1987.

Appendix. Personal Choices

Buenos Aires café: Las Violetas in Almagro. **Author:** Jorge Luis Borges. **Play:** Friedrich Dürrenmatt, *The Physicists.* **Winter hike in the Catskills:** Slide Mountain. **Hike in the Hudson Valley:** Breakneck Ridge. **Films:** Ingmar Bergman, *The Seventh Seal*; Cacá Diegues, *Deus É Brasileiro.* **Artists:** René Magritte, Jim Alford, Thomas Cole. **Philosopher:** Leibniz. **Mathematician:** Euler. **Cuisine:** Indian vegetarian. **Musical:** *Fermat's Last Tango.* **Hike in Rio:** o pico da Tijuca. **Opera:** an opulent Jacques Offenbach, *Les contes d'Hoffmann*, at the Teatro Colón. **Music:** Dmitri Shostakovich's first violin concerto with David Oistrakh; Johann Sebastian Bach, *Die Kunst der Fuge*, played by Glenn Gould; Ravi Shankar. **Winter hike in the White Mountains:** Lonesome Lake. **Novel:** Aldous Huxley, *The Genius and the Goddess.* **Science fiction:** Fred Hoyle, *Ossian's Ride*; Fred Hoyle, *October the First is Too Late.* **Popular music:** Bossa Nova, Olodum. **Brazilian singers:** Caetano Veloso, Adriana Calcanhotto. **TV series:** *Mahabharat* (1988–1990). **Sculptors:** Kenneth Snelson, Franz Weissmann. **Photographers:** Lucien Clergue, Ralph Gibson. **Engraver:** Gilvan Samico. **Popular science:** Einstein, Infeld, *The Evolution of Physics*; Feynman, *The Character of Physical Law*; Hardy, *A Mathematician's Apology*; Pólya, *How to Solve It*; Dantzig, *Number: The Language of Science*; Bell, *Men of Mathematics*; Nagel, Newman, *Gödel's Proof*; Einstein, *Ideas and Opinions*; Newman, *The World of Mathematics.* **Textbooks:** Hardy, *A Course of Pure Mathematics*; Hardy, Wright, *An Introduction to the Theory of Numbers*; Feller, *An Introduction to Probability Theory and Its Applications*; Courant, Robbins, *What is Mathematics?*; Pólya, *Mathematics and Plausible Reasoning*; Kac, *Statistical Independence in Probability, Analysis and Number Theory*; Kac, *Probability and Related Topics in the Physical Sciences.* **Research books:** von Neumann, Morgenstern, *Theory of Games and Economic Behavior*; Shannon, Weaver, *The Mathematical Theory of Communication*; Davis, *The Undecidable: Basic Papers on Undecidable Propositions, Unsolvable Problems and Computable*

Functions; von Neumann, *Theory of Self-Reproducing Automata*; von Neumann, *The Computer and the Brain*; Weyl, *The Open World*; Weyl, *Philosophy of Mathematics and Natural Science*; Popper, *The Logic of Scientific Discovery*; Leibniz, *Discours de métaphysique.*

Contents

Foreword vii

Preface xi

About the Editors xv

About Dr. Gregory Chaitin xvii

1. An Experiment in Autobiography 1
Gregory Chaitin

2. Unknowability in Mathematics, Biology and Physics 15
Gregory Chaitin

3. Conceptual Complexity and Algorithmic Information 21
Gregory Chaitin

4. Chaitin, The Thinker 41
Richardo Espinoza Lolas

5. Dropping In on Gottfried Leibniz 49
Stephen Wolfram

6. Literature and Mathematics: Jorge Luis Borges' Library and Gregory Chaitin's Omega 83

Arturo Sangalli

7. Selva Selvaggia 91

Francisco Antonio Doria

8. Rules, Networks, and Evolution: On Chaitin's Algorithmic Information Theory and the Threefold Complexity of the Law 107

Ugo Pagallo

9. Learning the Undecidable from Networked Systems 131

Felipe S. Abrahão, Ítala M. L. D'Ottaviano, Klaus Wehmuth, Francisco Antonio Doria and Artur Ziviani

10. Limits of Computation and Physical Reality 181

Apostolos Syropoulos

11. Compression is Comprehension and the Unreasonable Effectiveness of Digital Computation in the Natural World 201

Hector Zenil

12. On Information, Quanta, and Context 239

J. Acacio de Barros

13. A New Bifurcation in the Universe 271

A. E. S. Hartmann and M. Novello

14. From the Origins of Life to the Nature of Intelligence 289
Sara Imari Walker and Paul C. W. Davies

15. Meta+phenomenology: Primer Towards a Phenomenology Formally Based on Algorithmic Information Theory and Metabiology 317
Michael Winter

16. Metabiology, Interdisciplinarity and the Human Self-image 335
Virginia Maria Fontes Gonçalves Chaitin

17. Incompleteness and Complexity in Economic Theory 345
J. Barkley Rosser, Jr.

18. The Incompletability of Markets 369
Sami Al-Suwailem

Gallery 403

Index 411

Chapter 1

An Experiment in Autobiography*

Gregory Chaitin[†,‡]

Mathematics is easy, real life is complicated

— John von Neumann

Logic is too sharp a knife to cut reality

— Sergio Luiz de Castilho Fernandes

Dedicated to *Jacob T. ("Jack") Schwartz* (1930–2009),[a] mathematician,[b] Leibnizian polymath,[c] and good friend.[d]

1. Algorithmic Information Theory

Algorithmic information theory uses the idea of the computer, particularly the size of computer programs, to study the limits of knowledge, in other words, what we can know and how. This theory can be

*With apologies to H. G. Wells!

†University of Buenos Aires, Argentina; gjchaitin@gmail.com.

‡https://independent.academia.edu/GregoryChaitin; https://arxiv.org/search/?query=g.+j.+chaitin&searchtype=author.

[a]"In memory of Jacob Schwartz," *AMS Notices* **62**: 473–490 (2015).

[b]https://ams.org/notices/201505/rnoti-p473.pdf.

[c]Gian-Carlo Rota, *Indiscrete Thoughts*, Chapter 2.

[d]In later years, I had many delightful dinners with him at hole-in-the-wall ethnic restaurants in Greenwich Village in Manhattan, and he encouraged me to continue thinking about biology, which was one of his many interests. He showed me the immense textbook *Molecular Biology of the Cell* and invited me to visit him at the Cold Spring Harbor Laboratory in Long Island and see the automatic equipment for gene sequencing and other wonders.

traced back to Leibniz in 1686, and it features a place in pure mathematics where there is absolutely no structure, none at all, namely the bits of the halting probability Ω.

There are related bodies of work by other people with other applications, but in my case, the emphasis is on using the idea of algorithmic complexity to obtain incompleteness results. I became interested in this as a teenager and have worked on it ever since, taking advantage of the freedom granted to me by the fact that this was always my hobby, never my profession.

Let me tell you that story. History is extremely complicated. What makes my account simple is the unity of purpose imposed on a field that is a personal creation, that has a central spine, that pulls a single thread. What did it feel like to do that? In fact, it's not something I did. It's as if the ideas wanted to be expressed through me.

It is an overwhelming experience to feel possessed by a tsunami of promising new ideas. As I said, this happened to me as a teenager, and I have spent the rest of my life — more than half a century — trying to develop these ideas. There are many connections with crucial concepts in other fields: complexity of physical theories, complexity of biological organisms, measuring consciousness, etc. Let me try to remember...

2. Incompleteness

Gödel discovered incompleteness in 1931 using a version of the liar paradox, "This statement is unprovable." I was fascinated by Gödel's work. I devoured Nagel and Newman, *Gödel's Proof* — I was 11 years old when it was published in 1958, and had already read the 1956 *Scientific American* article by the same authors containing an unforgettably stark photo of Gödel.

I was also fascinated by computers and by computer as a mathematical concept. In 1936, Turing derived incompleteness from uncomputability. My work follows in Turing's footsteps, not Gödel's, but adds the idea of looking at the size of computer programs.

For example, let's call a program Q **elegant** if no program written in the same language that is smaller than Q produces the same output. Can we prove that individual programs are elegant? In general, no. Any given formal axiomatic system can only enable us to show that finitely many programs are elegant.

It's easy to see that this must be so. Just consider a program P that calculates the output of the first provably elegant program that is larger than P. P runs through all the possible proofs in the formal axiomatic system until it finds the first proof that an individual program Q larger than P is elegant, and then P runs Q and returns Q's output as its (P's) output.

If you assume that only true theorems can be proved in your formal axiomatic system, then P is too small to be able to produce the same output as Q. If P actually succeeds in finding the program Q, then we have a contradiction. Therefore, Q is never found, which means that no program that is bigger than P can be proven to be elegant.

So, how big is P? Well, it must include a big subroutine for running through all the possible proofs of the formal axiomatic system. The rest of P, the main program, is rather small; P is mostly that big subroutine. That's the key thing: to focus on the number of bits in that subroutine.

So, let's define the **algorithmic complexity of a formal axiomatic system** to be the size in bits of the smallest program for running through all the proofs and producing all the theorems. Then we can state what we just proved like this: You can't prove that a program is elegant if its size is substantially larger than the algorithmic complexity of the formal axiomatic system that you are using.

Instead of saying "a formal axiomatic system of algorithmic complexity N," I'll just say "N bits of axioms." So, if you have N bits of axioms, then no program larger than $N + c$ bits in size can be proven to be elegant. That's the result we just proved.

A more sophisticated example is the number I call Ω, which is the **halting probability** of a computer running a program produced one bit at a time by repeatedly tossing a coin. Because it is a probability, this number has to be between zero and one. Imagine writing it out in binary:

$$\Omega = 0.011100\ldots$$

These bits are peculiar, they are **irreducible mathematical information**. This means that a formal axiomatic system with N bits of axioms can enable you to determine at most $N + c$ bits of Ω. Essentially, the only way to determine bits of Ω is to add that information directly to your axioms. Even though Ω is a single well-defined real number (once you fix the programming language), its bits have no structure, no pattern, none at all, they are irredundant, irreducible mathematical information.

In other words, the bits of Ω are mathematical facts that are **true for no reason**, no reason simpler than themselves.

So, that's the basic idea, and those are my two favorite results, but the devil is in the details. You can spend your life on those details, and I did.

3. Randomness in Arithmetic

Andy Warhol has said dismissively, "Everyone has their ten minutes of fame!" Let me tell you the story of the most colorful chapter in my career, the **randomness in arithmetic** episode.

It started innocently enough, with a letter from Cambridge University Press. It turned out they were going to launch a book series *Cambridge Tracts in Theoretical Computer Science*, and they wished to show that computer science was intellectually significant, not mere software engineering. And to make this point as effectively as possible, they invited me to write the first book in their series!

I was working in an engineering group at the IBM Watson Lab in New York at that time, the mid-1980s, and I showed the letter from CUP to the head of IBM Research, the legendary Ralph Gomory, who happened to have started his career as a mathematician. He immediately picked up the phone, called the head of the physics department, and I was granted a one-year sabbatical in the theoretical physics group so that I could write my book!

For the book, it occurred to me that it might be fun to give a more concrete example of irreducible complexity in pure math than the bits of Ω, so I dressed up Ω as an exponential diophantine equation. It had 17,000 variables, was 200 pages long, with a parameter K, and was constructed so that fixing the value of the parameter K, the equation would have a finite or an infinite number of unsigned whole number (non-negative integer) solutions depending on whether the Kth bit of Ω is, respectively, a 0 or a 1.

In fact, this equation was an arithmetization of a LISP interpreter (for a toy version of LISP), into which one inserted a LISP program for approximating Ω in the limit from below. Ω can be approximated in the limit from below very, very slowly, but of course there can be no computable regulator of convergence or Ω would be computable instead of irreducible. In other words, one can never know how close the approximation is to the correct value of Ω.

My book, *Algorithmic Information Theory*, duly appeared as volume one of *Cambridge Tracts in Theoretical Computer Science*. And it featured a foreword by my late friend Jacob T. Schwartz, who at City College in New York had been a student of Emil Post. Who, you may ask, is Emil Post?[e] In fact, he is the unjustly largely forgotten equal of both Gödel and Turing.

Events now began to accelerate. I discovered a review of my book written by Ian Stewart in the highly visible and extremely prestigious journal *Nature*. In fact, it was more than a book review, it was a news item entitled "The Ultimate in Undecidability."

The *Nature* piece was quickly echoed in less prestigious but also well-known magazines such as *Scientific American*, the *New Scientist*, and *La Recherche*.

[e]See Pierre Cassou-Noguès, *Les Démons de Gödel. Logique et folie* [logic and madness], Le Seuil, 2007, a book which is half about Gödel and half about Post.

Of course, it all blew over in a couple of years, but I was left with two charming souvenirs: a fascinating literary work, and two musical compositions.

First let me tell you about the novel *Pythagoras' Revenge* by Arturo Sangalli, published by Princeton University Press and reprinted in a number of other languages. I enjoyed his novel immensely, although it was a bit threatening. It's a marvelous tale revolving around a mysterious Pythagorean secret society that kills a mathematician who — building on my work — shows that there is **randomness everywhere** in pure mathematics. Murdering him is a terrible mistake, since he is in fact a reincarnation of the sect's founder, the philosopher–mathematician Pythagoras!

The conceptual composer Michael Winter's beautiful piece *approximating omega* sonifies the LISP program for approximating Ω in the limit from below. His piece *for Gregory Chaitin* includes the 64 known bits of a particular Ω number; it is rather like a meditation or a religious ceremony based on Ω. Michael uses mathematical ideas in a great many of his compositions.

This might be a good place to include some Ω aphorisms:

- Ω is a diamond to mathematical truth's carbon.
- Ω is the concentrated essence of mathematical creativity.
- Ω is the DNA for pure mathematics.

One final thought: Physicists still hope to find a *Theory of Everything*, a small set of simple equations that explains the entire universe. Biology is a very different kettle of fish; everything is extremely complicated. What about pure mathematics? Well, Ω shows that **pure math is more biological than biology itself**, because biology is extremely complicated, but Ω is infinitely complicated, provably so.

4. Metabiology and Beyond

After being awarded the Leibniz medal by Wolfram Research in 2007 — one side commemorates Leibniz's discovery of binary arithmetic and the other commemorates the discovery of Ω, in both cases in Latin verse — I turned from incompleteness to what I call **metabiology**, an attempt to characterize **life as evolving software**.

Biology is based on the idea of software, which was discovered by Nature eons before humanity did. It is software which accounts for the plasticity of computer technology and the biosphere. DNA is procedural digital software for constructing and running an organism.

This idea appears for the first time in a remarkable 1951 paper by the mathematician John von Neumann — another of my heroes — "The General and Logical Theory of Automata," where he talks about *natural automata* and *artificial automata*. Artificial automata are computers and natural automata are biological organisms. One of the creators of molecular biology, Sydney Brenner, who for many years shared an office at Cambridge with Francis Crick, was inspired by von Neumann's paper, and not, as was more often the case, by Erwin Schrödinger's 1944 book *What is Life?*, which Brenner regarded as sadly misguided!

Be that as it may, metabiology studies randomly mutating computer programs, not randomly mutating DNA. So our space of all possible organisms is a software space, and random evolution becomes **a random walk in software space**. In our model of evolution, mutations are accepted only if they increase the fitness of the organism, so it actually is **a hill-climbing random walk in software space**.

Instead of **design without a designer** this is **programming without a programmer**. A crucial feature of this model is that we allow *global algorithmic mutations* and there is a natural *a priori* probability distribution on the space of all possible mutations: a mutation that is a K-bit algorithm has probability 2^{-K}. And we can define the **mutation distance** between two organisms A and B as the size in bits of the smallest program for converting A into B, which is almost precisely the same as $-\log_2$ of the probability that a random mutation will convert A into B.

The original purpose of metabiology was to give a mathematical proof that Darwin was right, that random evolution plus natural selection can account for the myriad of sophisticated lifeforms that surround us. What we have actually been able to prove is that random evolution plus natural selection will, with overwhelming probability, be open-ended and produce software organisms of increasing conceptual complexity. And it is possible to characterize the rate of increase of the complexity of these organisms.

Furthermore, metabiology turns incompleteness on its head. Instead of being bad it is good, it is the first step in the direction of **a mathematical theory of creativity**...

That's about where I got with these ideas. My wife Virginia and my former Ph.D. student Felipe Abrahão, the two other members of the metabiology team here in Brazil, continue working on related questions. I prefer to think about the brain and consciousness.

My thoughts on the brain have been outlined in the paper "Conceptual complexity and algorithmic information" published in *La Nuova Critica* and reprinted in this volume.

And my latest work has been a critique of David Chalmers' 1996 *The Conscious Mind: In Search of a Fundamental Theory.* I reappraise Chalmers' panpsychic proposal that any physical system that represents and processes N bits of information possesses N bits of consciousness.

I find Chalmers panpsychism more compelling than the currently fashionable and elaborate integrated information theory of consciousness (Tononi's Φ).

Nevertheless, I feel compelled to ask Chalmers some questions: Is the immune system conscious? Is the entire universe conscious? And which measure of information are we talking about: Shannon information, algorithmic information or quantum information (qubits)?

Chalmers panpsychism fits in beautifully with our final topic, *digital philosophy.*

Pythagoras asserted that

All is number, God is a mathematician.

Digital philosophy, originally due to Leibniz, asserts that

All is algorithm, God is a programmer.

In fact, digital philosophy postulates that the universe is built out of information, algorithmic information.

Furthermore, Stephen Wolfram and Edward Fredkin have proposed digital philosophies based on cellular automata. And there are two beautiful books in Italian about all of this: *Bit Bang: The Birth of Digital Philosophy* (2013) co-authored by a theologian, Andrea Vaccaro, and an information theorist, Giuseppe Longo; and *Introduction to Digital Philosophy: From Leibniz to Chaitin* (2005) by a philosopher, Ugo Pagallo.

However, in the meantime, physicists have become enamored and are pursuing vigorously the idea that the universe is built out of quantum information. See for example Erik Verlinde's spectacular work on space–time emerging from qubit entanglement with gravity as an entropic force rather than a fundamental interaction, a theory of gravity which by modifying the inverse-square law may also eliminate any need for dark matter...

It's all been quite a hobby! And now my wife Virginia and I are delighted to have a new project, our little baby boy! Meanwhile I continue following research such as David Glanzman's challenging the current paradigm of how the brain works, and tracking progress on Andrea Rossi's *energy catalyzer* (E-Cat) and Randell Mills' attempts to imitate the energy source in the solar corona by converting hydrogen into *hydrinos* — a form of hydrogen below the conventional ground state, and which Mills believes to be the dark matter.[f]

The future should be a lot of fun! Perhaps, our little boy will emigrate to Elon Musk's Mars colony. I would love to be there!

[f] "RNA from trained *Aplysia* can induce an epigenetic engram for long-term sensitization in untrained *Aplysia*," David L. Glanzman *et al. eNeuro*, 14 May 2018; *An Impossible Invention: The True Story of the Energy Source that Could Change the World*, Mats Lewan, 2014; *Randell Mills and the Search for Hydrino Energy*, Brett Holverstott, 2016.

5. A Letter to Crete

Dear Mr. Christos Gikas

You live in Crete? Small world: My wife's father was a Brazilian diplomat posted in Europe who loved ancient Greece, and every summer they would visit Crete and the palace at Knossos.

I don't think you will like my answers to your questions; I am an iconoclast. But here goes:

Q1. As far as I am concerned, you are a mathematician. What was your initial inspiration to make you study mathematics and not any other field of science?

As a child, I was interested in deeply hidden mysteries like general relativity and quantum mechanics. Then I discovered Gödel's incompleteness theorem and dropped physics and astronomy. As I think Plato said, in mathematics one thinks like the Gods! I was always interested in the Good, the True, and the Beautiful.

Q2. I suppose you have encountered difficulties during your career concerning your work. Could you describe the toughest one and the way you managed to overcome it?

From an early age I was interested in developing a mathematical theory of life and evolution. Already in my twenties I had developed the right toolkit, but it took me until my sixties to realize that! I could have done metabiology in my twenties, not my sixties, but one gets distracted, one's attention is diverted.

Q3. Scientists struggle to explain the importance of their work, seeking acknowledgement or money. Does money come with acknowledgement and vice versa? Apart from being a good scientist, do you think it is of vital importance to be a good marketeer in order not to have your work neglected during one's lifetime?

A true scientist is a heroic figure, a mortal who behaves as if he were immortal. A professional scientist, however, is mostly a politician. Fields of science are power structures and do not welcome new ideas, especially not from outsiders. When there were few scientists, it was easier to do good science. Now science is a big business with a giant administrative bureaucracy. That explains why there have been no fundamental advances in physics since the 1920s. The shut up and calculate dogma has killed physics.

Q4. Ordinary people are generally interested in thinking what to do on Saturday night instead of reading a book or discussing the past and the future of humanity. What would you propose in order to change such a mentality and turn more people into admiring natural science and science in general?

Current societies want mere consumer citizens, who spend money they don't have to buy things they don't need. Why are we here? Not to be consumers, but to understand and to create...

Q5. If you could turn back time, then would you choose the same field of study, mathematics? If so, then what would you suggest to newcomers to do in order to gain acknowledgement for their work?

I have been very fortunate to have done work on all the questions in mathematics that interested me, fundamental philosophical questions that most mathematicians have no interest in at all. So if I could start over I would be a mountain climber or a Martian colonist, not a mathematician.

And being a creative person who honors the human mind by advancing understanding is quite different from being good at scientific public relations and politics. My hero Leibniz wrote mostly for his drawer. Some of his best ideas were discovered by chance among his posthumous papers centuries after his death.

Furthermore, if your work attracts too much attention you will be hated by your colleagues! And most scientific advances are attributed to the wrong persons — if it happens to you it can be very upsetting, but as Jacob T. Schwartz told me, "In a million years, what difference will it make?" Real history is extremely complicated... Real science is a lonely solitary pursuit, between you and God, not for anyone else!

Q6. One last question about your path through life: How did you manage to "oppose" these modern power structures of knowledge and to become creative in your own way, the way you wanted?

That is a very good question, Chris. I did not oppose those modern power structures. That is impossible. I did manage to ignore them, though, for a while.

I did this in many different ways.

First of all, I am from Argentina and now living in Brazil, two countries which are not so closely coupled with the international academic mafias.

Second of all, I am an autodidact: self-taught. My doctorates are honorary doctorates, my professorships are honorary professorships. So I did not have to work on a project for my thesis advisor. I did not have to work on the fashionable topics.

Third, I started research and publishing very young, as a teenager. My first major publication was in the *ACM Journal* in 1966 when I was 19. At that time the field of theoretical computer science did not yet exist, the theoretical computer science mafia had not yet solidified. I was one of the creators of complexity theory, and at that time the *ACM Journal* was the only theoretical computer science magazine. My work was rushed into print because the United States was competing with what was then a rather substantial Soviet scientific effort originally under Stalin's personal protection because of military applications. The Americans were scared of the Russian scientists, they knew that Russian mathematicians were absolutely top-level.

Fourth, for many years, I had a position as a research staff member in what was then the premier computer-related research center, the IBM Watson Research Center in New York. I did practical computer engineering (software and hardware) design work for them; my theoretical research was my hobby, not what they paid me for. This position at an IBM industrial lab protected me from academic politics for a while.

Fifth, my work went around the mafia, I did not oppose it on its own turf. My work attracted attention among the scientifically educated public, but it was not viewed with approval by logicians and mathematicians. And my South American personality was out of place in a country where down-to-earth pragmatists are preferred to European-style intellectuals like Jorge Luis Borges, Argentina's greatest writer.

People would invite me to give talks because I was a strange creature, a kind of entertainer. Nevertheless giving these talks helped me to develop my ideas.

Sixth, complexity became a fashionable topic in theoretical physics, and physicists were interested in my work. I received innumerable invitations to physics meetings, but none to math, logic or computer science meetings. I did not compete with physicists, since I was not a physicist. They enjoyed my work criticizing math and saying it was not that different from an empirical science, but mathematicians hated my ideas.

Now complexity is not a hot topic in the physics community, it has been replaced by quantum information and computation.

As physicists lost interest in what I was doing, some philosophers began to be interested, especially because of my great admiration for Leibniz...

New ideas, new points of view, always encounter opposition. During my career the immune systems of the math, logic and computer science communities reacted strongly. But I did the work, I was able to pursue my dream, which is what counts. The rest is unimportant, it's just politics...

Anyway, as you can see this life path worked for me, but I could not repeat it now. Industrial labs no longer also want to be known for their basic research. Basic research, blue sky work, is no longer tolerated, it is considered a waste of money...

I do not believe in institutions, I believe in creativity and exceptional individuals, like yourself!

Best of luck in your endeavors, and much thanks for the interesting, thought-provoking questions!

Chapter 2

Unknowability in Mathematics, Biology and Physics*

Gregory Chaitin†

It's a privilege and a pleasure for me to be here at The New School today. A pleasure, because one of my favorite books was written here during World War II: *The Psychology of Invention in the Mathematical Field* by Jacques Hadamard, a refugee from Hitler's Europe. It's a fascinating book, so for me this is like visiting a sacred shrine.

I'm going to discuss unknowability in mathematics, in biology, and in physics.

There are three fields in mathematics that deal with unknowability: in chronological order, they are probability theory, metamathematics, and algorithmic information theory.

The oldest is the theory of probabilities. This started as letters between Pascal and Fermat. One of the first books in this field is *A Philosophical Essay on Probabilities* by Pierre–Simon Laplace, which is still very much worth reading. I would love to own a first edition of

---

*Unknowability in science and mathematics was discussed at The New School in New York on the morning of April 5, 2019. Chaitin spoke for 20 minutes on probability theory, metamathematics, algorithmic information theory, metabiology, cold fusion and dark matter. This is an edited transcript of his remarks.

†University of Buenos Aires, Argentina; gjchaitin@gmail.com.

this work, but instead I only have an early edition of his *Exposition du Système du Monde*, a wonderful book that is not about probabilities.

The great physicist Boltzmann declared that the most important scientific work of the 19th century was Darwin's theory of evolution, because for the first time it gave randomness a fundamental role in the sciences. Darwin's theory inspired Boltzmann to become one of the creators of statistical mechanics, whose methods are now pervasive in theoretical physics.

The theory of probability is about indeterminate events, but it is not about mathematical facts, which are always black or white, not gray, and which are necessary, not contingent, truths.

However, the field of metamathematics is actually about unknowability within mathematics itself. This is a fascinating piece of intellectual history associated with the names of David Hilbert, Kurt Gödel and Alan Turing. These three gentlemen have already been mentioned several times in this meeting, so I will not dwell on this.

The newest mathematical field dealing with unknowability is algorithmic information theory. The star of this theory is a real number, the halting probability Omega, which is defined to be the probability that a self-contained computer program that is chosen at random will eventually halt.

If all programs halted, the halting probability would be one. If no program halted, the halting probability would be zero. And since some programs eventually halt and others don't, Omega is actually between zero and one.

Omega has a rather straightforward mathematical definition, and it is a single well-defined real number. Yet its numerical value is maximally unknowable. The base-two bits of the numerical value of the halting probability Omega provide a perfect simulation within pure mathematics of independent tosses of a fair coin. Whether an individual bit is a 0 or a 1 is, to God or to an infinite mind, a single well-determined individual mathematical truth, but down here, to us, it looks maximally unknowable and very much like the result of independent tosses of a fair coin.

Hilbert had hoped that there would be a theory of everything for pure mathematics, but Omega shows that pure math is more like biology, the domain of the irreducibly complex, than it is like fundamental physics, where there is still hope for finding a small set of simple equations that determine the universe.

So that brings us to biology, a very messy, complicated field. I have a wife and a child, so I love messy biology, but from a mathematical point of view biology looks rather intractable, it looks rather opaque. Will we ever be able to prove mathematically that life must evolve? Does Darwin's theory provably account for the spectacular biological creativity that led to human beings, to you and me? Can one prove that random mutations and natural selection are enough to account for the richness and diversity of our biosphere?

There is a highly developed field called systems biology, dealing with computer simulations of complex biological systems, but there are very few mathematical proofs in biology. What can be done about this?

Well, in a remarkable 1948 lecture published in 1951 which is remarkably forgotten, *The General and Logical Theory of Automata*, John von Neumann already identified the fundamental mathematical idea hidden in biology. That's the idea of software, which accounts both for the plasticity of computer technology and its overwhelming success, and for the plasticity of the biosphere with its spectacular diversity of forms.

To put it bluntly, nature invented software millennia before humanity did. Biology deals with extremely ancient software. Evodevo is a kind of software archeology.

The new field I call metabiology recognizes the impossibility of a fundamental mathematical theory dealing with the effect of random mutations on natural software, on DNA, so instead it proposes studying the effect of random mutations on artificial software, on computer programs. That in fact turns out to be tractable mathematically, and it is probably as close as one can get to a theoretical biology as deeply impregnated with mathematics as theoretical physics already is.

When I was a visiting professor at the Federal University of Rio de Janeiro, I sketched out these ideas, and they were published in a little book in English which has also appeared in Chinese, Japanese, Italian and Spanish. Teaching this material in Rio helped me to organize my thoughts and write the book, and for a while we even had a small metabiology research group including my wife, Virginia, who studied metabiology's epistemological foundations and implications, and Felipe Abrahão, who did his Ph.D. on extending the mathematical framework.

After my visiting professorship expired there was a hiatus, but I am now happy to say that these ideas have been taken up by Hector Zenil and his collaborators in Europe. And some of their work on metabiology has recently been covered in an excellent article in *Quanta* magazine, *Mathematical Simplicity May Drive Evolution's Speed.*

So perhaps the mathematical unknowability of biology is decreasing slightly.

For my next topic I would like to turn to physics. A week or so ago, the end of March, marked the thirtieth anniversary of a notorious 1989 press conference at the University of Utah, announcing a discovery by Martin Fleischmann, an extremely distinguished electrochemist, and his younger colleague Stanley Pons. They claimed they had an electrochemical cell that was generating so much excess energy, so much heat, that the source had to be nuclear energy — it was far too big to be chemical in origin.

For a month or so there was a media frenzy and many replication attempts. The cover of *Time* magazine proclaimed that nuclear fusion had been achieved in a test tube. But this quickly turned into a fiasco.

The final verdict was that the two scientists were incompetent or worse. They resigned their professorships and fled the country. Fleischmann returned to England, which is not such a bad place to live, although I love hiking in Utah. And Pons fled to France and even resigned his American citizenship.

Just mentioning "cold fusion" can destroy a scientist's career. It is an unmentionable topic and has been so for three decades.

That's in most of the world. But strangely enough there are places where people think differently.

In Japan there is now a national project to save the world from global warming using a technology based on a much more sophisticated version of Pons and Fleischmann's discovery. Remember, the Japanese have no choice. They have no petroleum, and Fukushima was a catastrophe. In Japan there is tremendous political and economic backing for what they wisely refer to as the "new hydrogen energy" and not as "cold fusion."

A press release in English strangely invisible outside of Japan has in fact recently announced a collaboration between Mitsubishi, a Japanese foundation called Clean Planet, and Tohoku University, where there is even a university department devoted to developing this new hydrogen energy source.

I think that the fact that what is unmentionable in much of the world is a national imperative in Japan is a fascinating example of the sociology of science.

Continuing in this vein, for my last topic I would like to talk about Randell Mills' work on hydrinos and dark matter, the mysterious 90% of the mass of the universe that no one can see and nobody knows what it is.

It is a curious fact that the solar corona is much, much hotter than the surface of the sun. What can the source of all this energy be?

Well, according to Mills, who has his own private research institute, the energy in the corona is released when atomic hydrogen turns into what he calls hydrinos, a more compact form of hydrogen than is allowed by conventional quantum mechanics. The electron in a hydrino atom is closer to the proton than normally thought possible, because it forms what Mills calls an orbisphere, that is some kind of a surface, not a point.

According to Mills, in fact, hydrinos are the dark matter.

And Mills has some intriguing experiments in which he claims he creates a piece of the solar corona in his laboratory. According to Mills, the brilliant spectrum the device emits in the deep UV is similar to the deep UV part of the spectrum of the solar corona.

What, then, are we to make of "new hydrogen energy" and "hydrinos?" Whether or not they eventually save humanity from global warming, whatever the final verdict, there is no doubt in my mind that they constitute fascinating chapters in the history of ideas and in the sociology of science.

Chapter 3

Conceptual Complexity and Algorithmic Information*

Gregory Chaitin[†,‡]

1. Introduction

In this essay, we propose that the fundamental philosophical concept of *conceptual complexity* is captured mathematically by the notion of *algorithmic information content*, and we discuss the complexity of physical and mathematical theories, the complexity of biological mutations, and the most complex system in biology, the human brain. These are steps in the direction of a mathematical philosophy, by which we mean a mathematical approach to philosophical questions, not a philosophy of mathematics. For as Leibniz said:

> Without mathematics we cannot penetrate deeply into philosophy.
> Without philosophy we cannot penetrate deeply into mathematics.
> Without both we cannot penetrate deeply into anything.

*Reprinted from *La Nuova Critica* **61–62** (2015) 9–28 with the kind permission of Prof. Arturo Carsetti. Invited talk at *IACAP 14: Conference of the International Association for Computing and Philosophy*, Thessaloniki, July 2–4, 2014.
†Federal University of Rio de Janeiro, Brazil; gjchaitin@gmail.com.
‡https://ufrj.academia.edu/GregoryChaitin.

Sans les mathématiques on ne pénètre point au fond de la philosophie.
Sans la philosophie on ne pénètre point au fond des mathématiques.
Sans les deux on ne pénètre au fond de rien.

2. Origins of the Concept of Complexity in the Philosophy of Science and in Biology

2.1. *Complexity in the work of Leibniz*

We begin the intellectual history of the concept of complexity with Leibniz's brilliant *Discours de métaphysique* (1686). In particular, we are interested in Sections V and VI of the *Discours*, in which Leibniz explains why science is possible and what it means for facts to be governed by law rather than be random. Let's paraphrase Sections V and VI of the *Discours*.

God has created the best of all possible worlds, in that the richness and diversity — including us! — that we see in the world around us is the consequence of a small, simple set of concepts and laws. God *maximizes* the richness of the world and at the same time *minimizes* the complexity of the laws that determine this world. In other words, the world is comprehensible, science is possible, there are elegant theories. God's perfection resides in the fact that He uses the smallest possible number of bricks, the simplest possible tools, in order to create the world.

By the way, this excludes miracles, which would be like amendments to the constitution, exceptions that have to be added to the constitution and unnecessarily complicate it, and analogously would unnecessarily complicate the fundamental laws of physics.

Furthermore, what is a law, how can we distinguish the lawless from the lawful? Well, said Leibniz, if we have a finite set of points on a graph representing the behavior of a physical system, there is **always** an equation passing precisely through those points. Hence the mere existence of a mathematical law is not enough. Real laws have to be simple, because anything can be explained if one allows sufficiently complicated, *ad hoc* laws.[a]

In our modern reading of Leibniz, Sections V and VI both assert that the essence of explanation is compression. An explanation has to be much simpler, more compact, more concise, than what it explains.

2.2. *Complexity and evolution*

Leibniz is a tremendously deep thinker. We now turn to biology, a more obvious source of the notion of complexity. Please see Peter Bowler's book *Darwin Deleted* for an analysis of the many threads in the development of ideas on evolution and progress in Nature.

One of these threads was Lamarck, who proposed the inheritance of acquired characteristics in order to explain the evolution, the increase in the complexity of organisms, that was becoming increasingly apparent in the fossil record. Another precursor of Darwin was his own grandfather, Erasmus Darwin. Even before Charles Darwin, many realized that evolution was taking place. The problem was to find a mechanism, to explain how evolution works. That it was taking place was not at issue. Many even assumed that Nature's goal was to progress, a notion that is rather submerged — but not entirely eliminated — in Darwin's theory.

2.3. *Complexity and spontaneous generation*

A further impetus to the crystallization of the concept of complexity in biology was provided by the dispute over spontaneous generation. Spontaneous generation is immediately seen to be impossible once the enormous complexity of an individual cell was properly grasped.

2.4. *The immensely complicated human brain*

Later in this essay, we shall attempt to analyze human intelligence and the brain. That's also connected with complexity, because the

[a]As Hermann Weyl put it, if arbitrarily complex laws are permitted, then the concept of law becomes vacuous because there is always a law. See Weyl, *The Open World* (1932).

human brain is the most complicated thing there is in biology. Indeed, our brain is presumably the goal of biological evolution, at least for those who believe that evolution has a goal. Not according to Darwin! For others, however, evolution is matter's way of creating mind.

Fundamental ideas that have inspired us may be discredited and slip out of sight, but they have a way of reemerging years later in more modern dress. Bowler sees this process at work in Darwinism and it certainly applies to our next topic, the search for the perfect language.

3. The Search for the Perfect Language

In Section 2, we discussed four important applications of the concept of complexity but neglected to specify what complexity is. The time has come to define complexity rigorously with mathematical precision.

To show how that's done, let's review the saga of *The Search for the Perfect Language*. Umberto Eco's book with this title reminds us that in the Middle Ages it was thought that knowing the language used by God to create the universe — perhaps Hebrew — would give us a way to analyze ideas into their basic conceptual components and provide a path to all truths. An early effort in this direction was the *Ars magna* of Ramon Llull. Inspired by Llull, Leibniz converted this project into the search for a *characteristica universalis*.

This project is mocked by Jonathan Swift in *Gulliver's Travels*, in his account of Gulliver's visit to Laputa. A later version of this dream is Hilbert's program for a single consistent, complete formal system for all of mathematics which would solve the *entscheidungsproblem* by providing an algorithm for deciding if any mathematical statement were true or false, merely by running through all possible proofs in the formal system until finding a proof of correctness or a refutation.

Thanks to Gödel and Turing we know that this cannot be done: There is no universal language for formalizing all possible mathematical proofs. However, Turing also showed that there are universal languages for formalizing all possible mathematical algorithms, and algorithmic information theory tells us which are the most concise, the most expressive such languages.

Clearly these would be the languages of choice for creating the world! In this essay we define the *conceptual complexity* of an object X to be the size in bits of the most compact program for calculating X, presupposing that we have picked as our complexity standard a particular fixed, maximally compact, concise universal programming language U. This is technically known as the *algorithmic information content* of the object X, denoted $H_U(X)$, or simply $H(X)$ since U is assumed fixed. In medieval terms, $H(X)$ is the minimum number of yes/no decisions that God would have to make to create X.

As Leibniz said, "As God calculates, so the world is made," "*Cum Deus calculat, fit mundus.*"[b]

4. Mathematical Definition of Conceptual Complexity

Whence cometh this definition? Who invented the necessary mathematics? Here is a brief summary. If we sum

Alan Turing (1936) + Claude Shannon (1948)
computation **information**

this gives us

algorithmic information
[R. Solomonoff, A. N. Kolmogorov, G. J. Chaitin (1960s)]
[G. J. Chaitin, L. A. Levin (1970s)]

And the concept of algorithmic information in turn gives us a handle on the deep notion of conceptual complexity, which is fundamental

[b]The actual statement by Leibniz seems to be "*Cum Deus calculat et cogitationem exercet, fit mundus*" (When God thinks things through and calculates, the world is made). But the abbreviated version is more to the point, or at least more to *our* point.

in epistemology. (In contrast, I think of the more practical topic of *time complexity* as a branch of software engineering.) To illustrate the power of this new intellectual toolkit, we shall now discuss applications in metaphysics, in metamathematics, and in a new research program that I have dubbed *metabiology*.

5. Conceptual Complexity in Software Models of Physics, Mathematics and Biology

Now you know how to define conceptual complexity mathematically and a little about its history. In this section, we start by examining the complexity of physical theories, take a detour to consider the complexity of mathematical theories, and then finally arrive at a model of biological evolution. In later sections we shall present some thoughts on the brain and consciousness.

5.1. *Complexity of a* physical theory

A physical theory T is a computer program that calculates and exactly reproduces a finite set of experimental data by using the equations of physics and the initial conditions of the physical system in question. Then T halts. The data must be reproduced precisely; no errors are allowed.

Thus we can only deal with ***deterministic*** *physical theories in this model, not with* ***statistical*** *theories, which necessarily cannot exactly reproduce the experimental data.*

The conceptual complexity $H(T)$ of the theory T is the size in bits of the program T.

Why is this definition of complexity of any value? Because it enables us to formalize and study mathematically the idea that the best theory is the simplest theory.

Following Ray Solomonoff, there are numerous practical applications of this criterion in machine intelligence, in Bayesian statistics, and in data mining, but in this essay I am interested in philosophy and not in practical applications.

5.2. *Complexity of a* formal axiomatic theory

A formal axiomatic theory T is an unending computer program that calculates one by one the infinite set of all theorems that follow from the axioms by applying the rules of symbolic logic [David Hilbert, Emil Post]. T does this by systematically running through all possible proofs.

The conceptual complexity $H(T)$ of the theory T is the size in bits of the program T.

The surprising fact that you can almost never prove that you have the best — the simplest possible — physical theory, gives us a new kind of Gödel incompleteness, an information-theoretic incompleteness theorem in fact.

To be able to prove that an N-bit physical theory is the most concise computer program that calculates the output that it does (which I like to refer to as an "elegant" physical theory or an "elegant" program), you need to use a formal axiomatic theory T whose conceptual complexity $H(T)$ is at least N bits.

Proof Sketch: Given a formal axiomatic theory T with conceptual complexity $H(T)$, consider the paradoxical program P that calculates the output of the first large provably elegant program Q produced by T — the first such Q you find when systematically running through all possible proofs that programs are elegant in theory T. More precisely, P searches for a provably elegant Q with the property that the size in bits of Q is greater than the size in bits of P. P calculates the same output as Q, which is supposed to be the smallest program that produces the output that it does, but P is smaller than Q!

This easily shows that in a given formal axiomatic theory T a provably elegant computer program Q cannot be larger than the size in bits of P. And what is the size of P? One can show that it's $H(T)+c$ bits, where c is a constant independent of T and is negligible if $H(T)$ is large.

5.3. *Complexity of an* algorithmic mutation

An algorithmic mutation M is a computer program that calculates the new organism from the old organism.

The conceptual complexity $H(M)$ of the mutation M is the size in bits of the program M.

Note that in the toy model of evolution given below the organisms are also software, but we are not too concerned with their conceptual complexity. We concentrate on the conceptual complexity of the mutations.

Defining a mutation in this very general and somewhat abstract way is the key idea that we shall use to model Darwinian evolution mathematically. It's crucial for the success of our model. It's what enables us to prove that Darwinian evolution works in this model and yields open-ended, endless evolution.

6. Modeling Darwinian Evolution

Since this is a little-known, new area of application of the notions of conceptual complexity and algorithmic information, we shall sketch here some of the basic features of our highly-simplified toy model of evolution.

6.1. *How do we pick a random mutation?*

As we stated before, in our model an algorithmic mutation M is a computable function that takes as input the original organism A and produces as output the mutated organism A':

$$A' = M(A).$$

Here M, A and A' all consist of binary software, of bit strings:

$$M = \text{algorithmic mutation},$$

$$A = \text{original organism},$$

$$A' = \text{mutated organism}.$$

But what is the probability of the mutation M? What probability distribution shall we use over this very rich space of possible mutations?

In our model, simple mutations are much more probable than complex mutations. In fact, if the conceptual complexity of the mutation M is $H(M)$, then its probability is $2^{-H(M)}$. In other words, if the binary software for the mutation M has K bits, then its probability is $1/2^K$, for each bit of the program for M is chosen by using an independent toss of a fair coin.

If M is a K-bit program, then probability of $M = 2^{-K}$.

If the *conceptual complexity* of M is K bits, then
the probability of $M = 2^{-K}$.

This is a very natural way of assigning probabilities to programs and it has been used in theoretical computer science since the 1970s.

(Although there is an important technical proviso: The program for M must be "self-delimiting" — otherwise the total probability for all possible mutations is infinite instead of being less than one as a probability should be. We shall encounter self-delimiting programs again later when discussing the halting probability Ω.)

Please note that for any A and A' there is a non-zero probability that the mutation M will transform A into A'. But most mutations are extremely unlikely. The fact that we allow arbitrary *algorithmic* mutations M has some surprising consequences. For example, the global change M that consists of inverting each bit of a program A yielding the complementary bit string A' is a very simple and therefore a highly probable mutation.

Furthermore, M's probability does not depend on the size of A and A', since the same program M will work for inverting arbitrarily large bit strings A. Unfortunately M will probably completely wreck the program A. This is not a useful mutation, even though it is a highly probable one.

We've explained the mutational model we're using. We are now ready to talk about the evolution of mutating software.

6.2. *Life as evolving software*

Now let's explain the rest of our evolutionary model. It's just a *hill-climbing random walk in software space*, for our organisms are programs too, and a mutation is accepted if and only if it increases the fitness of the organism.

In fact, our organisms are programs that calculate a single positive integer, and the bigger the integer, the fitter the program. In other words, the well-known computer science Busy Beaver problem consisting of naming the largest possible integer is the challenge that will force our organisms to evolve.

There is just a single software organism A at a time, and we try mutating it at random using algorithmic mutations as previously described, until we obtain a fitter organism A', one that calculates a bigger number. Then everything continues as before but with this new fitter organism A' replacing the previous software organism A.

Darwinian evolution has been described as

"Design without a Designer!"

Here what we have instead is

"Programming without a Programmer!"

In other words, our organisms consist only of hereditary material, they do not have bodies or a metabolism. And in our model, instead of randomly mutating DNA, which is the actual software of life, we are mutating computer programs that are working on an extremely difficult computational problem, the Busy Beaver problem.

Simple as it is, this model exhibits some of the features of Darwinian evolution, in fact provably so. If it weren't so simple, we would not be able to prove anything. One important concept that emerges from this analysis is *mutation distance.*

6.3. *Mutation distance*

We are modeling evolution as a random walk in software space. How far does one have to walk to get from the organism A to the organism A'? Is there a distance measure associated with this model? Yes indeed, the mutation distance. What is mutation distance? Well, there are several different but ultimately equivalent ways of defining this important concept:

Mutation distance $= H(A'|A)$
= the *relative information content* of A' given A
= the number of bits of algorithmic information needed to transform A into A'
= the *conceptual complexity* of the mapping from A to A'
= $-\log_2$ of the probability that a random mutation M carries A into A'.

For more information about this model of evolution and its remarkable properties, please see the book *Proving Darwin* or my article "Life as evolving software" in Hector Zenil's Turing centennial volume *A Computable Universe.* Now let's change topic.

7. Designing a Brain

We have discussed applications of the concept of conceptual complexity in metaphysics, in metamathematics, and in a toy model for Darwinian evolution. Now let's turn to the most complicated object there is, the human brain.

What we present here may or may not be how our brain works, but I believe it is a possible design for a brain.

7.1. *Clues for solving the puzzle*

Our analysis of the brain will follow John von Neumann's posthumous classic *The Computer and the Brain* (1958) and will be based primarily on considering *processing power* and *memory capacity*, the two key parameters in a computer.

And the RNA/DNA molecular biology level has much greater memory capacity and processing power than the neuronal level, so why not use it?

We also take into account some intriguing experiments with planaria that were formerly dismissed but which have recently received a certain amount of confirmation.

There is a deep remark in von Neumann, *The Computer and the Brain*, that in computers the technology used for memory and for logic is always different, so maybe the same is true in the brain.

Von Neumann also makes this anti-connectionist remark in the question and answer discussion at the end of his paper "The general and logical theory of automata" (presented 1948, published 1951), the paper in which he proposes thinking of DNA as software.

Following von Neumann, we shall explore anti-connectionist models for a brain, ones in which logic may be encoded in the connections of neurons but perhaps a different encoding is used for memory, especially long-term memory, which has a huge capacity, as illustrated by von Neumann himself, who reportedly had a photographic memory.

In a nutshell, we propose that

$$\textit{logic} = \textit{neurons} = \textit{conscious mind},$$

while

$$\textit{memory} = \textit{RNA/DNA level} = \textit{unconscious mind}.$$

This proposal is also inspired by some old but intriguing planaria learning through cannibalism experiments which were greeted with skepticism by the scientific community but have just received some confirmation. Please see the following papers:

"Planaria: Memory transfer through cannibalism reexamined"
Science **146**, 9 October 1964: 274–275.

"An automated training paradigm reveals long-term memory in planarians and its persistence through head regeneration"
J. Exp. Biol. **15**, October 2013: 3799–3810.

7.2. *Is the brain a two-level system?*

Here then is our two-level proposal:

- Conscious, Rational, Serial, Sensual Front-End Mind:
  Neurons (Fast)
- Unconscious, Intuitive, Parallel, Combinatorial Back-End Mind:
  Molecular Biology (Slow)
  (much greater compute and memory capacity)

This proposal was also inspired by three other works, which we now list.

On the hugeness of human memory capacity and the fact that people seem to remember everything even if they can't recall it, see *Borges and Memory* by Rodrigo Quian Quiroga.

See also *Dissertatio de arte combinatoria*, an early work by Leibniz (1666). Following Llull, the *Dissertatio* considers intelligence and the power of invention as the ability to quickly search through all possible combinations of a set of ideas.

And see Jacques Hadamard, *The Psychology of Invention in the Mathematical Field* (1945). This book emphasizes the fundamental role of the unconscious in mathematical creation. Having a problem in your research? Sleep on it! Consult with your pillow!

On the basis of these readings, we tentatively conclude that the processing power and memory capacity of the unconscious mind is much greater than that of the conscious mind.

7.3. *Discussion of the model*

After all, the immune system does information processing at the molecular level. If the brain worked only at the neuronal level, for example by storing one bit per neuron, it would have roughly the capacity of a pen drive, far too low to account for human intelligence. But at the RNA/DNA molecular biology level, the total information capacity is quite immense.

In the life of a research mathematician it is frequently the case that one works fruitlessly on a problem for hours then wakes up the next morning with many new ideas. The intuitive mind has much, much greater information processing capacity than the rational mind. Indeed, it seems capable of exponential search.

We can connect the two levels postulated here by having a unique molecular "name" correspond to each neuron, for example to the proverbial "grandmother cell." In other words, we postulate that the unconscious "mirrors" the associations represented in the connections between neurons. Connections at the upper conscious level correspond at the lower unconscious level to enzymes that transform the molecular name of one neuron into the molecular name of another. In this way, a chemical soup can perform massive parallel searches through chains of associations, something that cannot be done at the conscious level.

When enough of the chemical name for a particular neuron forms and accumulates in the unconscious, that neuron is stimulated and fires, bringing the idea into the conscious mind.

And long-chain molecules can represent memories or sequences of words or ideas, i.e., thoughts.

To end this discussion, I should repeat that whether or not the human brain actually works this way, it seems to me that this is a design for a brain that just possibly might be made to work. At any rate, I think it is important to escape from the current cul-de-sac and propose new lines of research.

8. The Mystery of Consciousness

We talked about consciousness above. Now we discuss some ideas about consciousness put forth by other authors, ideas that are totally unrelated to — and perhaps even contradict — the brain design speculations that were just presented.

I once fell asleep at the wheel while driving, then woke up and wondered where I was and "who" had continued driving while my conscious mind was asleep. Fortunately *someone* had continued driving!

According to Christof Koch's book *Consciousness*, the level of my brain that continued driving was **also** conscious. Similarly, Koch believes the immune system must be conscious. In fact, he believes that everything is conscious; this is called *panpsychism*. What's important is the degree of consciousness.

According to Giulio Tononi's book *PHI*, consciousness can be measured in terms of the integrated information Φ. (Tononi's Φ is also discussed in Koch's book.) The greater a system's Φ, the more integrated and conscious it is. A binary switch has one bit of consciousness.

I suspect Φ has something to do with what in algorithmic information theory is called *mutual information*

$$H(X:Y) = H(X) + H(Y) - H(X,Y),$$

which is the extent to which X and Y are simpler when seen together than when seen separately.

In a 1979 paper "Toward a mathematical definition of 'life'," I discuss using the mutual information of the parts of a system to determine the extent to which the parts are integrated into a whole.

These ideas might also help with the question of "the creation of the self:" what it means to isolate a certain physical system from the rest of the world and declare it to be a unity. How do we decompose the world into things? How do I distinguish you from me? And how can we define this mathematically?

9. Digital Ontology

We have discussed epistemology (of physics and mathematics), evolution, the brain and consciousness. Our final topic is ontology.

9.1. *The new Pythagoreans*

The digital, algorithmic approach inspired by the computer suggests that everything is discrete, everything is 0's and 1's. The continuum is deemphasized, dethroned. Is the world a computer?

Instead of having a world built of number (Pythagoras) we now have a world built of information, a **new substance**. This view of the world is also known as *digital philosophy*, and other practitioners are Stephen Wolfram and Edward Fredkin. See Longo and Vaccaro, *Bit Bang: La nascita della filosofia digitale.*

9.2. *Digital physics*

There are also hints from physics.

The so-called "holographic principle" has emerged in the field of quantum gravity, as a consequence of results on the thermodynamics and entropy of black holes. According to this principle, the total number of bits of information in any physical system is finite, and moreover grows only as the surface area of its boundary, not with its volume.

These tentative results — quantum gravity is a relatively new field — suggest both that the physical world is in some sense discrete, and that perhaps at some level space is more two-dimensional than three-dimensional. See Lee Smolin, *Three Roads to Quantum Gravity.*

9.3. *The halting probability Ω*

In order to continue our discussion of ontology we have to define the number Ω and say a few words about why it is so interesting. Ω is the total probability of all the self-contained (without any input) programs p that eventually halt, assuming that the bits of p are picked using independent tosses of a fair coin:

$$\Omega = \sum_{p \text{ halts}} 2^{-|p|} = \sum_{U(p) \text{ halts}} 2^{-(\text{size in bits of program } p)}.$$

This infinite sum defines Ω, but does not enable us to calculate its numerical value, because in fact Ω is wildly, extravagantly uncomputable.

Technical Note: For this infinite sum to converge and be between zero and one it is essential that we stipulate that the programs p for our universal computer U be "self-delimiting" so that U knows by itself where to stop reading the bits of p and no extension p' of a p that halts is ever included in the sum for Ω. Otherwise this sum will diverge to infinity. For a full explanation of this please consult any treatise on algorithmic information theory.

Imagine writing the numerical value of Ω in base-two notation. This would give us its binary expansion, for example,

$$\Omega = 0.01110\ldots$$

Knowing the first N bits of this expansion would enable us to discover all the programs up to N bits in size that ever halt, see what output they produce before halting, and output something different. Therefore the algorithmic information content of the first N bits of the binary expansion of Ω is greater than $N - c$:

$$H(\text{first } N \text{ bits of } \Omega) > N - c.$$

Leibniz in his *Monadology*, Sections 33–35, says that mathematical proof consists in analyzing a complex statement into the consequence of simpler statements, continuing this analysis until one reaches statements that are so simple that their truth is self-evident and no longer requires any proof — otherwise there is an infinite regress.

However, it takes an N-bit theory — a formal axiomatic theory whose conceptual complexity is N — for us to be able to determine N bits of Ω. To be able to prove what are the precise values of N bits of the binary expansion of Ω, we must use a formal axiomatic theory which itself has at least N bits of complexity.

Therefore the precise values of the successive bits of Ω are *irreducible mathematical facts*, mathematical facts that are true for no reason, more precisely, true for no reason simpler than themselves. They thus violate Leibniz's *principle of sufficient reason* and seem to be *contingent* rather than *necessary* truths.

Does the number Ω exist? Does anything with infinite complexity exist? If we can never know something, why should we believe it exists? Well, we **can** prove lovely theorems about Ω. For example, Ω is Borel normal, which means that in the binary expansion for

Ω, in the limit we encounter 0's and 1's with exactly the same relative frequency. And hopefully Ω gives us some intuition about more down-to-earth situations and how things may behave in somewhat more realistic contexts. Remember, the world of pure mathematics is much simpler and much more understandable than the messy real world!

All by itself, Ω shows that pure mathematics is biological. Because one of the defining characteristics of the field of biology — as opposed to the field of theoretical physics which is based on equations — is that there are few simple organizing principles in biology. In other words, biology is very complicated. That may be so, but Ω is infinitely complicated, and therefore, in a sense, more biological than biology itself.

Furthermore, the Platonic world of pure mathematics has infinite conceptual complexity, whilst if Hilbert had been correct, the world of pure mathematics would only have finite complexity, namely the complexity of the universal consistent, complete formal axiomatic theory that Hilbert was searching for. Mathematics is open-ended, creative, not closed as Hilbert thought. And this makes mathematical truth more tentative and not as totally black or white as Hilbert thought it would be.

This complexity-based point of view also emphasizes the similarities between pure mathematics and theoretical physics rather than the differences, and suggests, as I have argued elsewhere, a *quasi-empirical* view of mathematics. Mathematics and physics may be different, but they are not **that** different — at least not from an information-theoretic point of view.

10. Summary

We have used the program size $H(T)$ to measure the *conceptual complexity* of a physical or a mathematical theory T. The fact that you can never prove that you have the simplest physical theory yields a new version of Gödel's theorem on the limitations of formal axiomatic mathematical theories.

In modeling Darwinian evolution, we have used *relative* program size $H(A'|A)$ to measure the conceptual complexity of an $A \to A'$ mutation. We can prove that our model evolves. We have a Pythagorean world in which life provably evolves.

What about the brain and consciousness? Following von Neumann, maybe a computer engineering approach is relevant. And following Tononi and Koch, maybe *mutual* program size complexity $H(X:Y)$ is related to *integrated information* as measured by Φ.

We now have a new fundamental substance, information, that comes together with a digital world-view.

And — most ontological of all — perhaps with the aid of these concepts we can begin again to view the world as consisting of both mind and matter. The notion of mind that perhaps begins to emerge from these musings is mathematically quantified, which is why we declared at the start that this essay pretends to take additional steps in the direction of a mathematical form of philosophy.

The eventual goal is a more precise, quantitative analysis of the concept of "mind." Can one measure the power of a mind like one measures the power of a computer?

Acknowledgment

The author is grateful to the Brazilian government for supporting this research with a CAPES PVE grant.

Further Reading

1. G. J. Chaitin, *Meta Math!*, Pantheon, 2005.
2. G. J. Chaitin, *Thinking about Gödel and Turing*, World Scientific, 2007.
3. N. Shubin, *Your Inner Fish*, Pantheon, 2008.
4. G. J. Chaitin, *Matemáticas, Complejidad y Filosofía / Mathematics, Complexity and Philosophy* (bilingual Spanish/English edition), Midas, 2011.

5. G. Chaitin, N. da Costa, F. A. Doria, *Gödel's Way*, CRC Press, 2012.
6. G. J. Chaitin, *Proving Darwin*, Pantheon, 2012.
7. S. C. Meyer, *Darwin's Doubt*, HarperOne, 2013.
8. P. J. Bowler, *Darwin Deleted*, Chicago University Press, 2013.
9. J. C. Venter, *Life at the Speed of Light*, Viking Adult, 2013.
10. G. O. Longo, A. Vaccaro, *Bit Bang. La nascita della filosofia digitale*, Maggioli, 2013.
11. G. Chaitin, V. M. F. G. Chaitin, F. S. Abrahão, "Metabiología: los orígenes de la creatividad biológica," *Investigación y Ciencia* **448**, January 2014: 74–80.
12. S. Pääbo, *Neanderthal Man*, Basic Books, 2014.

Chapter 4

Chaitin, The Thinker

Richardo Espinoza Lolas*

"Without mathematics we cannot penetrate deeply into philosophy. Without philosophy we cannot penetrate deeply into mathematics. Without both we cannot penetrate deeply into anything"

— Leibniz

"The following sentence is false. The preceding sentence is true"

— Russell

Today, more than ever, we are in the presence of a paradox: not any paradox, but "the" paradox. Humankind cannot progress along a clear and distinct pathway, as Descartes, Laplace and the entire rational European Modernity dreamed of; there is not a single destiny predetermined forever, there is no equation that bluntly determines the sense of reality; we are paralyzed in the presence of the paradox, be it in a disciplinary manner, since sciences, humanities, arts cannot do research in full homogeneous fields; reality is fissured, with a high porosity, how can one think about Russell's paradox? Is it possible to think about the paradox? What is thinking? Who is a thinker?... Chaitin is a thinker!

The paradox is written in terms of Lewis Caroll's literature and it is philosophically thought, for instance, from Deleuze as: "When I say 'Alice becomes larger,' I mean that she becomes larger than she was. By the same token, however, she becomes smaller than she is

*Pontificia Universidad Católica de Valparaíso, Chile; ricardo.espinoza@pucv.cl.

now. Certainly, she is not bigger and smaller at the same time. She is larger now; she was smaller before. But it is at the same moment that one becomes larger than one was and smaller than one becomes" (Deleuze, *The Logic of Sense*). How can one think, mathematically, the paradox?

We are in waters of the becoming, of the event, of evolution and its processes; we are in the flow and we move in the flow. How can we think this? In this becoming, in this event, we elude the present, the steady, the dogmatic position, the belief in itself, in the typical abstract representation of the outside of the system, in its most extreme coming into presence of indifferent showing up. How can one elude the present of a scientific research, though in systems other than continuous systems, with tools other than the ones of the clear and distinct scope of the aged European science? Scientific research is done in the residues and the edges of the present (far away from the equilibrium, far away from an abstract fixed position); in its powers, this flow cannot remain under the eaves of the simple homogenizing linearity of before and after. Not only Greece and Aristotle failed with their metaphysical project, but also sameness from the universal homogeneity, that has ruled for centuries, has been perforated by pure events, by potentialities, by folds, by fissures and cracks, by crossings and ligatures. The paradox in the paradox itself is the difference of the discrete that constitutes the continuous as continuous that has always been the one that has ruled and has globalized everything the empire of continuous sameness has fallen!... Reality is porous and has folds...

In *The Logic of Sense*, Deleuze reminds us of Lewis and his Alice: "It pertains to the essence of becoming to move and to pull in both directions at once: Alice does not grow without shrinking and vice versa". That is, falseness only "falsifies" in the truth; truth only "truthifies" in falseness. Power of the falseness!... This crossing of the paradox, that in order to be solved pulls in both directions at once, is arises and qualitativeness comes into presence; this causes the leap to start once again, but differently, and, for instance, allows evolution because allows the becoming of different species.

Moonlight! That's what we are, a moment that takes place, a differential that is expressed in a manner extremely different from that of the homogeneous animality. How can one think, metaphorically speaking, that beautiful power of the falseness of the moonlight itself? There is always the good sense, that is, the dogmatic sense of the realm of sameness and unilateralism tells us that it is this or that, but never this and that "at the same time"; that's why, according to Nietzsche, good sense knows nothing of humankind as "the sense of the earth", that is, of a bodily men, with intensities, feelings, projects that don't let themselves get caught nor let themselves get chained to aged European rationalities. The paradox is the affirmation of the two senses at the same time, like Escher's (1945) hands that shake one another's hands.

But, how can pure being and pure nothing be, therefore, the same? wondered Hegel at the beginning of his book *Science of Logic*. That's another paradox that scandalizes non-Deleuzian Hegelian purist and non-Hegelian Deleuzian purists. If we play a game, it would go like this. "pure Hegel's thought and pure Deleuze's thought is, therefore, the same", that is, the pure No of the dead Hegelian's dialectics and the pure Yes of the affirmed Deleuzian's life is, therefore, movement, flow, but how to think this event?

Is there any way to think this paradox, to think life itself that moves forward, both upwards and, at the same time, downwards?... Eternal Yes of being! Uttered Nietzsche in one of his dithyrambs. How can we find a kind of instrument to measure, articulate, organize, relate what cannot be tamed, because it is the "wildest horse" that we will find? How can one think that which forces us to think, that which in its fugacity, in its randomness, requires us with necessity? How can we think that which is irreducible to thought, that which is incompressible since all compression escapes through the fissures, through the holes, through the monads that cannot be captured? How can we think what is "de suyo" unthinkable, what is "de suyo" incompressible? How can we think the "outside" itself of very thought? That's what Gödel thought in mathematics. He thought the "outside" itself of mathematics in his paradox *par excellence*.

Given a very personal, human, all too human matter, Heidegger indicated us in 1952 that science does not think, science only calculates, relates, measures. Science, as a mere norm establishes to each its own and sorts it out in the shelf of what is socio-historical at that time, that is why science cannot think change, innovation, difference, qualitativeness, emergence, what is different, the differential itself which potentiates every difference that raises from the horizon of continuity. Science used to work as the dispositive power that has everything regulated in its showing up (according to Foucault, the presence of the dispositive, its empowerment, is related to the institution) that is why science cannot think change, the leap into otherness; science is always moving in what is steady, in what is continuous homogeneous where everything is already determined. How can science think? Apparently, more than ever, these Heideggerian words make sense to us, since in science there is nothing thought anymore, even less in the paradox itself, because everything has become applied science, under application protocols with fast results financed by the institution in office that works in the political apparatus of the moment; apparatus that urges us, demands us, controls us, watches over our researches; the apparatus that keeps our research all to themselves. It is true that science as an institution is not thought anymore, but is merely calculations.

We think that Heidegger's statements regarding sciences, as he knew them at his time, were right; however, mathematics is different because there is thought in mathematics, it is not just a matter of mere calculations. Gödel (1931) indicated it, Turing (1936) confirmed it and Chaitin by the mid-60s with his "Algorithmic Information Theory" places us on the pathway of mathematics thought (and in the best part of science), on the pathway of mathematics, that by thinking itself it thinks itself from the unthoughtness itself, from its paradox, from its "outside". I've said "it thinks itself?" but, when did we leap to think the paradox?

But the event has already taken place; the paradox flows, the hands "shake hands" with one another, and the paradox is becoming in the becoming of this thinking from the "outside" itself of its thinking. That are just a few who can think and not merely do calculations. The mathematical thinker is not just a mathematician, but is formally a thinker. Because they have to think beyond the grounds of their discipline, they have to think from the paradox that constitutes their knowledge. That is why the thinker who thinks their discipline always new territory not yet discovered. The thinker is always creator, is a dancer who dances above the homogeneous territory of the already worn discipline and that give nothing of itself anymore, because it is exhausted. In this sense, the thinker is always denigrated, unloved, is always banished from the territory of the discipline, he is an exiled, an outcast looking for a new homeland where he can create and develop his thought. Hence, the thinker is a loner who seeks to be heard, who travels, who is always on the move, because he wishes to find the place where he can be recognized.

Chaitin thinks!... Is this clear enough?... How can one compress what "de suyo" in incompressible in Chaitin's thought? How to think, from what has not yet been thought from the discipline itself? How can the circle be squared? How can the piece or wood be ironed? Is this a matter only for mathematicians and their mathematics? I don't think so. This is a matter of thinkers and their thoughts, of those scarce thinkers around the world that, as passersby and with only one eye, like Wotan, can see beyond what anyone can see within the confines of their view, that is, paradoxically, in and due to the paradox itself, they can see beyond than any simple mortal can see and those men are the "Centaurs creators and dancer stars". Those are the ones who are in the crossings and flow ligatures in the discrete monads of their thoughts; their thoughts are like monads that compress information and when in this compressing they can stand on the shoulders of other thinkers and their systems of thought.

Chaitin is one of those unusual beings; he is one of those few thinkers. Chaitin, the thinker! The OMEGA number is his Centaur. How many Centaur creators do you know? Maybe favored by the gods. I have know a few ones; just a few, of those who are Atlases supporting the mountains of sameness, of what is homogeneous. Chaitin is one of them. Chaitin is one of those great titans such as Boltzmann and his works in thermodynamics. ENTROPY was Boltzmann's Centaur.

Chaitin has taught us to, what in his words is, "measure computational complexity", but deep onside he has taught us to think. In his search to measure programs, measure computational program sizes, or better, in his attempt to measure information, measure complexity, Chaitin comes across Zarathustra at an altitude above 6000 feet beyond time and beyond the man merely man of the indifferent abstract City of what is universal, dominant and ruling, in that present city that has became institution where nothing takes place, from where nothing emerges, where there is no innovation. How much information is needed for such and such program to be executed? How much history is needed to know if this is true? Is it possible to get to know this? How do we know this, how do we think it? How to think freedom? — the paradox of paradoxes or we might as well change Chaitin's question to the Nietzsche-Greek question, how can you become such as you are? Is it possible to think it? Is it possible to think it right now?

What is it gas at room temperature? What is it with the size of a computer program? How can we know this? How can we think this? What program could "think" where all the gas atoms are? It is madness! We know what happened knowing it is unthinkable, it is part of the paradox, because it cannot be compressed, theorized, regulated, thought.

That is the question! That is the number! That is the OMEGA! That is the incompressible!, That is the unthinkable! The paradox itself of the becoming of the number since numbers for being such are already understood, compressed, thought; but are on the run, adrift, residually, in blurriness, on the ground, not on heavens anymore. There's randomness!

Stating the above sentence is the paradox itself, the incompressible it does compress! Thought, like a pragmatism, like a deictic *a posteriori*. Like a performative gesture, but how can this happen in the *a priori* mathematics? This is outrageous to many! But this is the life of numbers; they are living beings, Tarkovsky's mirrors. Numbers in experience, numbers becoming, flowing, "towards", according to Zubiri, a scandal?...

Chaitin, the scandalizer! Chaitin, the mathematics heretic! Chaitin, the guilty one! Chaitin, the demolisher! Chaitin, the new Socrates of mathematics! Greg, watch out the community of the court of the scientists of the contemporaneous Athens, they might chase you, and they surely do, because they fear you. They will look to have you exiled, to have you drink hemlock, but they won't succeed, as a matter of fact they haven't you are still there!...

Chaitin, like every thinker, becomes nomad and lives in rhizomes, that's his "Moira" and what has come to him for being a Centaur creator.

The complexity of your thought, which has looked for measuring thinking the "wild horse", the randomness, chases you. If the complexity of something is measured, is thought, in terms of the minimum size of a computer's program that allows calculate it, then how can we know what is the minimum program, that is, how can we indicate here and now its inherent complexity?... My words won't reach you, we might need a kind of shot of your Yasujiro Ozu, which you are very fond of, in order to measure your inherent complexity, is it possible to indicate the degree of brightness of your thought? "Well, No!... it is impossible!" can only indicate it.

There's Gregory Chaitin... The thinker!

Chapter 5

Dropping In on Gottfried Leibniz

Stephen Wolfram*

I've been curious about Gottfried Leibniz for years, not least because he seems to have wanted to build something like *Mathematica* and Wolfram|Alpha, and perhaps *A New Kind of Science* as well — though three centuries too early. So, when I took a trip recently to Germany, I was excited to be able to visit his archive in Hanover.

Leafing through his yellowed (but still robust enough for me to touch) pages of notes, I felt a certain connection — as I tried to imagine what he was thinking when he wrote them and tried to relate what I saw in them to what we now know after three more centuries:

*Wolfram Research, USA; sw-staff@wolfram.com.

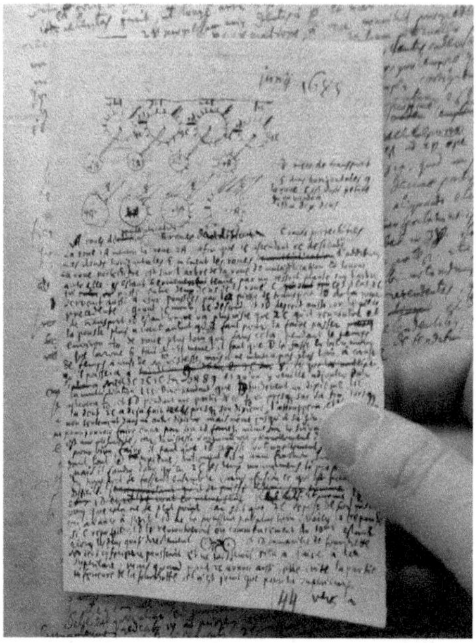

Some things, especially in mathematics, are quite timeless. Like here's Leibniz writing down an infinite series for $\sqrt{2}$ (the text is in Latin):

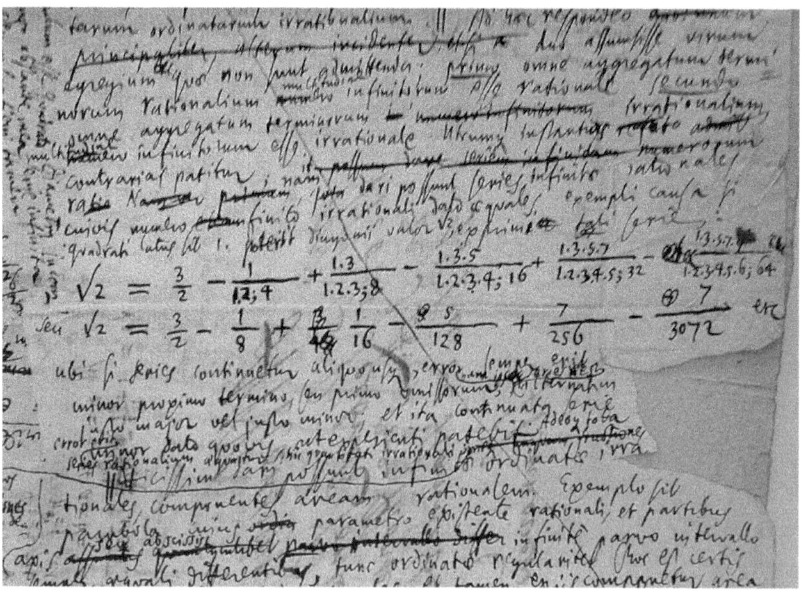

Or here's Leibniz tried to calculate a continued fraction — though he got the arithmetic wrong, even though he wrote it all out (the Π was his earlier version of an equal sign):

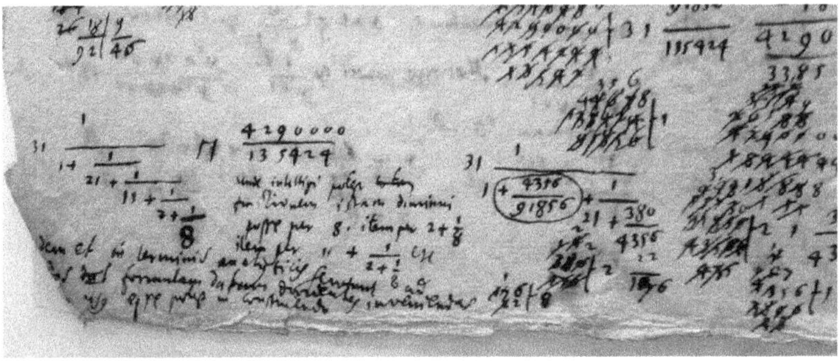

Or here's a little summary of calculus, that could almost be in a modern textbook:

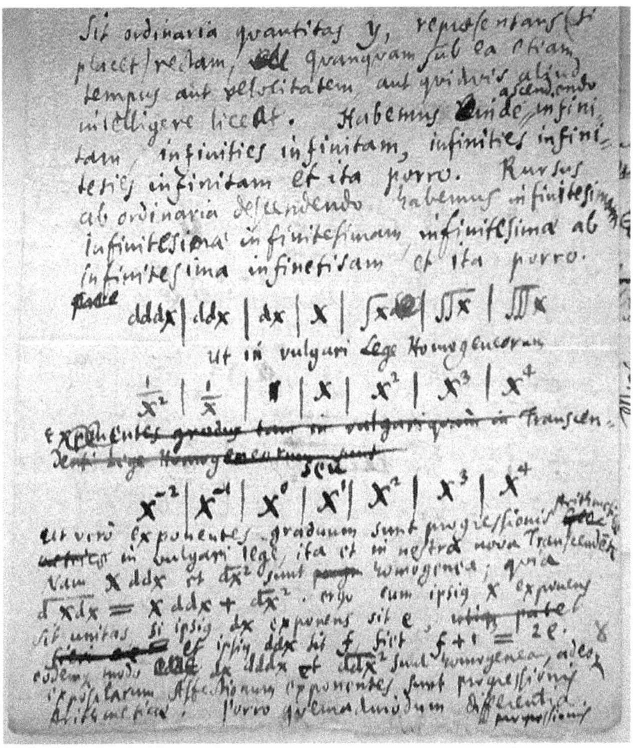

But what was everything else about? What was the larger story of his work and thinking?

I have always found Leibniz a somewhat confusing figure. He did many seemingly disparate and unrelated things — in philosophy, mathematics, theology, law, physics, history, and more. And he described what he was doing in what seem to us now as strange 17th century terms.

But as I've learned more, and gotten a better feeling for Leibniz as a person, I've realized that underneath much of what he did was a core intellectual direction that is curiously close to the modern computational one that I, for example, have followed.

Gottfried Leibniz was born in Leipzig in what's now Germany in 1646 (four years after Galileo died and four years after Newton was born). His father was a professor of philosophy; his mother's family was in the book trade. Leibniz's father died when Leibniz was 6 — and after a 2-year deliberation on its suitability for one so young, Leibniz was allowed into his father's library and began to read his way through its diverse collection of books. He went to the local university at age 15, studying philosophy and law, and graduated in both of them at age 20.

Even as a teenager, Leibniz seems to have been interested in systematization and formalization of knowledge. There had been vague ideas for a long time — for example, in the semi-mystical *Ars Magna* of Ramon Llull from the 1300s — that one might be able to set up some kind of universal system in which all knowledge could be derived from combinations of signs drawn from a suitable (as Descartes called it) "alphabet of human thought". And for his philosophy graduation thesis, Leibniz tried to pursue this idea. He used some basic combinatorial mathematics to count possibilities. He talked about decomposing ideas into simple components on which a "logic of invention" could operate. And, for good measure, he put in an argument that purported to prove the existence of God.

As Leibniz himself said in later years, this thesis — written at age 20 — was in many ways naive. But I think it began to define Leibniz's lifelong way of thinking about all sorts of things. And so, for example, Leibniz's law graduation thesis about "perplexing legal cases" was all about how such cases could potentially be resolved by reducing them to logic and combinatorics.

Leibniz was on a track to become a professor, but instead he decided to embark on a life working as an advisor for various courts and political rulers. Some of what he did for them was scholarship, tracking down abstruse — but politically important — genealogy and history. Some of it was organization and systematization — of legal codes, libraries and so on. Some of it was practical engineering — like trying to work out better ways to keep water out of silver mines. And some of it — particularly in earlier years — was "on the ground" intellectual support for political maneuvering.

One such activity in 1672 took Leibniz to Paris for four years — during which time he interacted with many leading intellectual lights. Before then, Leibniz's knowledge of mathematics had been fairly basic. But in Paris he had the opportunity to learn all the latest ideas and methods. And for example, he sought out Christiaan Huygens, who agreed to teach Leibniz mathematics — after he succeeded in passing the test of finding the sum of the reciprocals of the triangular numbers.

Over the years, Leibniz refined his ideas about the systematization and formalization of knowledge, imagining a whole architecture for how knowledge would — in modern terms — be made computational. He saw the first step as being the development of an *ars characteristica* — a methodology for assigning signs or symbolic representations to things, and in effect creating a uniform "alphabet of thought". And he then imagined — in remarkable resonance with what we now know about computation — that from this uniform representation it would be possible to find "truths of reason in any field... through a calculus, as in arithmetic or algebra".

He talked about his ideas under a variety of rather ambitious names like *scientia generalis* ("general method of knowledge"), *lingua philosophica* ("philosophical language"), *mathematique universelle* ("universal mathematics"), *characteristica universalis* ("universal system") and *calculus ratiocinator* ("calculus of thought").

He imagined applications ultimately in all areas — science, law, medicine, engineering, theology and more. But the one area in which he had clear success quite quickly was mathematics.

To me, it's remarkable how rarely in the history of mathematics notation has been viewed as a central issue. It happened at the beginning of modern mathematical logic in the late 1800s with the work of people like Gottlob Frege and Giuseppe Peano. And in recent times it's happened with me in my efforts to create *Mathematica* and the Wolfram Language. But it also happened three centuries ago with Leibniz. And I suspect that Leibniz's successes in mathematics were in no small part due to the effort he put into notation, and the clarity of reasoning about mathematical structures and processes that it brought.

When one looks at Leibniz's papers, it's interesting to see his notation and its development. Many things look quite modern. Though there are charming dashes of the 17th century, like the occasional use of alchemical or planetary symbols for algebraic variables:

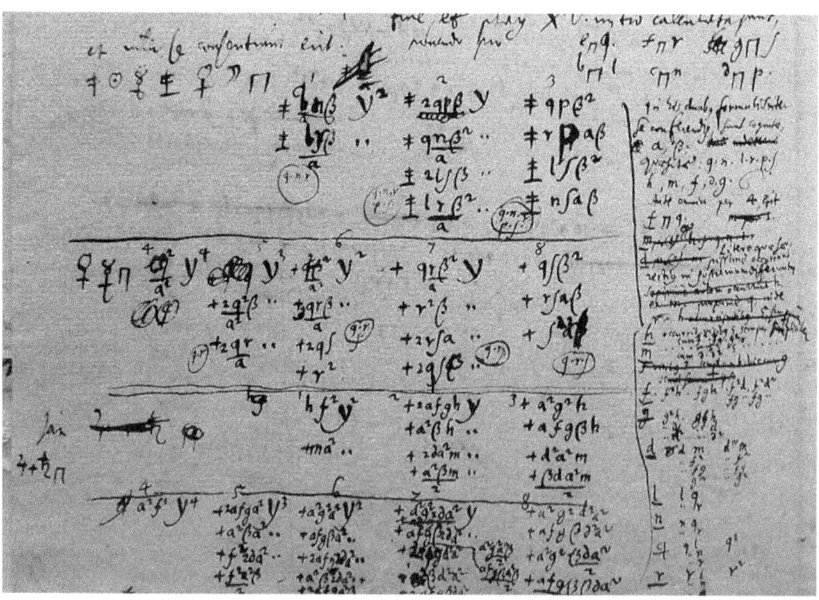

There's Π as an equals sign instead of =, with the slightly hacky idea of having it be like a balance, with a longer leg on one side or the other indicating less than ("<") or greater than (">"):

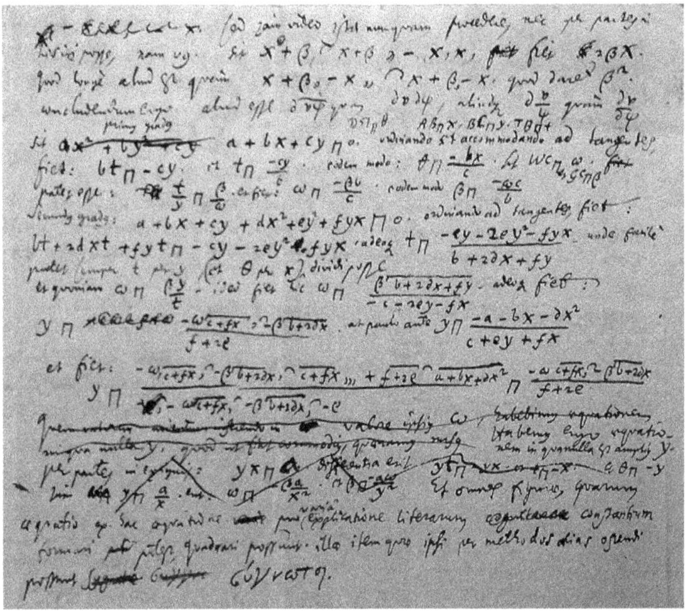

There are overbars to indicate grouping of terms — arguably a better idea than parentheses, though harder to type, and typeset:

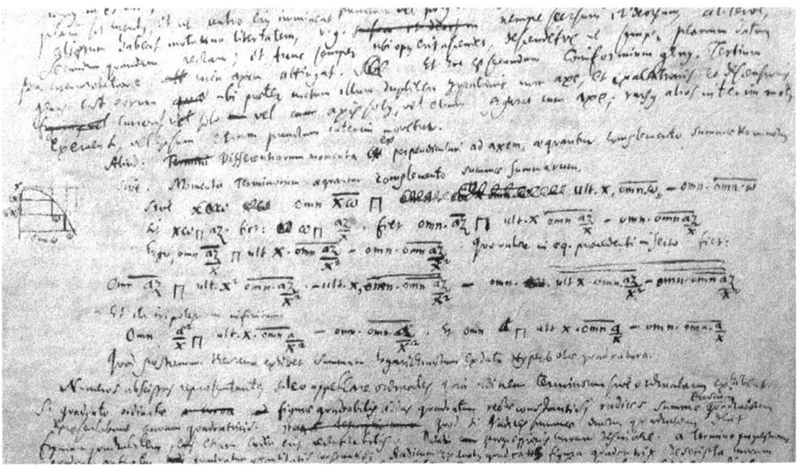

We do use overbars for roots today. But Leibniz wanted to use them in integrals too. Along with the rather nice "tailed d", which reminds me of the double-struck "differential d" that we invented for representing integrals in *Mathematica*.

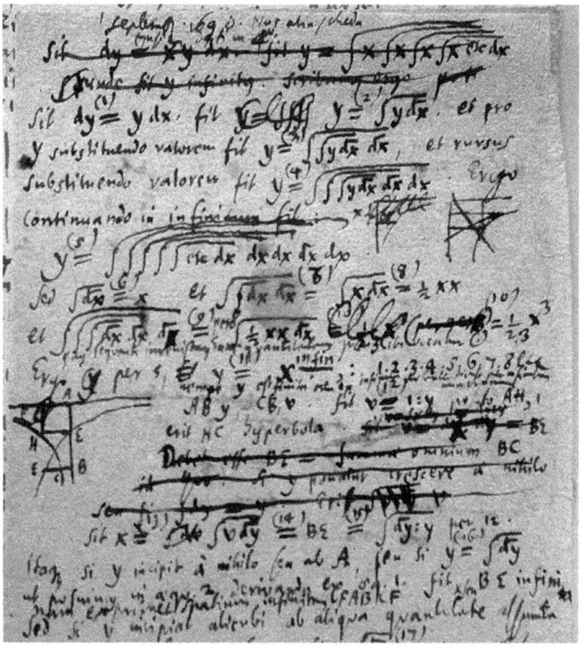

Particularly in solving equations, it's quite common to want to use $\pm$, and it's always confusing how the grouping is supposed to work, say in $a \pm b \pm c$. Well, Leibniz seems to have found it confusing too, but he invented a notation to handle it — which we actually should consider using today too:

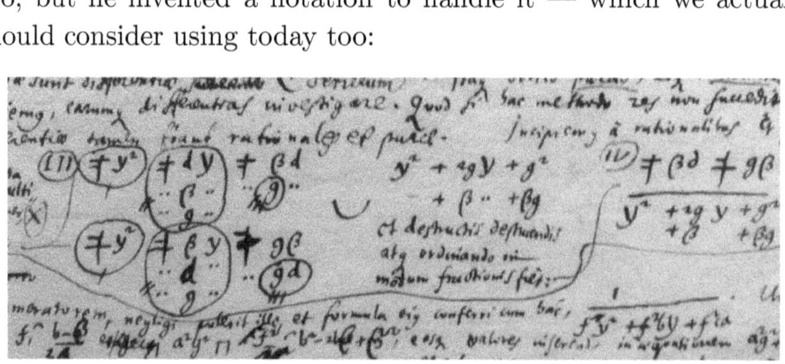

I'm not sure what some of Leibniz's notation means. Though those overtildes are rather nice-looking:

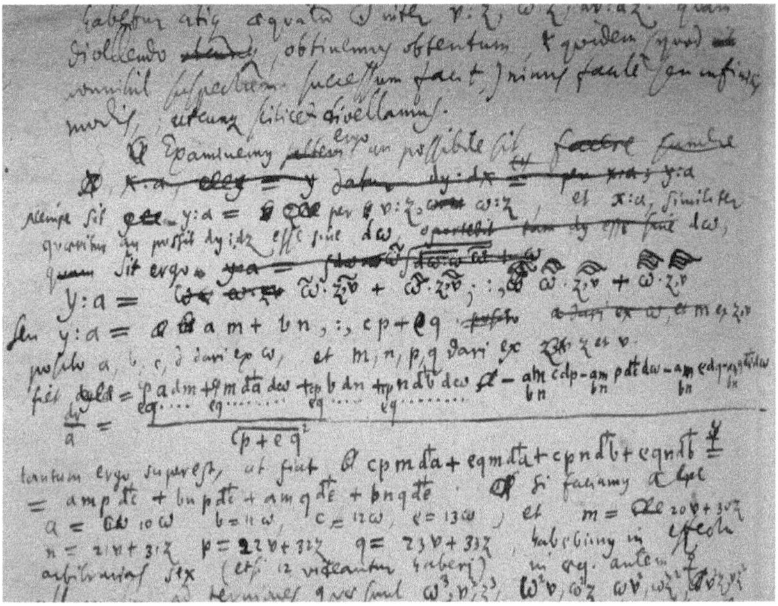

As are these things with dots:

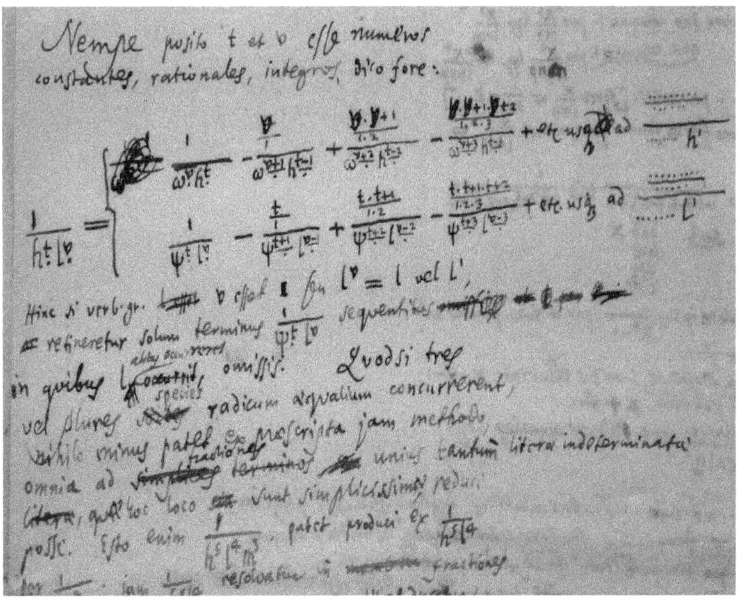

Or this interesting-looking diagrammatic form:

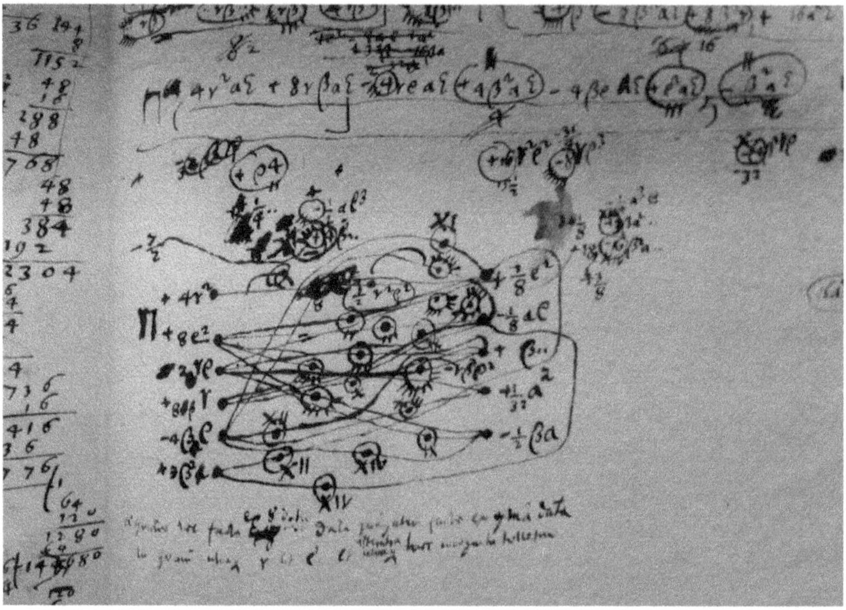

Of course, Leibniz's most famous notations are his integral sign (long "s" for "summa") and d, here summarized in the margin for the first time, on November 11th, 1675 (the "5" in "1675" was changed to a "3" after the fact, perhaps by Leibniz):

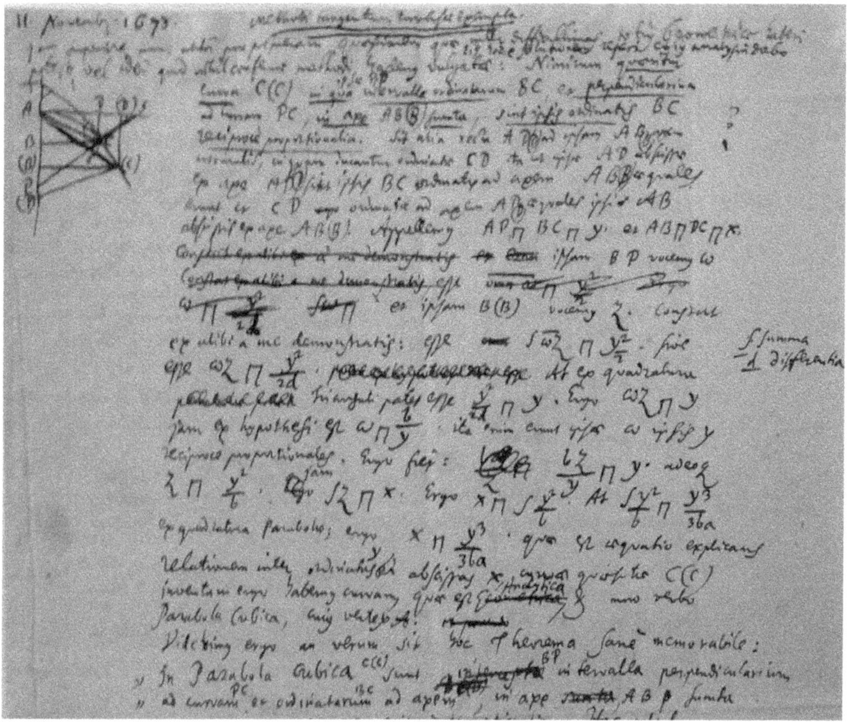

I find it interesting that despite all his notation for "calculational" operations, Leibniz apparently did not invent similar notation for logical operations. "Or" was just the Latin word *vel*, "and" was *et*, and so on. And when he came up with the idea of quantifiers (modern ∀ and ∃), he just represented them by the Latin abbreviations U.A. and P.A.:

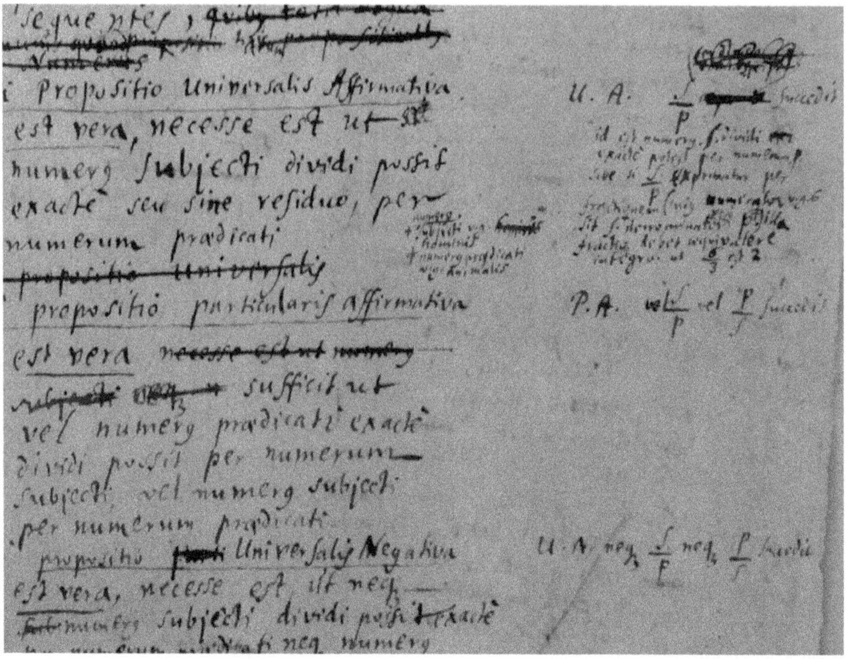

It's always struck me as a remarkable anomaly in the history of thought that it took until the 1930s for the idea of universal computation to emerge. And I've often wondered if lurking in the writings of Leibniz there might be an early version of universal computation — maybe even a diagram that we could now interpret as a system like a Turing machine. But with more exposure to Leibniz, it's become clearer to me why that's probably not the case.

One big piece, I suspect, is that he didn't take discrete systems quite seriously enough. He referred to results in combinatorics as "self-evident", presumably because he considered them directly verifiable by methods like arithmetic. And it was only "geometrical", or continuous, mathematics that he felt needed to have a calculus developed for it. In describing things like properties of curves, Leibniz came up with something like continuous functions. But he never seems to have applied the idea of functions to discrete mathematics — which might for example have led him to think about universal elements for building up functions.

Leibniz recognized the success of his infinitesimal calculus, and was keen to come up with similar "calculi" for other things. And in another "near miss" with universal computation, Leibniz had the idea of encoding logical properties using numbers. He thought about associating every possible attribute of a thing with a different prime number, then characterizing the thing by the product of the primes for its attributes — and then representing logical inference by arithmetic operations. But he only considered static attributes — and never got to an idea like Gödel numbering where operations are also encoded in numbers.

But even though Leibniz did not get to the idea of universal computation, he did understand the notion that computation is in a sense mechanical. And indeed quite early in life he seems to have resolved to build an actual mechanical calculator for doing arithmetic. Perhaps in part it was because he wanted to use it himself (always a good reason to build a piece of technology!). For despite his prowess at algebra and the like, his papers are charmingly full of basic (and sometimes incorrect) school-level arithmetic calculations written out in the margin — and now preserved for posterity:

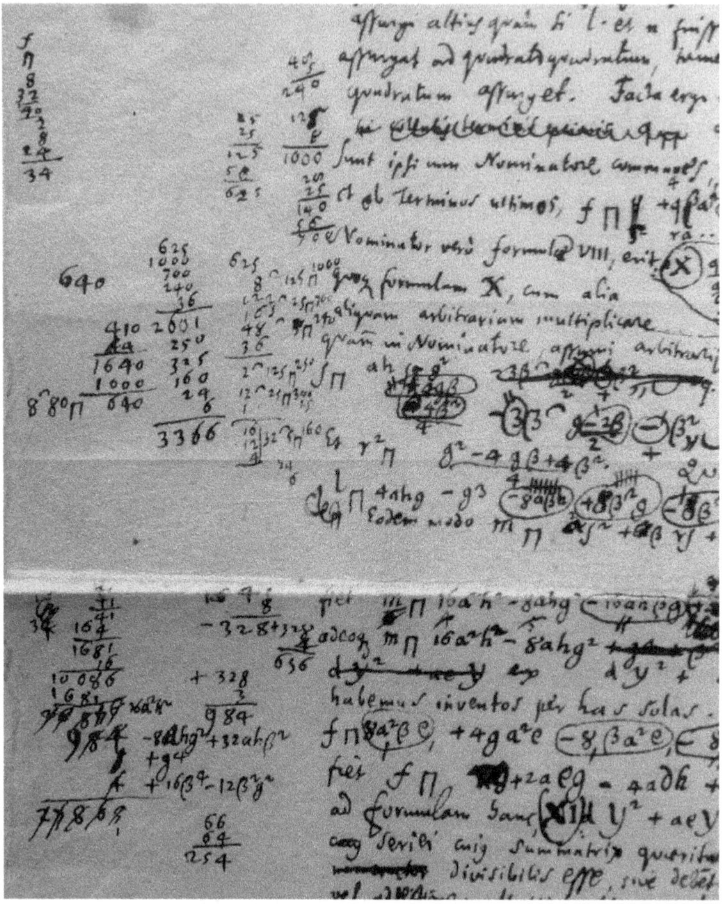

There were scattered examples of mechanical calculators being built in Leibniz's time, and when he was in Paris, Leibniz no doubt saw the addition calculator that had been built by Blaise Pascal in 1642. But Leibniz resolved to make a "universal" calculator, that could for the first time do all four basic functions of arithmetic with a single machine. And he wanted to give it a simple "user interface", where one would for example turn a handle one way for multiplication and the opposite way for division.

In Leibniz's papers there are all sorts of diagrams about how the machine should work:

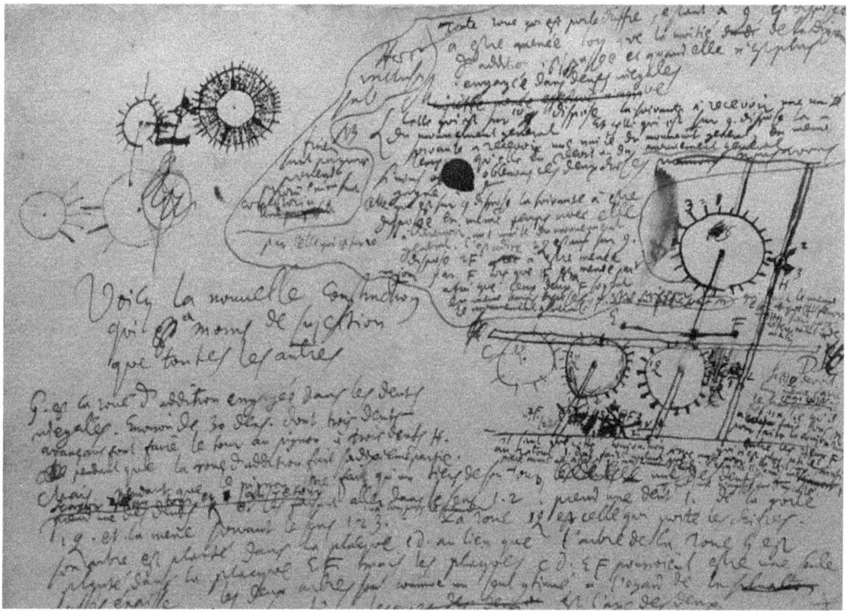

Leibniz imagined that his calculator would be of great practical utility — and indeed he seems to have hoped that he would be able to turn it into a successful business. But in practice, Leibniz struggled to get the calculator to work at all reliably. For like other mechanical calculators of its time, it was basically a glorified odometer. And just like in Charles Babbage's machines nearly 200 years later, it was mechanically difficult to make many wheels move at once when a cascade of carries occurred.

Leibniz at first had a wooden prototype of his machine built, intended to handle just 3 or 4 digits. But when he demoed this to people like Robert Hooke during a visit to London in 1673, it didn't go very well. But he kept on thinking he'd figured everything out — for example in 1679 writing (in French) of the "last correction to the arithmetic machine":

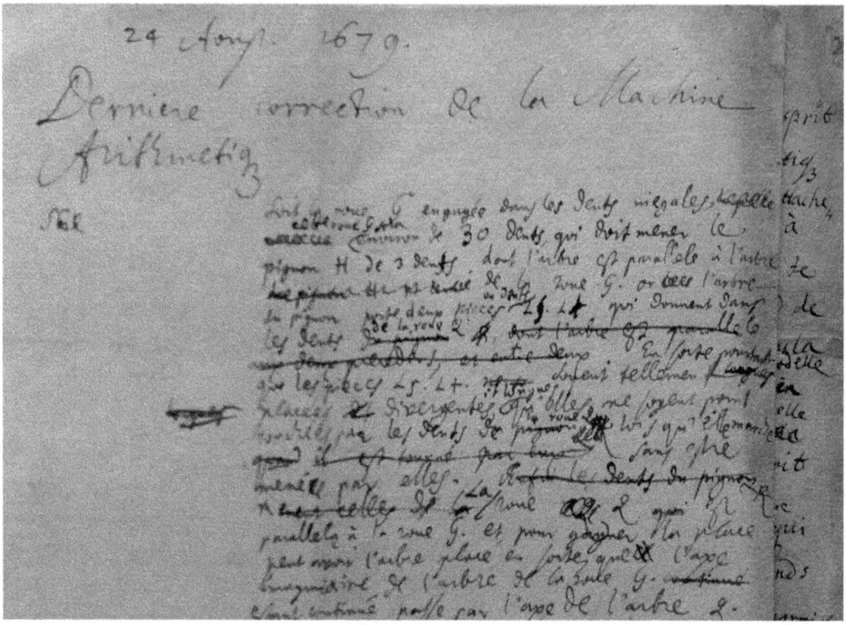

Notes from 1682 suggest that there were more problems, however:

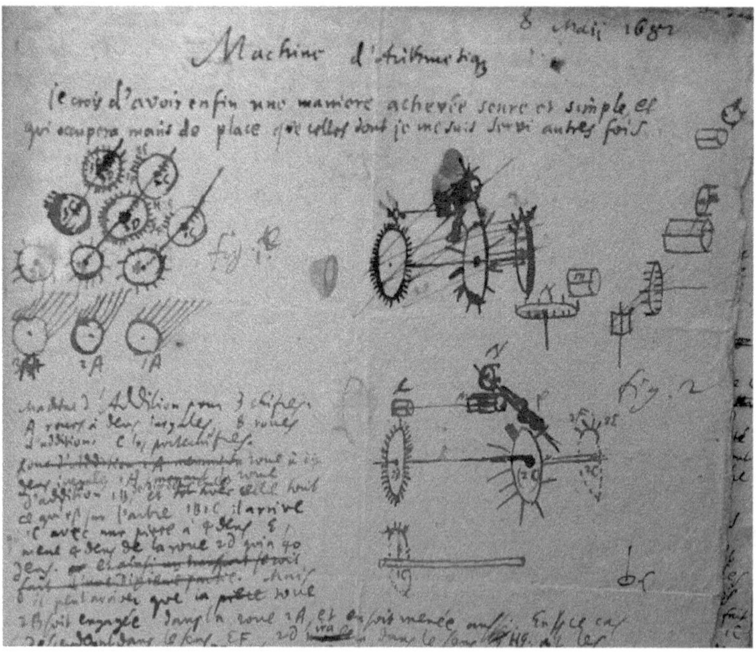

But Leibniz had plans drafted up from his notes and contracted an engineer to build a brass version with more digits:

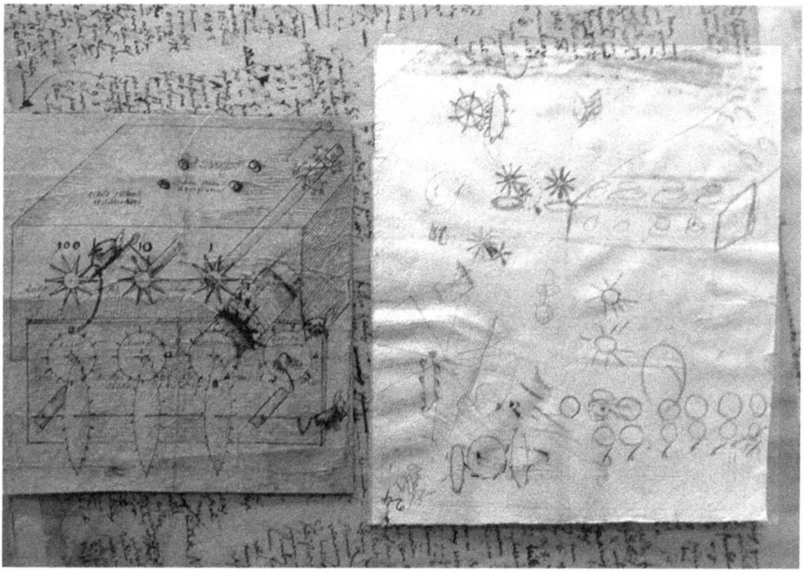

It's fun to see Leibniz's "marketing material" for the machine:

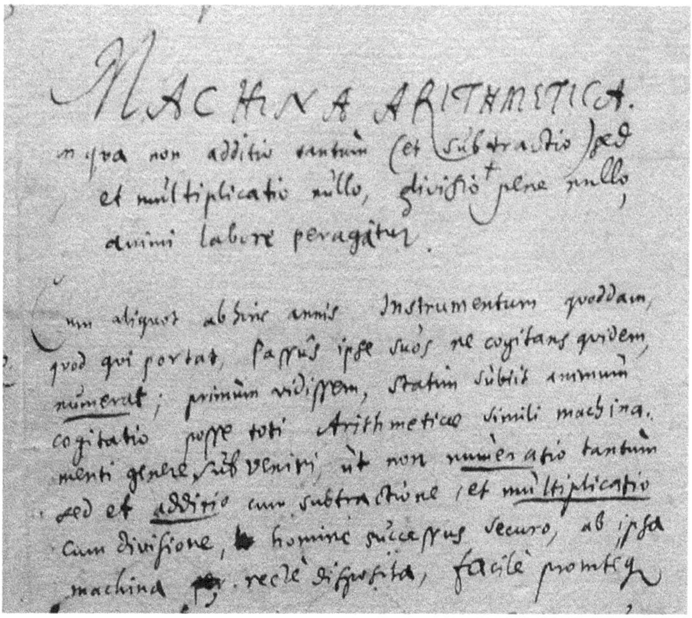

MACHINA ARITHMETICA.
in qua non additio tantum et subtractio sed
et multiplicatio nullo, divisio vero paene nullo,
animi labore peragatur.

Cum aliquot abhinc annis Instrumentum quoddam,
quod qui portat, passus ipse suos ne cogitans quidem,
numerat; primum vidissem, statim subiit animum
cogitatio posse toti Arithmeticae simili machina.
generi subveniri, ut non numeratio tantum
sed et additio cum subtractione et multiplicatio
cum divisione, homine successus securo, ab ipsa
machina recte disposita, facile promte

As well as parts of the "manual" (with 365 × 24 as a "worked example"):

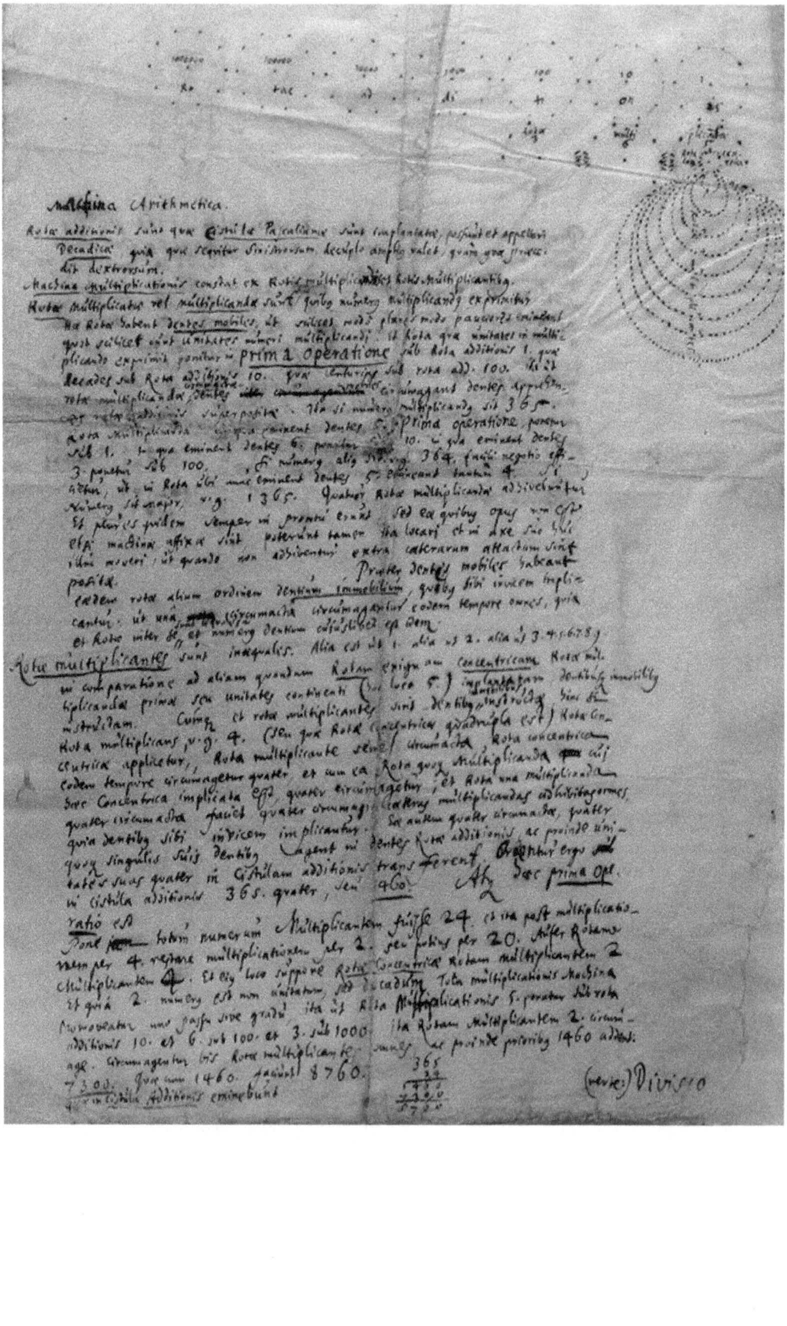

Complete with detailed usage diagrams:

But despite all this effort, problems with the calculator continued. And in fact, for more than 40 years, Leibniz kept on tweaking his calculator — probably altogether spending (in today's currency) more than a million dollars on it.

So, what actually happened to the physical calculator? When I visited Leibniz's archive, I had to ask. "Well", my hosts said, "we can show you". And there in a vault, along with shelves of boxes, was Leibniz's calculator, looking as good as new in a glass case — here captured by me in a strange juxtaposition of ancient and modern:

All the pieces are there. Including a convenient wooden carrying box. Complete with a cranking handle. And, if it worked right, the ability to do any basic arithmetic operation with a few minutes of cranking:

Leibniz clearly viewed his calculator as a practical project. But he still wanted to generalize from it, for example trying to make a general "logic" to describe geometries of mechanical linkages. And he also thought about the nature of numbers and arithmetic. And was particularly struck by binary numbers.

Bases other than 10 had been used in recreational mathematics for several centuries. But Leibniz latched on to base 2 as having particular significance — and perhaps being a key bridge between philosophy, theology and mathematics. And he was encouraged in this by his realization that binary numbers were at the core of the *I Ching*, which he'd heard about from missionaries to China, and viewed as related in spirit to his *characteristica universalis*.

Leibniz worked out that it would be possible to build a calculator based on binary. But he appears to have thought that only base 10 could actually be useful.

It's strange to read what Leibniz wrote about binary numbers. Some of it is clear and practical — and still seems perfectly modern. But some of it is very 17th century — talking for example about how binary proves that everything can be made from nothing, with 1 being identified with God, and 0 with nothing.

Almost nothing was done with binary for a couple of centuries after Leibniz: in fact, until the rise of digital computing in the last few decades. So when one looks at Leibniz's papers, his calculations in binary are probably what seem most "out of his time":

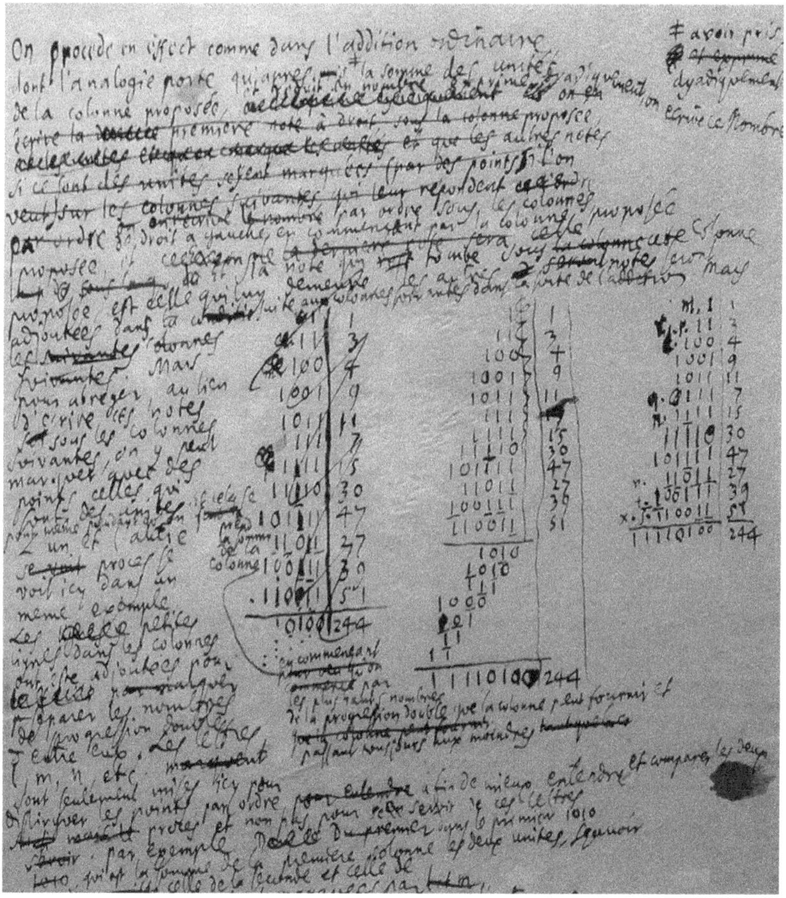

With binary, Leibniz was in a sense seeking the simplest possible underlying structure. And no doubt he was doing something similar when he talked about what he called "monads". I have to say that I've never really understood monads. And usually when I think I almost have, there's some mention of souls that just throws me completely off.

Still, I've always found it tantalizing that Leibniz seemed to conclude that the "best of all possible worlds" is the one "having the greatest variety of phenomena from the smallest number of principles". And indeed, in the prehistory of my work on *A New Kind of Science*, when I first started formulating and studying one-dimensional cellular automata in 1981, I considered naming them "polymones" — but at the last minute got cold feet when I got confused again about monads.

There's always been a certain mystique around Leibniz and his papers. Kurt Gödel — perhaps displaying his paranoia — seemed convinced that Leibniz had discovered great truths that had been suppressed for centuries. But while it is true that Leibniz's papers were sealed when he died, it was his work on topics like history and genealogy — and the state secrets they might entail — that was the concern.

Leibniz's papers were unsealed long ago, and after three centuries one might assume that every aspect of them would have been well studied. But the fact is that even after all this time, nobody has actually gone through all of the papers in full detail. It's not that there are so many of them. Altogether there are only about 200,000 pages — filling perhaps a dozen shelving units (and only a little larger than my own personal archive from just the 1980s). But the problem is the diversity of material. Not only lots of subjects but also lots of overlapping drafts, notes and letters, with unclear relationships between them.

Leibniz's archive contains a bewildering array of documents. From the very large:

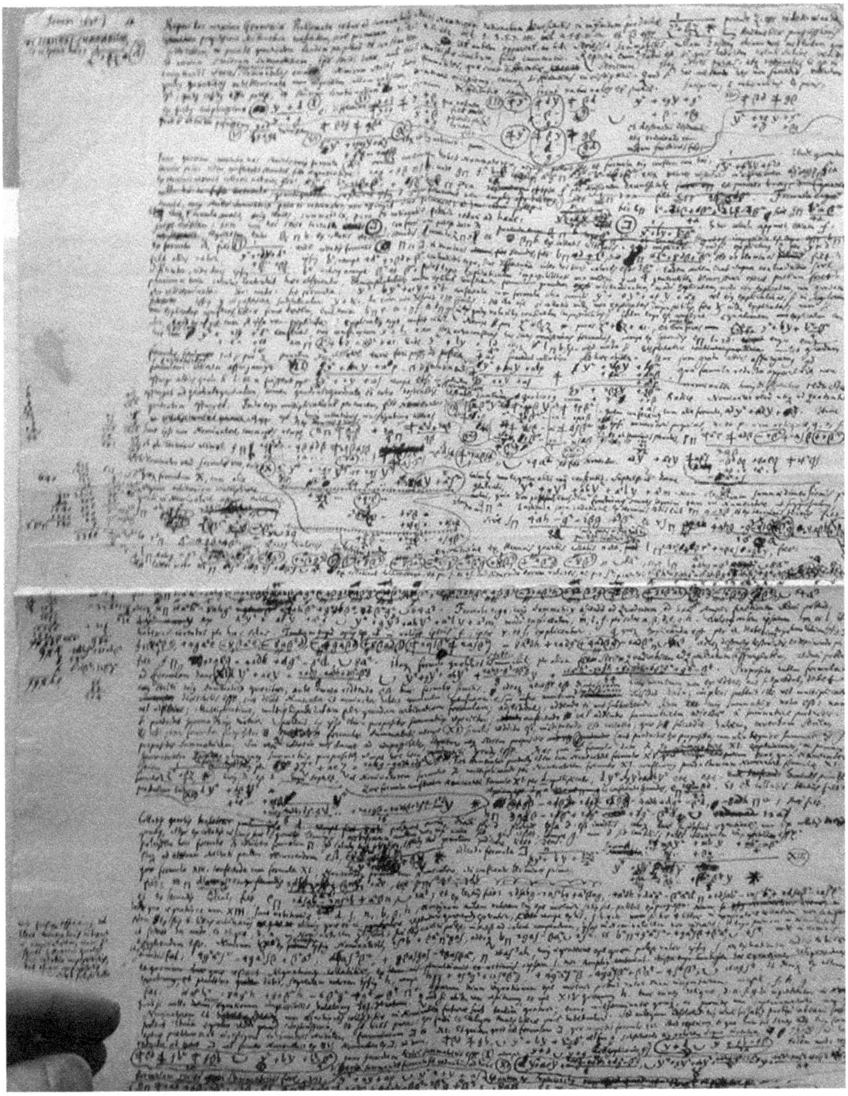

To the very small (Leibniz's writing got smaller as he got older and more near-sighted):

Most of the documents in the archive seem very serious and studious. But despite the high cost of paper in Leibniz's time, one still finds preserved for posterity the occasional doodle (is that Spinoza, by any chance?):

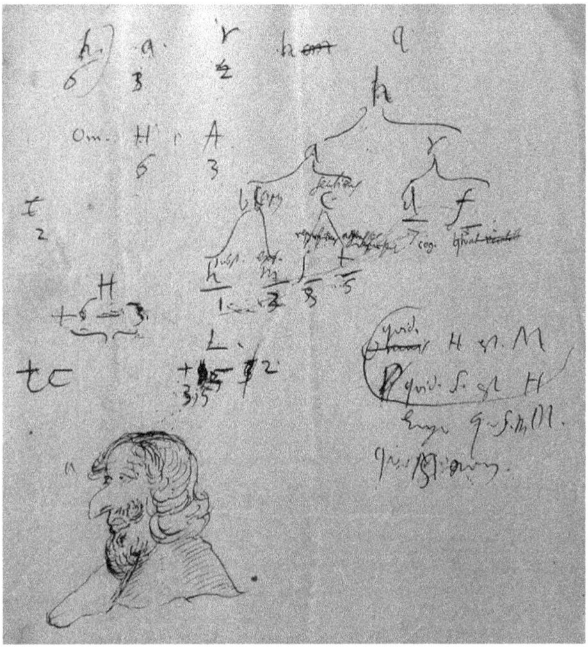

Leibniz exchanged mail with hundreds of people — famous and not-so-famous — all over Europe. So now, 300 years later, one can find in his archive "random letters" from the likes of Jacob Bernoulli:

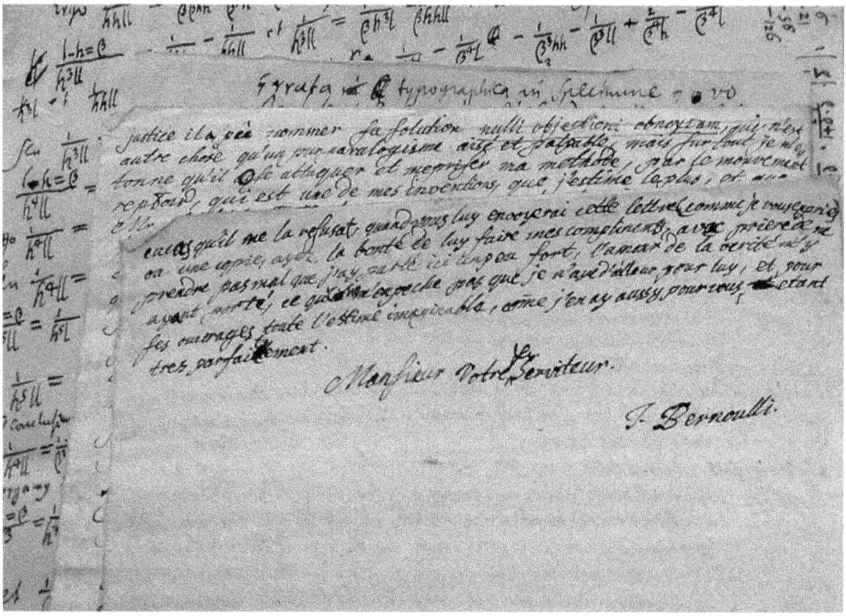
Monsieur Votre Serviteur.

J. Bernoulli.

What did Leibniz look like? Here he is, both in an official portrait, and without his rather oversized wig (that was mocked even in his time), that he presumably wore to cover up a large cyst on his head:

As a person, Leibniz seems to have been polite, courtierly and even tempered. In some ways, he may have come across as something of a nerd, expounding at great depth on all manner of topics. He seems to have taken great pains — as he did in his letters — to adapt to whoever he was talking to, emphasizing theology when he was talking to a theologian, and so on. Like quite a few intellectuals of his time, Leibniz never married, though he seems to have been something of a favorite with women at court.

In his career as a courtier, Leibniz was keen to climb the ladder. But not being into hunting or drinking, he never quite fit in with the inner circles of the rulers he worked for. Late in his life, when George I of Hanover became king of England, it would have been natural for Leibniz to join his court. But Leibniz was told that before he could go, he had to start writing up a history project he'd supposedly been working on for 30 years. Had he done so before he died, he might well have gone to England and had a very different kind of interaction with Newton.

At Leibniz's archive, there are lots of papers, his mechanical calculator, and one more thing: a folding chair that he took with him when he traveled, and that he had suspended in carriages so he could continue to write as the carriage moved:

Leibniz was quite concerned about status (he often styled himself "Gottfried von Leibniz", though nobody quite knew where the "von" came from). And as a form of recognition for his discoveries, he wanted to have a medal created to commemorate binary numbers. He came up with a detailed design, complete with the tag line *omnibus ex nihilo ducendis; sufficit unum* ("everything can be derived from nothing; all that is needed is 1"). But nobody ever made the medal for him.

In 2007, though, I wanted to come up with a 60th birthday gift for my friend Greg Chaitin, who has been a long-time Leibniz enthusiast. And so I thought: why not actually make Leibniz's medal? So we did. Though on the back, instead of the picture of a duke that Leibniz proposed, we put a Latin inscription about Greg's work.

And when I visited the Leibniz archive, I made sure to bring a copy of the medal, so I could finally put a real medal next to Leibniz's design:

It would have been interesting to know what pithy statement Leibniz might have had on his grave. But as it was, when Leibniz died at the age of 70, his political fates were at a low ebb, and no elaborate memorial was constructed. Still, when I was in Hanover, I was keen to see his grave — which turns out to carry just the simple Latin inscription "bones of Leibniz":

Across town, however, there's another commemoration of a sort — an outlet store for cookies that carry the name "Leibniz" in his honor:

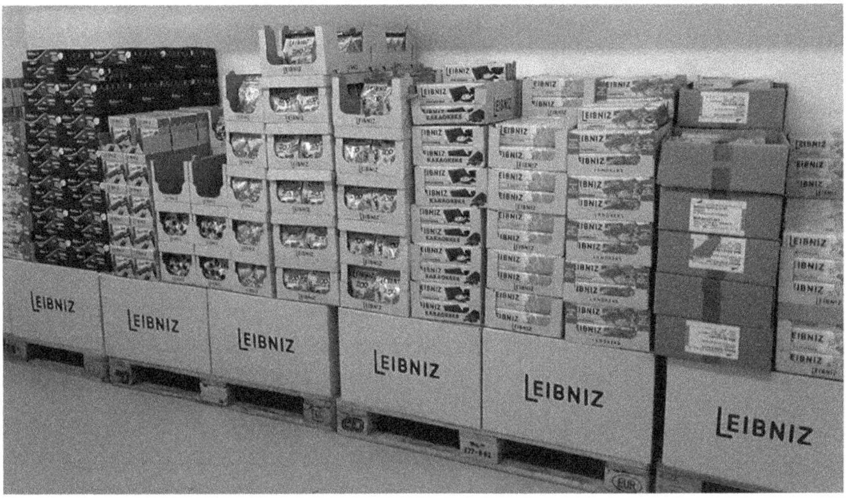

So what should we make of Leibniz in the end? Had history developed differently, there would probably be a direct line from Leibniz to modern computation. But as it is, much of what Leibniz tried to do stands isolated — to be understood mostly by projecting backward from modern computational thinking to the 17th century.

And with what we know now, it is fairly clear what Leibniz understood, and what he did not. He grasped the concept of having formal, symbolic, representations for a wide range of different kinds of things. And he suspected that there might be universal elements (maybe even just 0 and 1) from which these representations could be built. And he understood that from a formal symbolic representation of knowledge, it should be possible to compute its consequences in mechanical ways — and perhaps create new knowledge by an enumeration of possibilities.

Some of what Leibniz wrote was abstract and philosophical — sometimes maddeningly so. But at some level Leibniz was also quite practical. And he had sufficient technical prowess to often be able to make real progress. His typical approach seems to have been to start by trying to create a formal structure to clarify things — with formal notation if possible. And after that his goal was to create some kind of "calculus" from which conclusions could systematically be drawn.

Realistically he only had true success with this in one specific area: continuous "geometrical" mathematics. It's a pity he never tried more seriously in discrete mathematics, because I think he might have been able to make progress, and might conceivably even have reached the idea of universal computation. He might well also have ended up starting to enumerate possible systems in the kind of way I have done in the computational universe.

One area where he did try his approach was with law. But in this he was surely far too early, and it is only now — 300 years later — that computational law is beginning to seem realistic.

Leibniz also tried thinking about physics. But while he made progress with some specific concepts (like kinetic energy), he never managed to come up with any sort of large-scale "system of the world", of the kind that Newton in effect did in his *Principia.*

In some ways, I think Leibniz failed to make more progress because he was trying too hard to be practical, and — like Newton — to decode the operation of actual physics, rather than just looking at related formal structures. For had Leibniz tried to do at least the basic kinds of explorations that I did in *A New Kind of Science*, I don't think he would have had any technical difficulty — but I think the history of science could have been very different.

And I have come to realize that when Newton won the PR war against Leibniz over the invention of calculus, it was not just credit that was at stake; it was a way of thinking about science. Newton was in a sense quintessentially practical: he invented tools then showed how these could be used to compute practical results about the physical world. But Leibniz had a broader and more philosophical view, and saw calculus not just as a specific tool in itself, but as an example that should inspire efforts at other kinds of formalization and other kinds of universal tools.

I have often thought that the modern computational way of thinking that I follow is somehow obvious — and somehow an inevitable feature of thinking about things in formal, structured, ways. But it has never been very clear to me whether this apparent obviousness is just the result of modern times, and of our experience with modern practical computer technology. But looking at Leibniz, we get some perspective. And indeed what we see is that some core of modern computational thinking was possible even long before modern times. But the ambient technology and understanding of past centuries put definite limits on how far the thinking could go.

And of course this leads to a sobering question for us today: how much are we failing to realize from the core computational way of thinking because we do not have the ambient technology of the distant future? For me, looking at Leibniz has put this question in sharper focus. And at least one thing seems fairly clear.

In Leibniz's whole life, he basically saw less than a handful of computers, and all they did was basic arithmetic. Today there are billions of computers in the world, and they do all sorts of things. But in the future there will surely be far far more computers (made easier to create by the Principle of Computational Equivalence). And no doubt we'll get to the point where basically everything we make will explicitly be made of computers at every level. And the result is that absolutely everything will be programmable, down to atoms. Of course, biology has in a sense already achieved a restricted version of this. But we will be able to do it completely and everywhere.

At some level we can already see that this implies some merger of computational and physical processes. But just how may be as difficult for us to imagine as things like *Mathematica* and Wolfram|Alpha would have been for Leibniz.

Leibniz died on November 14, 1716. In 2016 that'll be 300 years ago. And it'll be a good opportunity to make sure everything we have from Leibniz has finally been gone through — and to celebrate after three centuries how many aspects of Leibniz's core vision are finally coming to fruition, albeit in ways he could never have imagined.

Chapter 6

Literature and Mathematics: Jorge Luis Borges' Library and Gregory Chaitin's Omega

Arturo Sangalli*

Literature and mathematics stand on opposite sides of the art/science divide. The freedom to create fictional realities and the absence of definite rules are at the heart of literature, in stark contrast with the search for universal truths and the strict adherence to the rules of logic that are characteristic of mathematics. Nor is there much in common — apart from both requiring creativity and imagination — between the work of mathematicians and that of writers; mathematicians prove theorems, writers tell stories.

In spite of all this, their paths occasionally cross. There are biographies of famous mathematicians; novels in which characters mention or discuss mathematical concepts and results, literary essays about mathematics, mathematical poetry, and works of fiction where mathematics plays a central role. However, it is not our purpose to discuss

*Formerly affiliated to Champlain Regional College, Sherbrooke, Quebec, Canada; asangall@hotmail.com.

mathematics *in* literature, but rather to apply certain mathematical ideas *to* literature considered essentially as a book-producing activity — all books, good, bad, mediocre, and sublime.

In that perspective, let us first take a look at mathematics as a theorem-producing activity and ask the following question: Can the mathematician be replaced — at least in principle — by a machine, a sort of robot that would automatically produce, one after the other, all mathematical theorems. The short answer is "yes, but."

Let me explain. The robot we have in mind is of course an ideal one: it doesn't require any energy to operate, it has an unlimited supply of paper and ink at its disposal to write down the theorems, and it can work as long as it takes to give its answers without breaking down. We know that, in principle, such a theorem-generating machine could be built, in the sense that it is possible to write a computer program that would achieve the desired outcome. But there is a problem. The robot, given enough time, will not only write down every theorem, i.e., every true mathematical proposition, but *all* mathematical propositions, including those such as "$2 + 2 = 5$" that are false. An easy fix would then be to have a second robot check the output from the first one and separate the grain from the chaff, that is, eliminate the false propositions and keep only the theorems. Alas, such a "theorem-checking" robot cannot be built, not even in theory: the problem has no solution. So mathematicians have to prove their theorems the hard way, using brainpower — no machine can do the job for them.

Moving over to literature, we may ask a similar question: Can the writer be replaced by a machine? The short answer is again "yes, but," an answer that requires an explanation and has some surprising consequences.

Our ideal writing robot will have stored in its memory all the symbols: letters, numerals, punctuation signs, the blank space, etc., needed to write a text in English or in any other language. What is important here is that the set A of all these symbols or characters — the alphabet — be finite. The robot then operates in the following way: Every second, say, it chooses one symbol at random from the alphabet and types it out. The physical details are unimportant, so we may imagine the symbols being typed out linearly, one after the other, on a paper tape. For example, the first few symbols may be: psk5&,a)xxc]df$;0?if...

We are going to assume that the robot has been typing characters at random for a very long time. A representation taken from Buddhist cosmology may help to convey the notion of extremely long periods of time. In this cosmology, a *kalpa* is a unit of time whose duration varies according to the particular Buddhist tradition, but the following analogy gives an idea of its magnitude. Imagine a mountain made of stone and twice as high as Mount Everest. Once every one hundred years, a bird flies by and lightly brushes the tip of the mountain with a small piece of silk. The mountain will be completely worn out by the repetition of this ritual before one kalpa ends.

Suppose then that the robot has kept typing for a very long time. By a stretch of the imagination (a sort of thought experiment), it is possible to conceive that the robot has been at work for an infinitely long period of time. What can be said about the infinite sequence s of symbols written on the (infinite) tape? Enter mathematics, or more precisely, the Borel–Cantelli lemma. The following version of the lemma is due to Émile Borel [1]:

Let $T_1, T_2, \ldots, T_n, \ldots$ be an infinite sequence of random "trials", each of which has one of two possible outcomes: Success or Failure. Let p_n be the probability of T_n resulting in Success. If the infinite series $p_1 + p_2 + \cdots + p_n + \cdots$ converges, then the probability for Success to occur infinitely many times is equal to 0; if the series diverges, this probability is equal to 1.

As a corollary of this lemma, *every* possible finite combination of symbols will appear somewhere on the tape with 100 percent probability — and not just once, but infinitely many times. Here's the proof:

Let $a_1, a_2, \ldots, a_k$ be a given finite sequence of symbols in A. We imagine the tape as divided up into sections of k consecutive symbols: the first section consists of the first k symbols, the second section is made up of the next k symbols, and so on. Let the nth trial T_n consist in looking at the nth section of k symbols. Success is defined as finding that this nth string of symbols is precisely the given sequence $a_1, a_2, \ldots, a_k$; otherwise, the trial results in Failure. The symbols having been typed at random, the probability of Success is $p_n = 1/m^k$, where m is the number of symbols in the alphabet A. Since the infinite series $p_1 + p_2 + \cdots + p_n + \cdots$ diverges, Success (i.e., the occurrence of the given finite sequence) will happen an infinite number of times. For example, the 26-symbol combination "Elementary, my dear Watson", or a much longer one, such as the complete adventures of Sherlock Holmes as written by Arthur Conan Doyle; or the complete adventures of Sherlock Holmes in which every occurrence of "Sherlock Holmes" has been replaced with "Barack Obama", the rest of the text remaining unchanged. Let me repeat and emphasize this fact: Even if practically the whole tape will consist of long chains of characters devoid of any discernible meaning, it is all but certain that *every conceivable text*, no matter how long, will appear on it sooner or later. Among the myriad of possibilities, we would find the various versions of the Bible, all masterpieces of world literature, and the chapter you are presently reading; after a sufficiently long period of time every book ever written or to be written in the future will appear on the tape. All these "meaningful" texts will be like very tiny islands in a vast linear sea of meaningless strings of symbols.

To sum up: Our robot typing at random can write every possible book, provided we are willing to wait a long time for this to happen. Does this mean that creativity can be replaced by randomness, that literature is just a combinatorial game? Writers, not to worry: "a long time" can be an extremely long time, longer than the age of the Universe, estimated at 13.7 billion years. Thus, the writing "by chance" of a complete book should be considered an impossible task on a human scale. The upshot of this is that we should be grateful to human authors for the books we so much enjoy reading. Perhaps one day intelligent robots may also be up to the task, but certainly not by typing at random.

This is not however the end of the story. The last act has not yet been played out. It features Gregory Chaitin's number Omega (Ω).

Introducing this number requires a detour through the theory of computation.

In 1935 Alan Turing, then a young Cambridge graduate, came up with the idea of a simple theoretical model of a computer known as a *Turing machine* [2]. Simply put, his virtual computer could mimic the operations of any (digital) computer, so that it could perform any computation that a real computer was capable of carrying out. Turing also proved that there is something that can never be computed: deciding in advance whether the execution of any given computer program will eventually terminate or is doomed to run forever — the so-called Halting Problem. Chaitin's Ω is the probability that a randomly chosen program for a Turing machine will eventually halt.

But what does Ω have to do with literature? To understand the connection, we have first to explain how a string of digits can be "decoded" or transformed into a string of characters. We begin by assigning to each letter or character a numerical "code". This can be done in many different ways, for instance: $a = 01$, $b = 02, \ldots, e = 05, \ldots, l = 12, \ldots, z = 26, \ldots, ? = 55$, and so forth. Thus, the string of digits 02011212, once decoded, becomes the word "ball". If a pair of digits does not correspond to the code of some character, we simply decode it as special symbol such as "#". The important thing is to realize that *every string of digits can be interpreted as a string of characters*, and therefore, in certain cases, as a word, or phrase, or even a long text in some language. In particular, the infinite string of decimals of $\pi = 3,141592\ldots$ can be so interpreted. The first four decimals decode as the word "no" ($14 = n$, $15 = o$). Further down the sequence of decimals, we find 0119 (at positions 493 to 496), which decodes as "as". If we continue even further, we may find other, longer words, perhaps even several words in succession — although, intuitively, it appears nearly impossible ever to come across something so complex as "Do not let the better be the enemy of the good", an epigram attributed to Winston Churchill.

But in fact, something even more counterintuitive is true: the sequence of decimals of Ω contains, in coded form, the whole of literature, past, present and future, and not only once but an infinite number of times. Therefore, we don't need an ideal robot typing at random to write all possible books, they already exist in the decimals of Ω.

The last statement is based on certain mathematical results concerning the so-called *normal numbers*, introduced by Borel at the beginning of the 20th century. Without going into the details, let us just say that the decimal part of a normal number contains every possible combination (of any given length) of the digits 0 to 9 an infinite number of times; and that Chaitin's Ω is normal [3] (and non-computable), so it would contain in coded form all possible books.

In *The Library of Babel* [4], the writer Jorge Luis Borges imagined a mysterious library that would contain all 410-page texts obtained by combining the letters of the alphabet, the point, the comma, and the blank space in all possible ways. The narrator of the story suspects that the library, a veritable allegory of literature or the Universe, is "unlimited and periodic", and he adds: "*If an eternal traveller would set out to explore it in any given direction, he would conclude after a journey of many centuries that the same books repeat themselves always in the same disorder — which, by its repetition, would become an order: The Order.*"

In a note added by Borges after the first publication of the story in 1941, he observed that his vast library could be replaced by a single book, of ordinary size but containing an infinite number of infinitely thin pages. It would have come as a surprise to him (or perhaps not) to learn that in fact one number suffices. The worlds of mathematics and literature might be far apart, but some rare and strange encounters between them should not be ruled out.

Bibliography

[1] É. Borel, *Le jeu, la chance* (Gallimard, Paris, 1941).

[2] A. Sangalli, *The Importance of Being Fuzzy: And Other Insights from the Border Between Math and Computers* (Princeton University Press, New Jersey, 1998), pp. 52–55.

[3] C. Calude, Borel normality and algorithmic randomness, in *Developments in Language Theory: At the Crossroads of Mathematics, Computer Science and Biology*, eds. G. Rozenberg and A. Salomaa (World Scientific, Singapore, 1994), pp. 113–119.

[4] J. L. Borges, *The Library of Babel*, in *Labyrinths: Selected Stories and Other Writings*, eds. D. A. Yates and J. E. Irby (Penguin, London, 1970).

Chapter 7

Selva Selvaggia

Francisco Antonio Doria*

Arturo Sangalli conjectured in his romance *Pythagoras' Revenge* that the collection of all mathematical objects — the "space of mathematics" — has a random structure. We call that conjecture "the Chaitin–Thorp theorem", as it originates in a result by Chaitin and is proved in the romance by fictional mathematician Norton Thorp. We prove in this chapter a version of the Chaitin–Throp theorem.

1. Unexpected Monsters?

Suppose that mathematics is a vast landscape which you can see from your mind's balcony. Someone guides you to that balcony, as you have your eyes closed. You are placed next to the railing by your guide — your eyes are kept closed. Before you open them, you draw in your mind the landscape you'll soon see. The landscape you expect, the space of mathematics. You expect a vast, clear city, with streets drawn in a gridiron pattern, with majestic buildings that seem to be organized in a clear hierarchical pattern. You expect order, clarity, rationality.

You open your eyes.

*Advanced Studies Research Group and Fuzzy Sets Laboratory, PIT, Production Engineering Program, COPPE, Federal University at Rio de Janeiro, Brazil; fadoria63@gmail.com.

You see a huge, monstrous, dark, impassable jungle. No order, no rationality, just a terrifying mess. Mathematics is the mind's wilderness.

> *O quanto a dir' qual era, è cosa dura,*
> *questa selva selvaggia ed aspra e forte,*
> *che nel pensier' rinova la paura.*[a]

Chaos in the place of order. But chaos, indeed?

Well, that's just a metaphor, albeit a surprising, unexpected one (with a helping hand from Dante). But is it so? Does it capture in an adequate way the landscape of mathematics?

In 1987, Chaitin published a beautiful, very elegant theorem [2, p. 132]. There is a family of exponential Diophantine equations:

$$Q(n, x_1, x_2, \ldots, x_m) = Q'(n, x_1, x_2, \ldots, x_m)$$

on the variables $x_1, \ldots, x_m$ and parameterized by n so that given the real number q in binary form by

- the nth digit of q is 0 if the corresponding equation has finitely many solutions;
- the nth digit of q is 1 if the corresponding equation has infinitely many solutions.

Then, we have the following proposition.

Proposition 1.1. *q is a random real.*

That is a big surprise: a family of equations which are recursively given leads to randomness in a key property! Randomness appears then in the depths of mathematics when we try to understand a simple property of a naïve-looking system of equations.

[a]To better describe it, it was a hard thing, that savage jungle, rough, impassable, that makes me fear it just by recalling...

1.1. *The Chaitin–Thorp theorem*

Since we began with a fantastic metaphor, let's now jump from mathematical reality to the side of fantasy. In 2009, Sangalli published a romance, *Pythagoras' Revenge* [21], with the subtitle *A Mathematical Mystery*. One of the subplots of that romance has to do with a great (fictional) mathematician called Norton Thorp who discovered an extension of Chaitin's result, which we call "the Chaitin–Thorp theorem" [21, p. 80]:

> "I've now found an irrefutable proof that, as Chaitin suspected, randomness is at the heart of mathematics," he announced, "and the implications of that fact are far-reaching. It means that we may be able to prove some theorems, answer some questions, but the vast majority of mathematical problems are essentially unsolvable. It means, as one of my colleagues has put it, that a few bits of math may follow from each other, but for most mathematical situations those connections won't exist because mathematics is full of accidental, reasonless truths. And if you can't make connections, you cannot solve or prove things. Solvable problems are like a small island in a vast sea of undecidable propositions."
>
> As he pronounced these last words, the aerial view of an island with "Solvable Problems" written on it appeared on the screen. Slowly the camera began to move away, revealing more and more water around the island until the water, with the inscription "Unsolvable Problems" floating in it, completely filled the screen and the island all but disappeared.

That's actually a revenge against Pythagoras: the world is described by mathematics, but mathematics itself is hyper-irrational. It is unfathomable.

1.2. *A generalized Rice theorem*

Well, the Chaitin–Thorp theorem (or better, metatheorem) may have been formulated in a romance, but it is mathematical fact. We are going to state it in a more acceptable language and sketch its proof.

Actually, we are going to state and very briefly sketch the proof of a Rice-like theorem that N. da Costa and the author have proved.

Roughly, if a nontrivial property P in a consistent theory T is such that it includes arithmetic, has a recursively enumerable set of theorems, and so that there are $a, b, a \neq b$ with $T \vdash P(a)$ and $T \vdash \neg P(b)$, then we get the following result:

Any such P is undecidable.

Everything which is of interest turns out to be undecidable in the general case. There is no general algorithm to test for theoremhood — o.k., we knew that for sure since Gödel proved his incompleteness theorems, but here we have something more specific. That is, we have an extension of Rice's theorem in computer science [18] the whole of classical analysis and, we can say, most of what we can call "standard mathematics."

2. The Halting Function in Extensions of Arithmetic

We call such function, or better, such functions the θ functions. θ functions are (roughly) a particular kind of image of Diophantine equations into the language of classical analysis. They are obtained through a map due to D. Richardson (see [3] on that); the corresponding map from the language of arithmetic into that of classical analysis can be recursively implemented, and zeroes are preserved through Richardson's map.

(If we start from a universal Diophantine equation, we get our image of the Halting Function.)

We deal here with elementary functions. Elementary functions are those that belong to an algebra $\mathcal{A}$ that contains the rational numbers, real-defined and real-valued polynomials, sines and cosines, exponentials, plus the number π, and is closed under sum, product and function composition. Richardson's main result translates expressions for Diophantine equations into expressions in the language of elementary functions. We'll see that innocent-looking problems in analysis are seen to be intractable because they are equivalent to the solution of an intractable Diophantine equation.

2.1. *A simple example*

Consider the interior of a square in the plane R^2 and consider two curves within that square described by some expression in the algebra of elementary functions. Is there a general algorithm that allows us to decide whether these curves meet? No. Moreover, if we formalize everything within ZFC, there will be curves s, s', so that the (formalized version of the) sentence "s and s' never intersect" can neither be proved nor disproved within ZFC, supposed consistent.

That is, if such a simple mathematical question cannot be decided in an arbitrary case, what are we to expect out of more complex queries?

2.2. *Another example*

We can present here another example of incompleteness that directly stems from the metamathematical properties of F, where F is a fast-growing total recursive function that can neither be proved total in our theory T nor can be disproved to be so. The so-called BGS set $\mathcal{S} = \langle \mathsf{M}_m, |x|^{F(n)} + F(n) \rangle$, $n = 0, 1, 2, \ldots$ has the following two properties, among many others of interest:

- It is a set of Turing machines M_m bound by a clock that stops it after $|x|^{F(n)} + F(n)$ computation steps, where x is the input to the machine and $|x|$ the binary length of x.
- $\mathcal{S}$ is a set of poly Turing machines in the standard model for arithmetic in our background theory T.
- The sentence "$\mathcal{P}$ is a set of Turing poly machines" is true of the standard model for the arithmetic portion of theory T.[b]
- But "$\mathcal{P}$ is a set of Turing poly machines" is undecidable in theory T.

[b]Of course, we mean the formal version of that sentence in the language of T.

2.3. *Chaitin–Thorp*

We can extend as follows Chaitin's theorem on randomness in the theory of exponential Diophantine equations, stated in Proposition 1.1: there are functions $R(n, x_1, x_2, \ldots, x_k)$ and $R'(n, x_1, x_2, \ldots, x_k)$ in the algebra of elementary functions, so that the binary number q defined as in Proposition 1.1 satisfies the following proposition.

Proposition 2.1. *q is a random real.*

Moreover, we can state, if we are dealing within ZFC which we suppose to have a model $\mathbf{N} \models$ ZFC that makes ZFC a theory with a sound arithmetic portion, the following proposition.

Proposition 2.2. *There are infinitely many R, R' so that the sentence "q is random" can neither be proved nor disproved in* ZFC, *but* $\mathbf{N} \models$ *"q is a random real."*

We may call that the main Chaitin–Thorp theorem.[c] Paraphrasing it, we may say: Portions of the algebra of elementary functions may exhibit randomness in real life — if we may say so of $\mathbf{N}$ — but yet, we can neither prove nor disprove that it exhibits it.

2.4. *Still another example*

Worse — how about the next theorem? Call it Chaitin–Thorp, another example.

Proposition 2.3. *There is a recursive family of recursive reals that are irrational in* $\mathbf{N}$, *but such that we can neither prove nor disprove it.*

That's only the beginning... (The proof requires those beautiful and intractable fast-growing recursive functions.)

Let's now explore a bit of the wild world of the θ functions.

[c]There will be others inspired by Chaitin's original result.

2.5. θ *functions, main results*

Now, our goal is to prove the following theorem, which we name "Chaitin–Thorp, again." Let Ψ be such that ZFC $\vdash \Psi(m_0)$, ZFC $\vdash \neg\Psi(n_0)$, $m_0 \neq n_0$; it is otherwise arbitrary.

Proposition 2.4 (Chaitin–Thorp, again). *Let* $\mathbf{N} \models$ ZFC *and make it sound. Then there is a given* q_0*, so that neither* ZFC $\vdash \Psi(q_0)$ *nor* ZFC $\vdash \neg\Psi(q_0)$.

Moreover, we have the following proposition.

Proposition 2.5. *There are infinitely many such* q_0.

It means that, given any such provable sentence ψ in ZFC, there are infinitely many sentences which essentially mean the same as ψ, but which can neither be proved nor disproved in ZFC. To prove Chaitin–Thorp, again, put:

$$q_0 = m_0\theta_a + n_0(1 - \theta_a).$$

$\theta_a = \theta(a)$ is a particular, undecidable, instance of an explicit expression for the Halting Function in computer science. Let's elaborate on that. θ is the Halting Function: given a universal Turing machine, $\theta(x_0) = 1$ if and only if the universal machine $\mathsf{U}(x_0)$ halts, and equals 0 otherwise. Since the set of values of θ that equal 1 is recursively enumerable but not recursive, then there is a value a so that $\theta(a)$ is 0 — it corresponds to a non-halting instance of U, that is, $\mathsf{U}(a)$ never halts — but we cannot algorithmically determine it. We use that particular instance (there are actually infinitely many such instances a, of course). So, the difficulty of the problem is of degree $\mathbf{0}'$ (it is recursively enumerable but not recursive). Can we go beyond that? Yes!

2.6. *More about the Halting Function*

From here onwards, we suppose that our formal constructions are made within a first-order axiomatic theory T which can be thought to include Zermelo–Fraenkel (ZF) set theory plus at least some portion of the Axiom of Choice.

In our version, the Richardson functor starts out of a polynomial p in q indeterminates over the integers Z. Richardson tells us how to explicitly and algorithmically construct two expressions out of that p:

- First, an expression for a q-variable function $f(p)$ which includes polynomial terms, sine functions and the number π. $f(p)$ satisfies the following conditions:
  (1) The Diophantine equation $p = 0$ has no solutions over the integers if and only if, for all values of its variables, $f(p) > 1$.
  (2) $p = 0$ has a solution over the integers if and only if $f(p)$ dips beyond 1 and has zero values.
- Second, an expression for a 1-variable function which can be explicit constructed with elementary functions and which has a similar behavior according to the existence of roots in $p = 0$.

So, the idea in Richardson's transform is that there is a kind of strip of finite width which is never crossed by the values of $f(p)$ if the Diophantine equation $p = 0$ has no solutions. If it does have solutions, that strip will be crossed at points depending on the Diophantine roots. If we now add an expression for the absolute value function $|\ldots|$ to our language, we can obtain a new function $c(f)$ with the following conditions:

- $c(f) = 0$ if and only if $p = 0$ has no solutions as a Diophantine equation.
- $c(f) > 0$ somewhere if and only if $p = 0$ does have integer solutions.

Richardson's interest seems to have been restricted to the construction of a few unsolvable problems in analysis. However, we realized that his two maps amounted to a true-blood functor from axiomatizable systems into classical analysis. In the case of our axiomatic system T, we can represent its proofs by a Turing machine M_T that halts whenever one of its theorems ξ (or a decidable sentence) is input; if we input an undecidable sentence ζ, $M_T(\zeta)$ will never halt over it.

If M_T halts over ξ, then we can explicitly obtain a Diophantine equation $p(m_\xi, x_1, \ldots) = 0$ which has solutions (m_ξ is a Gödel number for ξ); if M_T doesn't halt over ζ, then $p(m_\zeta, \ldots) = 0$ has no integer solutions. Therefore, with the help of Richardson's maps, we can code within the language of classical analysis the whole deductive machinery of an axiomatizable formal system.

The whole thing turns out to be wide-ranging. Out of a suggestion by Suppes, we proved a general undecidability and incompleteness theorem — a Rice-like theorem — within classical analysis (see Proposition 3.11). Let P be an arbitrary nontrivial[d] property in the language of analysis. The blueprint for our undecidable sentences out of those constructions is, informally, our extended version of Rice's theorem.

Proposition 2.6. *There is a term–expression ζ in the language of analysis such that neither $T \vdash P\zeta$ nor $T \vdash \neg P\zeta$.*

There is a corresponding undecidability result, and the associated decision problems can be made as high as one wishes in the arithmetical hierarchy, and even beyond [7] (Section 3).

3. Generalized Incompleteness

Notation here is standard. We follow [7] and some previous papers where these questions have been introduced and discussed such as [6]. We deal here (among other objects) with algorithmic functions. These are given by their programs coded in Gödel numbers e [20]. We will sometimes use Turing machines (noted by sans-serif letters with the Gödel number as index M_e) or partial recursive functions, noted $\{e\}$. Peano Arithmetic is noted PA. We require Russell's ι symbol: $\iota_x P(x)$ is, roughly, the x such that $P(x)$.

The *standard interpretation* for PA is: the variables $x, y, \ldots$ range over the natural numbers, and $\mathbf{0}$ and $\mathbf{1}$ are seen as, respectively, zero

[d]A property that isn't satisfied by either all or none of the objects in its domain.

and one. PA is strong enough to formally include Turing machine theory. Rigorously, for PA, we have the following definition.

Definition 3.1. A Turing machine of Gödel number e operating on x with output y, $\{e\}(x) = y$ is *representable* in PA if there is a formula $F_e(x, y)$ in the language of our arithmetic theory so that

(1) $\mathrm{PA} \vdash F_e(x, y) \wedge F_e(x, z) \to y = z$;
(2) for natural numbers a, b, $\{e\}(a) = b$ if and only if $\mathrm{PA} \vdash F_e(a, b)$.

Proposition 3.2. *Every Turing machine is representable in Peano Arithmetic. Moreover, there is an effective procedure that allows us to obtain F_e from the Gödel number e.*

A theory is *arithmetically sound* if it has a model with standard arithmetic for its arithmetical segment.

3.1. *Generalized incompleteness*

The following argument is valid for all theories that contain enough arithmetic, have a model where arithmetic is standard, and have a recursively enumerable set of theorems.

Suppose that our theory S has Russell's description symbol ι. Let P be a predicate symbol so that for closed terms ξ, ζ such that $S \vdash \xi \neq \zeta$, $S \vdash P(\xi)$ and $S \vdash \neg P(\zeta)$ (we call such P, nontrivial predicates). Then, for the term,

$$\eta = \iota_x[(x = \xi \wedge \alpha) \vee (x = \zeta \wedge \neg\alpha)],$$

where α is an undecidable sentence in S:

Proposition 3.3. $S \nvdash P(\eta)$ *and* $S \nvdash \neg P(\eta)$.

From now on, we will consider theories S, T, like the one above.

Our main tool here will be an explicit expression for the Halting Function, that is, the function that settles the halting problem. We have shown elsewhere that it can be constructed within the language of classical analysis.

Proposition 3.4. *If* $\{e\}(a) = b$, *for natural numbers* a, b, *then we can algorithmically construct a polynomial* p_e *over the natural numbers so that* $\{e\}(a) = b \leftrightarrow \exists x_1, x_2, \ldots, x_k \in \omega\, p_e(a, b, x_1, x_2, \ldots, x_k) = 0$.

Proposition 3.5. $a \in R_e$, *where* R_e *is a recursively enumerable set, if and only if there are* e *and* p *so that* $\exists x_1, x_2, \ldots, x_k \in \omega\, (p_e(a, x_1, x_2, \ldots, x_k) = 0)$.

3.2. *The halting function*

Remark 3.6. Let $\mathsf{M}_m(a) \downarrow$ mean: "Turing machine of Gödel number m stops over input a and gives some output." Similarly, $\mathsf{M}_m(a) \uparrow$ means, "Turing machine of Gödel number m enters an infinite loop over input a." Then we can define the Halting Function θ:

- $\theta(m, a) = 1$ if and only if $\mathsf{M}_m(a) \downarrow$;
- $\theta(m, a) = 0$ if and only if $\mathsf{M}_m(a) \uparrow$.

$\theta(m, a)$ is the Halting Function for M_m over input a.

θ isn't algorithmic, of course, that is, there is no Turing machine that computes it.

Then, if σ is the sign function, $\sigma(\pm x) = \pm 1$ and $\sigma(0) = 0$.

3.3. *Expressions for the Halting Function*

Proposition 3.7 (The Halting Function). *The Halting Function* $\theta(n, q)$ *is explicitly given by*

$$\theta(n, q) = \sigma(G_{n,q}),$$

$$G_{n,q} = \int_{-\infty}^{+\infty} C_{n,q}(x) e^{-x^2}\, dx,$$

$$C_{m,q}(x) = |F_{m,q}(x) - 1| - (F_{m,q}(x) - 1),$$

$$F_{n,q}(x) = \kappa_P p_{n,q}.$$

Here, $p_{n,q}$ is the two-parameter universal Diophantine polynomial and κ_P an adequate Richardson transform.

3.4. *More about undecidability and incompleteness in our theories*

Lemma 3.8. *There is a Diophantine set D so that*

$$m \in D \leftrightarrow \exists x_1, \ldots, x_n \in \omega p(m, x_1, \ldots, x_n) = 0,$$

p a Diophantine polynomial, and D is recursively enumerable but not recursive.

Corollary 3.9. *For an arbitrary $m \in \omega$, there is no general decision procedure to check whether $p(m, x_1, \ldots) = 0$ has a solution in the positive integers.*

3.5. *Main theorem*

Therefore, given such a p, and $F = \kappa_P(p)$, where κ_P is a Richardson transform [5], we have the following corollary.

Corollary 3.10. *For an arbitrary $m \in \omega$, there is no general decision procedure to check whether, for F and G adequate real-defined and real-valued functions,*

(1) *there are real numbers $x_1, \ldots, x_n$ such that $F(m, x_1, \ldots, x_n) = 0$;*
(2) *there is a real number x so that $G(m, x) < 1$;*
(3) *we have $\forall x \in \mathsf{R}\ \theta(m, x) = 0$ or $\forall x \in \mathsf{R}\ \theta(m, x) = 1$ over the reals.*
(4) *for an arbitrary $f(m, x)$, we have $f(m, x) \equiv \theta(m, x)$.*

Let $\mathcal{B}$ be a sufficiently large algebra of functions and let $P(x)$ be a nontrivial predicate. If ξ is any word in that language, we write $\|\xi\|$ for its complexity, as measured by the number of letters from ZFC's alphabet in ξ. We define the *complexity of a proof* $C_{\mathrm{ZFC}}(\xi)$ of ξ in the language of ZFC to be the minimum length that a deduction of ξ from the ZFC axioms can have, as measured by the total number of letters in the expressions that belong to the proof (this has been proved by da Costa and Doria in 1990).

Proposition 3.11. *If* ZFC *is arithmetically sound, then*

(1) *there is an* $h \in \mathcal{B}$ *so that neither* ZFC $\nvdash \neg P(h)$ *nor* ZFC $\nvdash P(h)$, *but* $\mathbf{N} \models P(h)$, *where* $\mathbf{N}$ *makes* ZFC *arithmetically sound;*
(2) *there is a denumerable set of functions* $h_m(x) \in \mathcal{B}$, $m \in \omega$, *such that there is no general decision procedure to ascertain, for an arbitrary* m, *whether* $P(h_m)$ *or* $\neg P(h_m)$ *is provable in* ZFC;
(3) *given the set* $K = \{m : \text{ZFC} \vdash \phi(\widehat{m})\}$, *and given an arbitrary total recursive function* $g : \omega \to \omega$, *there is an infinite number of values for* m *so that* $C_{\text{ZFC}}(P(\widehat{m})) > g(\|P(\widehat{m})\|)$.

Proof. Let θ be as above. Let f_0, g_0 satisfy our conditions on P, that is, ZFC $\vdash P(f_0)$ and ZFC $\vdash \neg P(g_0)$. Then, define

$$h(m, x) = \theta(m, x) f_0 + (1 - \theta(m, x)) g_0.$$

There goes (2). Now, let us specify θ so that the corresponding Diophantine equation $p = 0$ is never solvable in the standard model for arithmetic, while that fact can neither be proved nor disproved in ZFC. We then form, for such an indicator function,

$$h = \theta f_0 + (1 - \theta) g_0.$$

This settles (1). Finally, for (3), we note that as K is recursively enumerable but not recursive, it satisfies the conditions in the Gödel–Ehrenfeucht–Mycielski theorem about the length of proofs. □

4. Higher Degrees

Proposition 4.1. *If* T *is arithmetically sound, then we can explicitly and algorithmically construct in the language* $\mathcal{L}_T$ *of* T *an expression for the characteristic function of a subset of* ω *of degree* $\mathbf{0}''$.

Remark 4.2. We can obtain an expression in a recursive way, but such an expression isn't computable — every effort to compute it falls into an infinite loop.

That expression depends on recursive functions defined on ω and on elementary real-defined and real-valued functions plus the absolute value function, a quotient and an integration, or perhaps an infinite sum, as in the case of the β and θ functions associated to the halting problem.

4.1. *Higher arithmetic degrees*

We can explicitly obtain an expression for the $\theta^{(n)}$ function which is the characteristic function for a set of degree $\mathbf{0}^{(n)}$, and we can form a sequence:

$$\theta, \theta'', \theta''', \ldots, \theta^{(n)}, \ldots, \theta^{(\omega)},$$

where $\theta^{(\omega)}$ is obtained by diagonalization in the usual way. So, we can state:

Proposition 4.3 (More about Chaitin–Thorp). *Given the conditions of Proposition* 2.4, *any such* Ψ *gives rise to infinitely many problems of arbitrary arithmetic degree and even beyond the arithmetic hierarchy.*

And so it goes... It well fulfills the expectations stated in Thorp's address, as described by Arturo Sangalli in his novel.

For mathematical details, see [7].

5. Conclusion: Heraclitus

Heraclitus said something similar to the content of the Sangalli–Thorp conjecture, or to its answer in the Chaitin–Thorp theorem, in one of his most obscure utterances. Let's quote it; it is the Diels–Kranz fragment DK 22 B 124:

Fragment 124 :
Théophraste, Métaphysique, 15.

ἄλογον δὲ κἀκεῖνο δόξειεν ἂν, εἰ ὁ μὲν ὅλος οὐρανὸς καὶἕκαστα τῶν μερῶν ἅπαντ᾽ ἐν τάξει καὶ λόγῳ, καὶ μορφαῖς καὶ δυνάμεσιν καὶ περιόδοις, ἐν δὲ ταῖς ἀρχαῖς μηθὲν τοιοῦτον, ἀλλ᾽ ὥσπερ σάρμα εἰκῇ κεχυμένον ὁ κάλλιστος, φησὶν Ἡράκλειτος, [ὁ] κόσμος.

We now translate the final words in fragment 124:

> *Garbage accidentally having been thrown away, the most beautiful Order.*

Order and Disorder (to copy a technique much used by Martin Heidegger, the division of a word into its meaningful units, to emphasize how a new meaning arises out of the component submeanings, if we may say so), Order and Disorder are two sides of the same question. There is Order in the depths of mathematics, even if it is very very difficult to grasp it; it appears to us as an unfathomable, hellish, *selva selvaggia ed aspra e forte/che nel pensier' rinova la paura.* It frightens us, and it hides from us the hugely complex Order of the mathematical world.

Acknowledgments

The author acknowledges the support from CNPq-Brazil, Philosophy Program. He also wishes to thank C. A. Cosenza and F. Duarte at the Production Engineering Program, COPPE, UFRJ, for supporting this research project. Finally, he thanks A. Sangalli for a very helpful and interesting exchange on the Sangalli–Thorp conjecture.

Bibliography

[1] B. F. Caviness, *J. ACM* **17**, 385 (1970).
[2] G. J. Chaitin, *Algorithmic Information Theory*, Cambridge University Press (1987).
[3] N. C. A. da Costa and F. A. Doria, *Int. J. Theor. Phys.* **1041** (1991).

[4] N. C. A. da Costa and F. A. Doria, *Found. Phys. Lett.* **4**, 363–373 (1991).
[5] N. C. A. da Costa and F. A. Doria, in J. Echeverría *et al.* (eds.), *The Space of Mathematics*, pp. 168–191, Walter de Gruyter, Berlin (1992).
[6] N. C. A. da Costa and F. A. Doria, in P. Bystrov *et al.* (eds.), *Philosophical Logic and Logical Philosophy*, pp. 93–119, Kluwer (1996).
[7] N. C. A. da Costa and F. A. Doria, in C. Calude (ed.), *Randomness and Complexity: From Leibniz to Chaitin*, pp. 25–68, World Scientific (2007).
[8] N. C. A. da Costa, F. A. Doria and J. A. de Barros, *Int. J. Theor. Phys.* **29**, 935 (1990).
[9] M. Davis, H. Putnam and J. Robinson, *Ann. Math.* **74**, 425 (1961).
[10] M. Davis, *Amer. Math. Monthly* **80**, 233 (1973).
[11] M. Davis, Yu. Matijašević and J. Robinson, in F. E. Browder (ed.), *Mathematical Developments Arising from Hilbert Problems*, Proc. Symp. Pure Math., Vol. 28, American Mathematical Society (1976).
[12] F. A. Doria, in S. Wuppuluri and F. A. Doria (eds.), *The Map and the Territory*, Springer (2018).
[13] L. Jantscher, *Distributionen*, Walter de Gruyter (1971).
[14] J. P. Jones, *J. Symbol. Logic* **47**, 549 (1982).
[15] J. P. Jones and Yu. Matijašević, *J. Symbol. Logic* **49**, 818 (1984).
[16] Yu. I. Manin, *A Course in Mathematical Logic*, Springer (1977).
[17] E. Post, *Bull. Amer. Math. Soc.* **50**, 284 (1944).
[18] H. G. Rice, *Trans. Amer. Math. Soc.* **74**, 358 (1953).
[19] D. Richardson, *J. Symbol. Logic* **33**, 514 (1968).
[20] H. Rogers Jr., *Theory of Recursive Functions and Effective Computability*, McGraw-Hill, New York (1967).
[21] A. Sangalli, *Pythagoras' Revenge*, Princeton University Press (2009).
[22] B. Scarpellini, *Zeitschr. Math. Logik Grundlagen Math.* **9**, 265 (1963).
[23] J. Shoenfield, *Mathematical Logic*, Addison-Wesley (1967).
[24] Private comment by P. Suppes.
[25] P. Wang, *J. ACM* **21**, 586 (1974).
[26] K. Yosida, *Functional Analysis*, Springer (1980).

Chapter 8

Rules, Networks, and Evolution: On Chaitin's Algorithmic Information Theory and the Threefold Complexity of the Law

Ugo Pagallo*

In his work on complexity, information theory and randomness, Chaitin comes up with a remarkable idea on the equivalence between the rule of law and rule of reason. The more laws are complicated, the more they become arbitrary and incomprehensible. This chapter expands Chaitin's remark, according to a threefold perspective. First, the law as a set of rules can properly be conceived of as an algorithm: the application of Chaitin's ideas on complexity and the algorithmic information theory to the legal field follows as a natural result. This is what the chapter dubs as the algorithmic complexity of the law. Second, the complexity of the law can also be approached through network theory-methodologies and hence, formalized through a mathematical graph that consists of vertices and links: on this basis, we can properly distinguish between two magnitudes of complexity that regard the creation of the law (legislation), and its application (jurisprudence). This is what the chapter presents as the topological complexity of the legal phenomenon. Third, we should further distinguish between the complexity of the law and its own evolution through the emergence of unintentional orders: while human constructivism and political planning can only achieve certain amounts of complexity, spontaneous orders are not necessarily complex, but can comprise more particular facts than any brain could manipulate.

*University of Turin, Italy; ugo.pagallo@unito.it.

This is what the chapter defines as the ontological complexity of the law, which brings us back to Chaitin's ideas on mathematics, complexity, and randomness. Even in the realm of the law, it seems fair to admit that the world, that is, the legal phenomenon is more complex than its own language.

1. Introduction

In the appendix to Epistemology as *Information Theory: From Leibniz to* Ω, we find one of the rare pages in which — as far as I know — Gregory Chaitin deals with the law and the complexity of the legal phenomenon: "the rule of law is equivalent to the rule of reason, but if a law is sufficiently complicated, then it can in fact be completely arbitrary and incomprehensible." Although Chaitin draws this parallel almost en passant, he comes up with a remarkable idea. It merits to be expanded here, according to a threefold perspective. First, the law as a set of rules can properly be conceived of as an algorithm: the application of Chaitin's ideas on complexity and the algorithmic information theory to the legal field follows as a natural result. The law cannot only be understood in terms of information but more particularly, as a set of rules or instructions for the determination of every legal subject of the system. This sort of legal information 'for' reality goes hand-in-hand with a tenet of the rule of law, namely, the general and abstract nature of the legal rules and their standards. Once a statute, or an act, is simply tailored for the needs of an individual, that set of rules would not compress the complexity of life into the law. Rather, such statute or act should be declared unconstitutional. This is what happened with Article 1 of the Italian Act n. 51 from 7th April 2010 and the ruling n. 23 from 2011 of the Constitutional Court in Rome. On the one hand, the act concerned the 'legitimate impediment' of Mr. Berlusconi, namely, the legal reasons for which an accused party cannot show up in court. On the other hand, the Italian Supreme Court declared Article 1(3–4) to be invalid, for the provisions were set up in order to meet the necessities of the former national prime minister. This equivalence between the rule of law and the rule of reason is that which this chapter addresses in connection with the algorithmic complexity of the law.

Second, the complexity of the law can also be approached through the methodologies of network theory and hence, formalized through a mathematical graph that consists of vertices and links: on this basis, we can properly distinguish between two magnitudes of complexity that regard the creation of the law (legislation), and its application (jurisprudence). The differentiation is crucial because there often are concise acts, or statutes, that nevertheless spark disagreement among judges and scholars. Moreover, according to certain legal traditions, such as the case of American realism, the law appears as a sort of prophecy of what the courts will do in fact. Correspondingly, we should add a new kind of information to the previous one on legal information 'for' reality. By stressing the distinction between legislation and jurisprudence, between creation and application of the law, attention will be drawn to the law as legal information 'about' reality. While legal information 'for' reality refers to the signaling function of the law, legal information 'about' reality has to do with the semantic content of the information that refers to the states of the world and which is alethically qualifiable. This view is closely related to our ordinary experience every time we visit a lawyer's office or a clerk to the justices. All in all, what we hope is that such lawyer, clerk, or attorney, will speak the truth. Going back to the formalization of network theories, this second kind of complexity is what this chapter sums up as the topological complexity of the legal phenomenon.

Third, we should further distinguish between the complexity of the law and its own evolution through the emergence of unintentional orders: while human constructivism and political planning can only achieve certain amounts of complexity, spontaneous orders are not necessarily complex, but can comprise more particular facts than any brain could manipulate. Interestingly, over the past years, work on legal networks has shed light on this kind of phenomenon, i.e., the emergence of spontaneous orders that occurs in the case law of different courts, such as the US Supreme Court in Washington, D.C., the European Court of Justice in Luxembourg, or the aforementioned Italian Constitutional Court (ICC). That means we should add another kind of legal information to our previous ones on legal information 'for' reality (e.g., legislation), and 'about' reality (e.g., how courts will decide a case). This further kind of legal information can be grasped as legal information 'as' reality, so as to underline the fact that data can be meaningful regardless of an intelligent producer/informer. After the algorithmic complexity of the law and the topological complexity of the legal phenomenon, this new level of legal intricacy is presented as the ontological complexity of the law.

In light of this threefold level of complexity and of legal information, the overall aim of the chapter is to assess another profound insight of Chaitin's work. One of his claims in *Meta-Math!* is "that math does contain irreducible information, of which Ω is the prime example... so the world of mathematical truth has infinite complexity, even though any given FAS only has finite complexity" (Chaitin 2005, at 133 and 141). From an ontological viewpoint, this amounts to say that the world of math is more complex than its own language. Is the same true in the legal domain?

2. The Algorithmic Complexity of the Law

As a sort of Virgil in Dante's *Comedy*, let us assume a special guide for our tour in the realm of legal complexity, namely, the great German polymath Gottfried W. Leibniz. Both Chaitin and myself deeply admire Leibniz as the greatest of modern philosophers: after all, Leibniz was both a great mathematician and a jurist. As a mathematician, suffice it to mention that "Leibniz invented the calculus, invented binary arithmetic, a superb mechanical calculator, clearly envisioned symbolic logic, gave the name to topology (analysis situ) and combinatorics, discovered Wilson's theorem... etc." (Chaitin 2005: 57). As a jurist, Leibniz championed a logical and formal axiomatic program for which deductions could be performed mechanically even in the field of the law. Since his early work on jurisprudence — from his *magister* paper on *Specimen quaestionum philosophicarum ex jure collectarum* (1664), to his *Habilitation* research known as *Disputatio de arithmetica de complexionibus* (1665), followed by *De casibus perplexis in iure* (1666), and moreover the *Dissertio de arte combinatoria* (1666), down to the summa *Nova methodus* from 1667 — Leibniz's overall ambition was to turn legal arguments into computing through combinatorial analysis, probability calculus, and binary arithmetic (Pagallo 2005 and 2016). In the words of *de Arte Combinatoria*, "there would be no more need of disputation between two philosophers than between two accountants. For it would suffice to take their pencils in their hands, and say to each other: Let us calculate."

The convergence between Leibniz the mathematician and Leibniz the jurist can be found in Sections V and VI of his *Discourse on Metaphysics* from 1686. On the one hand, according to Section V of the *Discourse*, Leibniz's argument is that science is possible because all the richness and diversity that we see in Nature is actually the product of a small number of laws of physics. God is parsimonious, and the apparent complexity we see everywhere hides an inner simplicity. On the other hand, as Leibniz observes in Section VI, we have to distinguish a world governed by law from one which is lawless. Although "we should bear in mind that God does nothing that isn't orderly," Leibniz invites us to "suppose that someone haphazardly draws points on a page, like people who practise the ridiculous art of fortune-telling through geometrical figures. I say that it is possible to find a single formula that generates a geometrical line passing through all those points in the order in which they were drawn" (Leibniz ed. 2007).

Yet, if there is always a mathematical equation passing through any finite set of points, we have a problem, since this standpoint cannot enable us to distinguish between a lawful and a lawless system. As Hermann Weyl pointed out in his lectures on the metaphysical implications of science, "if arbitrarily complex theories are permitted then the concept of theory becomes vacuous because there is always a theory" (Weyl 1932). Therefore, both the concepts of law and simplicity require an accompanying concept of complexity in order to be meaningful. This is where Chaitin's algorithmic information theory

(AIT) has intervened, making Leibniz's ideas on law and simplicity more precise. Going back to Sections V and VI of the *Discourse on Metaphysics*, we can summarize Leibniz's theses with the following diagram (Chaitin 2005, at 64):

$$\text{ideas} \rightarrow \textbf{Mind of God} \rightarrow \text{universe.}$$

The diagram suggests that, according to Leibniz, although the universe appears very complicated, the ideas are simple: God minimizes the left-hand side of the diagram, and maximizes the right-hand size. In epistemological terms, empirical data is represented as points in a plane, i.e., pairs of real numbers, whereas the theory appears as a mathematical equation for a curve passing through those points. It goes without saying that the equation must be simple. AIT refines such ideas, by changing the context, namely, empirical data is understood as a finite string of data, while the theory is formalized as a software program that calculates exactly the empirical data. The number of bits of theory must be much smaller than the number of bits of data. In other words, the data is algorithmically compressed into the theory, as the following diagram illustrates:

$$\text{binary program} \rightarrow \textbf{Computer} \rightarrow \text{binary output.}$$

The fundamental idea of Chaitin is to define the complexity or algorithmic information content of something as the size in bits of the smallest program for calculating it. AIT solves the problem of the dependence of this concept on the programming language being employed, by identifying those languages that have the most concise programs. Accordingly, AIT considers only binary programs, bit strings p, and uses universal Turing machines U(p) with the property that for any other computer C there is a bit string π_C such that $U(\pi_C p) = C(p)$. As a result, the machine U can simulate the machine C by adding a fixed prefix π_C to each binary program p for C that indicates which computer to simulate.

In addition, the machine U should also be self-delimiting, namely, as U reads the program p, it does not have to read beyond the end of p. Rather, U gets by itself where p ends without a blank or other end-marker symbol. This has the important effect that the set of all possible computer programs attains a natural probability measure, for the probability of a program p is merely one over two raised to the size in bits of p. Accordingly, Leibniz's ideas on the "most perfect order," that is, as said in Section VI of the *Discourse*, "the order that is at once simplest in general rules and richest in phenomena," can be grasped as the smallest self-delimiting program that produces a given output. The size of that program in bits defines the complexity of the output (Chaitin 2005: 65).

Significantly, there have been many attempts to grasp on this basis how complex a legal act is. Since legal acts are increasingly translated into machine-readable versions of the natural language text, we may assess the complexity of the output, by taking into account the size of the corresponding compliance software program. One of the most popular legal fields in which scholars test these ideas has been, for obvious reasons, tax law. In the wording of Ruhl and Katz (2015: 196), "tax software companies are essentially selling the simplification of the Tax Code for the user, yet they must provide a product that accurately calculates the user's tax liability under any scenario and thus must somehow pack all of the Tax Code's substance into the software program. The program itself, however, has its own internal coding complexity, which software developers measure through various technical metrics and which software companies seek to control for business-cost purposes." For practical reasons, therefore, it would be in the interest of tax software companies to develop and sell programs that are no more intricate than that which is needed for accurate tax liability calculations, namely, the complexity in bits of reliable tax compliance software.

Still, there are several open issues with this kind of approach to legal complexity. The first kind of problem is theoretical, and has been stressed by Chaitin in *Meta-Math!* The question revolves around whether the model we adopt is the smallest possible program for producing the output that it does. Going back to the diagrams of this section, we can say that any given set of mathematical axioms can only prove that finitely many programs are the most concise possible for their output. Otherwise, if we could prove that a program is the smallest possible one, that would enable us to find a more concise program for the same output (Chaitin 2005: 126).

Second, there is no agreement in the legal community on the set of metrics, or standards, for measuring the complexity of a legal text, let alone the normative viewpoint on how complex a legal act should be. Going back to the example of tax law, some propose to measure its complexity vis-à-vis the number of tax rates and special provisions; others suggest the readability of the text; and so forth (Ruhl and Katz 2015, at 196).

Third, we should consider that the enforcement of the law presents a complex level of its own. This is a point, interestingly, that even Leibniz would concede. After his research on legal automation and computation, e.g., his arguments from the 1665 *Disputation on conditions*, he indeed mitigated these views. In 1669, for instance, Leibniz admitted in the *Doctrine of conditions* that in certain legal fields, such as inheritance, some cases hinge on equity, rather than strict law. Ten years later, in a text on evidence and legal inference, that is, *De legum interpretatione, rationibus, applicatione, systemate* from 1679–81, Leibniz declared that not always legal reasoning gets proofs but rather, depends on legal presumptions.

By drawing the attention to such limits of today's state-of-the-art in the legal domain,[a] is there any other way in which we may grasp the complexity of the law?

[a]Admittedly, there is another open issue, which regards randomness and the role it plays in mathematics, biology, or quantum mechanics. Going back to the diagrams of this section, this amounts to say that if the output is random, then no compression is possible, and the input has to be the same size of the output. Leibniz, of course, would reject this conclusion, since nothing happens without a reason. We can skip, however, this new divergence between Leibniz's ideas and Chaitin's AIT in this context.

Intriguingly, this way does exist and has been suggested, once again, by our Dante's Virgil and Chaitin's favorite modern thinker, Leibniz.

3. The Topological Complexity of the Legal Phenomenon

Scholars have attempted to clarify Leibniz's ideas on the "most perfect order," by tracing the structure of the universe back to some simple initial and boundary conditions (Smolin 1991; Barbour 1994). As instances of laws that directly determine structure, think about Darwinian evolution, or the dynamical self-organization of structure (e.g., Prigogine 1980), up to the inflationary scenario in modern cosmology (e.g., Albrecht 2003). In the case of Leibniz, the structure of a world which is the richest in phenomena can be understood in light of a basic tenet of his metaphysics, namely the principle of individuation, or the "identity of indiscernibles." In the wording of Section IX of the *Discourse*, "it is never true that two substances are entirely alike, differing only in being two rather than one... Moreover, each substance is like a whole world, and like a mirror of God, or indeed of the whole universe, which each substance expresses in its own fashion — rather as the same town looks different according to the position from which it is viewed. In a way, then, the universe is multiplied as many times as there are substances, and in the same way the glory of God is magnified by so many quite different representations of his work" (Leibniz 2007).

Some models have been proposed, in order to capture the Leibnizian essence of a world that is the richest in phenomena, namely, a world that presents the maximal variety of beings, or "monads." The model presented by Julian Barbour appears here particularly fruitful, for it brings together our investigation on Leibniz's metaphysics and the complexity of the law. As to Leibniz's ideas on the maximal variety of the world, Barbour formalizes them through a mathematical graph that consists of vertices and links (Barbour 2009). In this model, all what counts is whether any two vertices are connected, whereas, in accordance with the principle of individuation, only one link can hold between two different vertices. The maximal-variety graph is accordingly the best of all possible worlds, in which the vertices "are more varied, more readily distinguished, than in any other graph of *N* vertices" (*op. cit.*, at 198). Furthermore, once transformed into some kind of fundamental description of the universe, this model resembles biology, in that "all of the entities in a maximal-variety configuration are created in a kind of ecological balance between competing individuals. Each is trying to be as individualistic as possible, but in a curious way this selfish behavior is necessary if anything is to exist at all" (*op. cit.*, at 202).

This network approach to Leibniz's metaphysics has been curiously proposed over the past years, in order to grasp the complexity of the legal phenomenon. The analysis starts with the construction of the legal network, which requires the identification of the nodes, e.g., judicial decisions and scholarly citations, together with the explicit annotation of the references among them, namely, the edges of the network. By considering the flow of information among the source nodes, scholars have found out, time and again, a power law distribution in such legal networks (Pagallo 2006; Agnoloni and Pagallo 2014). This means that a small number of cases, such as the decisions of the courts, or scholarly papers, receive a large number of citations, while most cases and papers have few citations or none at all. Although some coefficients of the network, such as its diameter and average path length, may vary, the power law distribution of information is at work in such different networks as the case law of the US Supreme Court (Fowler and Jeon 2008), of the EU Court of Justice (Malmgren 2011), or of the Italian Constitutional Court (Agnoloni and Pagallo 2014). This small-world complexity is illustrated with Fig. 1.

A possible explanation of the phenomenon is simple and straight. Some features of complex networks, as their power laws and the presence of hubs, optimize the flow of information in the system. Remarkably, this was the suggestion of Herbert Simon in his seminal *The Sciences of the Artificial* (1969, ed. 1996). Simon's notion of "nearly

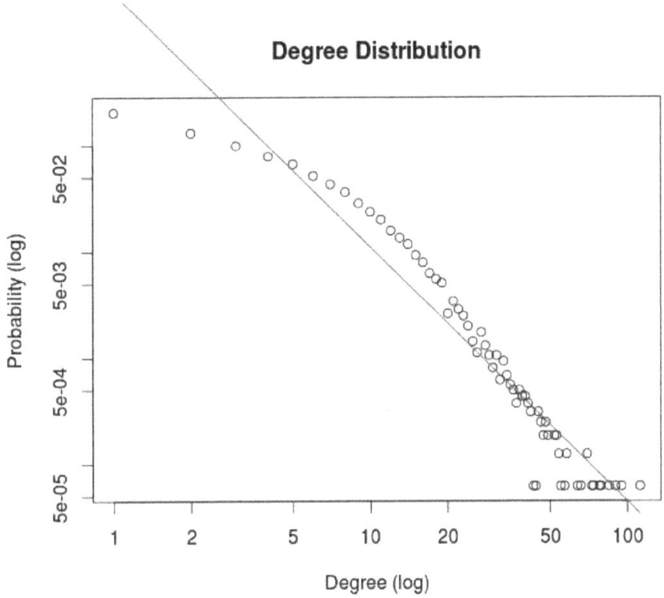

Fig. 1. Fitting of power law function with degree distribution in the case-law of the Italian Constitutional Court.

decomposable systems" proposes hierarchy as the clue for grasping the architecture of complexity, in that "most things are only weakly connected with most other things; for a tolerable description of reality only a tiny fraction of all possible interactions needs to be taken into account" (*op. cit.*) Additionally, Simon's "empty world hypothesis" can be properly grasped with the notion of hubs. It is this small fraction of nodes in the network with a much higher degree of connectivity than the average, which explains the clusters of dense interaction in the chart of informational exchange. Hubs offer the common connections mediating the short path lengths between the nodes of the network. These properties do not change vis-à-vis the legal technicalities of the system. For example, in the case of the Italian Constitutional Court, we can distinguish between the *in via incidentale* rulings, i.e., decisions on the constitutional legitimacy of norms raised by judges in a trial, and the *in via principale* rulings that regard the distribution of competences and powers between institutions. The

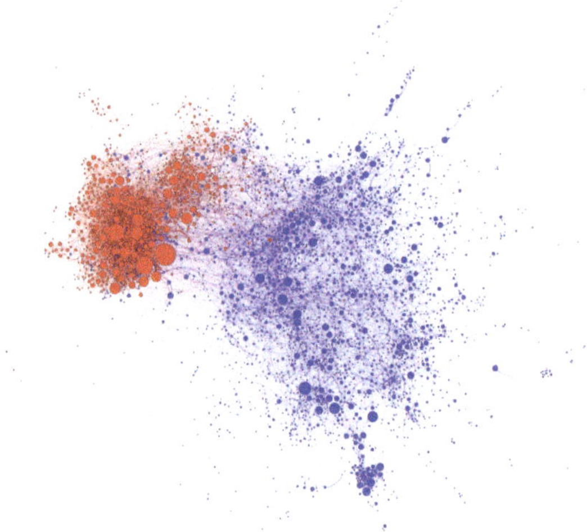

Fig. 2. The citation network of *in via principale* (red) and *in via incidentale* rulings (blue) of the ICC.

laws of informational flow remain the same, i.e., power-laws of distribution. Have a look at Fig. 2.

Once we have compressed the complexity of the law through its laws of informational flow, such a "family resemblance" between the case law of the US Supreme Court, of the EU Court of Justice, or of the Italian Constitutional Court (ICC), should not overlook the specificity of each legal system. All in all, as a matter of legal information "about" reality — rather than "for" reality or "as" reality — the differences between such networks can be illustrated with their semantic evolution. For instance, in the aforementioned 2008 work on the US Supreme Court jurisprudence, Fowler and Jeon showed that the most authoritative cases before the American civil war involved freedom of contract, namely the contract clause, whereas, after that war and until the end of the 1930s with the New Deal, the main core became balance of power in order to regulate commercial issues in a federal system. Then, since the late 1950s, the Supreme Court shifted its focus towards civil liberties and especially, freedom of speech. In the case of such legal network as the jurisprudence of the ICC, things are necessarily different. My work with Tommaso

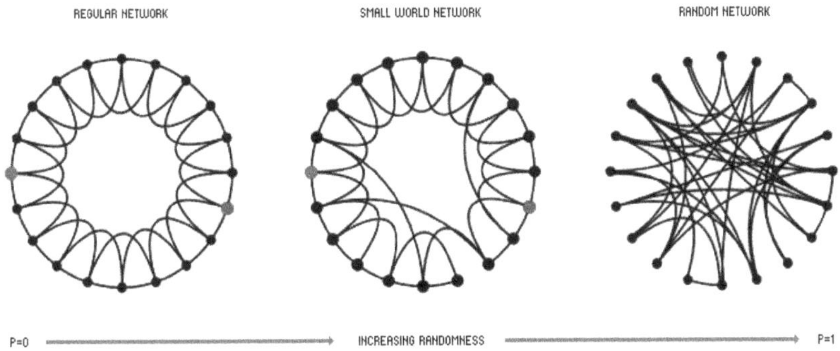

Fig. 3. Three models for networks complexity and randomness.

(Agnoloni and Pagallo 2014) has shown that over almost 60 years of evolution, the most authoritative cases of the ICC insist on two specific fields: 5 of the first 15 hubs in the network regard labour law, 3 criminal law (which become 6 if we take into account the first 25 hubs of the network). No Italian would be surprised by such results!

These outcomes do not mean that all legal systems should present small world-features, power laws of distribution, etc. In Boulet *et al.* (2010), for example, it is argued that the French environmental code presents a small world-structure and yet, it contrasts with the reference network of all the French legal codes. The latter shows "a rich club of ten codes very central to the whole... system," but this reference network has no small world properties at all. The result is particularly relevant in this context, because it stresses the differences between different models of (legal) networks in terms of randomness. In light of a new figure — see Fig. 3 — we should in fact distinguish between three toy models: regular, random, and in-between.

In a regular network, all of the nodes have the same number of links: this network has high clustering coefficients but a long diameter since the degree of separation between nodes is high. We may then refer to a random network with opposite features: it presents low clustering coefficients but a very short diameter. The explanation is that random links exponentially reduce the degree of separation between nodes in the network. In-between, there are the "small world-networks" which we found with the case law of the US Supreme Court, of the EU Court of Justice, of the ICC, etc. Their peculiarity depends on the apparent deviation from the properties of both regular and random networks. Like regular networks, small world-networks present high clustering coefficients, but they also share with random networks a short characteristic path length, i.e., the nodes of the network need few steps in order to reach each other. Therefore, once we have found the power law distribution of such legal networks, as in the case-law of Justices in Washington, in Luxembourg, in Rome, etc., we can accordingly flesh out the complexity of further legal networks — such as the reference network of all the French legal codes studied in (Boulet *et al.* 2010) — tracing them back to the tenets of Chaitin's AIT. The more such legal networks present random connections, the less we can compress the laws, according to which their information flows, the more such legal networks are complex.

On the other hand, the distinction between different legal networks sheds light on the kinds of legal information mentioned above in the introduction. More specifically, the discrepancies between the topological properties of some legal codes, as in e.g., France, and several case-law systems around the world, reminds us of the difference between legal information 'for' reality and legal information 'as' reality. In the first case, we referred to the signaling function of the law as a set of rules or instructions for the determination of every legal subject of the system. In the case of legal information 'as' reality, properties and regularities of (a specific domain of) the law are 'out there,' regardless of our awareness. This perspective has not only become popular nowadays in the field of evolutionary psychology, according to which basic human concepts would be wired in our brains thus grounding cultural evolution. In addition, this stance specifies what all the variables of the natural law tradition have in common, that is, the examination of being qua being, or its essence. This was in a way Leibniz's own stance, since old Roman law — in his view — would have captured most of the legal essence, paving the road to Leibniz's own calculations. Let us explore this third dimension of legal complexity, i.e., the ontological complexity of the law, separately in Section 4.

4. The Ontological Complexity of the Law

Legal information 'as' reality may refer to instructional information, namely, information that is meant to bring about something, and which regards the content of the information, i.e., semantic information. The assumption is that the law can be found in the nature of the individuals, of the world, of the things, i.e., the German "Natur der Sache," or in Nature as such, as in Kant's 1795 essay on the "perpetual peace" (Kant ed. 1996). An illustration of this stance on legal information 'as' reality is given by the Roman maxim, according to which "nobody is (legally) bound to that which is (physically) impossible." Key differences exist between modern theorizations of natural law and the state of nature, as in Locke, or in Kant, and the ancient school of Plato and Aristotle, to which Leibniz was indebted. Yet, despite these different instances of natural law, a common feature persists, namely, the idea that some basic principles and norms of the law could be deduced from an examination of being *qua* being.

A more recent illustration of this stance on legal information 'as' reality is given by Friedrich Hayek's philosophy of law. His ideas on the nature of the legal phenomenon and more particularly, his distinction between *kosmos* and *taxis*, i.e., between the emergence of spontaneous orders and human political planning, shed light on two crucial points. First of all, *pace* advocates of legal positivism — according to which, from Thomas Hobbes to Hans Kelsen, the only kind of legal information that matters is legal information 'for' reality — the law should be also understood in evolutionary terms, i.e., as an emergent property of social interaction. In the phrasing of *Law, Legislation and Liberty*, "man did not adopt new rules of conduct because he was intelligent. He became intelligent by submitting to new rules of conduct" (Hayek 1982: 163). Accordingly, we can grasp the law as a complex network, which adapts to the environment through learning and evolutionary processes, such as sophisticated signaling and information mechanisms. Collective behavior emerges from large networks of individual components, although no simple rules of operation direct them. Customs and social mores are examples of this kind of unintentional order. Further illustrations were given above in Section 3 with the formalization of three case-law networks. In all these cases, we deal with legal information 'as' reality and that which Hayek would call the *kosmos* side of the law.

Second, the distinction between *kosmos* and *taxis* matters in terms of complexity: "Spontaneous orders are not necessarily complex, but unlike deliberate human arrangements, they may achieve any degree of complexity. One of our main contentions will be that very complex orders, comprising more particular facts than any brain could ascertain or manipulate, can be brought about only through forces inducing the formation of spontaneous orders" (Hayek 1982: 38). This assumption entails two different meanings of complexity. One of them corresponds to the notion we have discussed so far in Chaitin's AIT terms of algorithmic compression: the more an order is complex, the less humans can ascertain or manipulate the set of facts comprised by such order. After all, no legislator can impose any custom, or mores, by decree. Still, we may wonder about the second meaning of complexity, that is, how a spontaneous order "achieves," in Hayek's terms, any degree of complexity. His examples on the market, or on legal rules, suggest that the achievements of spontaneous orders can be formalized through some elegant models. One of them was clarified in Section 3 with the case law of three courts. By fitting a power law function with the degree distribution of the network, the citation graph exhibits the properties of a scale free network. Furthermore, the topological properties of the network do not change, by restricting the analysis to a specific time period so as to follow the evolution of the network.

On this basis, we may suspect that acts and statutes of lawmakers are more complex than the case-law of courts and still, this hypothesis — which was discussed in Section 3 with the reference network of all the French legal codes studied in (Boulet *et al.* 2010) — does not contradict the further assumption that legal information 'as' reality, or the ontological dimension of the law, is more complex than *taxis* and, for that matter, than the formalization of legal language. This is the idea we found in Hayek's second meaning of complexity. Over the past years we started understanding how certain legal networks evolve (Pagallo 2006); and yet, it seems fair to admit that we are far away from obtaining satisfying models, in order to predict how a certain domain of the law may change. This attempt reminds me of Chaitin's work on making biology mathematical (e.g., Chaitin 2012), an attempt that can be further illustrated in the legal domain with my own research on scholarly work on the rulings of the ICC, i.e., the Italian Constitutional Court (Agnoloni and Pagallo 2015). What model can our stance propose for the evolution of this particular network on scholars, rather than judges, or justices?

An interesting outcome of this investigation has to do with the "scale free" features of the net of scholarly opinions, and the fact that some rulings of the Court with some relevant degree are not commented by scholars and experts at all. In particular, this is what occurred with 3 out of the first 30 top rank cases from the 2011 case law of the ICC; 5 out of the first 30 in 2012; 4 out of the first 30 in 2013; and so forth. This oblivion may be explained with the small world properties of the web of scholarly contributions and their in-degree values. Since most of the Italian experts seem to publish only on a restricted number of constitutional rulings, some of these rulings, although relevant, are simply overlooked. Hence, I proposed a prediction with Tommaso Agnoloni (and Pagallo 2015). We might suspect that the higher a case is ranked in the citation network of a court, the higher the probability that scholars will have to reflect on such case sooner, rather than later. The probability that a case will gain attention from scholars can be computed as the joint probability that a case has high ranking in the network of the court, and a low ranking in the web of scholarly opinions. This interplay can also be approximated in terms of incoming citations respectively from court cases and from scholars, as:

$$P_{\mathrm{SCHOLAR}} = (\mathit{CaseIndeg}/\mathit{MaxCaseIndeg}) \times (1 - \mathit{ScholarIndeg}/\mathit{MaxScholarIndeg}).$$

I admit that further work is needed to test the consistency of this conjecture. Finding out how a specific legal network, such as the web of scholarly opinions on the ICC rulings, may evolve, concerns a kind of exercise on legal information 'about' reality. Focus is indeed on the semantic content of the information that refers to the states of the world and is qualifiable in terms of truth. This kind of inquiry goes hand-in-hand with the increasing complexity of the events and environment that the legal language has to address and formalize. As mentioned above in Sections 2 and 3, such a relation between law 'as' information and the formalization of its language can explain many problems Leibniz had with his program on legal automation and computation. On the one hand, in the phrasing of a Leibniz's early essay on philosophy of law, "as to us, we hope that all cases can be decided on grounds of mere law" (in Artosi *et al.* 2013: 82). On the other hand, the method Leibniz had in mind, by means of which it would have been possible to move from the minimal agreement involved in the initial mutual tolerance of the parties to the maximal congruence achieved through the total agreement on the solution of the case, never materialized (Pagallo 2016). I suspect the reason is structural, so to speak, for the limits of Leibniz's project did not simply depend on what Bertrand Russell used to say about Leibniz squandering his time and immense energy, by meeting requests, curiosity and whims of princely employers (Russell 1900: 1). Rather, the limits of Leibniz's enterprise concern what we mentioned above in the introduction, with Chaitin's remarks on the complexity of the world of mathematical truth, which is infinite, and the finite complexity of any given FAS (Chaitin 2005, at 133 and 141). Regardless of whether the ontological complexity of the law is infinite or not, Chaitin's remarks on meta-math hold true in this context. In Hayek's phrasing, "the problem of the origin and function of these rules which govern both action and thought is therefore a problem wholly distinct from the problem of how they came to be articulated in verbal form. There is little doubt that even today the rules which have been thus articulated and can be communicated by language are only a part of the whole complex of rules that guide man's actions as a social being. I doubt whether anyone has yet succeeded in articulating all the rules which constitute 'fair play', for example" (Hayek 1982: 76).

Remarkably, we find 21 different definitions of fairness in today's computer science research (Narayanan 2018). Such classification mostly hinge on the legal meaning of the concept, as the prohibition to discriminate against an individual, or a group, on the basis of some criteria like sex, race, language, political or religious opinions, property and social origin, and so forth. Whereas the legal reference is Art. 7 of the 1948 Universal Declaration of Human Rights and, in Europe, the 1950 Convention on Human Rights, it is noteworthy how often current algorithmic definitions of fairness contradict each other (Kleinberg *et al.* 2016; Chouldechova 2017). From the limits of language and its formalization, however, it does not follow the arbitrariness of the results in this kind of research. Once again, Leibniz's example is instructive. Although he conceded that, at times, legal reasoning depends on presumptions, rather than proofs, on equity, rather than strict law, Leibniz never gave up his logical tour de force. Such limits stimulated his search for new kinds of logic, e.g., non-deductive forms of legal reasoning, that three centuries later would have been further developed by philosophical reflection and formalized with the study of logic and artificial intelligence, up to the logical analysis of legal arguments (Artosi and Sartor 2016; Pagallo 2016). Although the legal phenomenon appears more complex than its own language — as further confirmed by the incompleteness of every list of legal sources, namely, every fact or act which has or triggers legal effects — this sort of ontological incompleteness does not prevent us from increasing our understanding of the law as legal information 'for' reality, 'about' reality, and 'as' reality.

The time is ripe for the conclusions of this chapter.

5. Conclusions

Few people would doubt the law is something complex. The problem is how complex the law is and, from a normative perspective, how much complex the law should be. Too little of complexity — as occurs with some acts especially tailored for the needs of an individual — and such acts and norms should be deemed as against the rule of law. Too much of complexity, and the legal system would be incomprehensible. Correspondingly, both notions of law and complexity require a supplementary concept, in order to be meaningful. This is where Chaitin's AIT helped us in Section 2, suggesting to grasp the intricacy of the law in terms of algorithmic compression and the smallest self-delimiting program that produces a given output. On this basis, the next step of the analysis was to distinguish three different kinds of legal information and hence, three types of legal complexity. In addition to the algorithmic complexity of the legal phenomenon, Sections 3 and 4 focused on the topological and ontological complexity of the law, respectively.

In particular, Section 2 dwelt on the complexity of legal information 'for' reality and current work on the intricacy of reliable law-compliance software. We still lack common metrics, or standards, and a shared vision of justice, to determine the size of the program in bits that best defines the complexity of any output. By distinguishing between creation and application of the law, between legislation and jurisprudence, between legal information 'as' reality and 'about' reality, Section 3 thus explored a complementary approach through the methodologies of network theory. The properties of some relevant case-law systems — such as the case law of the US Supreme Court, of the EU Court of Justice, or of the ICC — shed light on the 'small world' properties of these legal networks. From the stance of this model, some legislative texts, as the reference network of all the French legal codes, appeared more complex than the small world networks of the US Supreme Court, of the EU Court of Justice, etc. The explanation had to do with the increasing randomness of the links, e.g., articles of the French civil code, in the system. This result fits like hand in glove with the further conclusion, illustrated in Section 4, according to which legal information 'as' reality or the ontological dimension of the law is far more complex than both the acts of lawmakers and every formalization of the legal sources. Two further insights of Chaitin's AIT helped in this context. First, as occurs with the world of mathematical truths, also the legal phenomenon appears more complex than its own language. Second, once again in correspondence with the world of math, this ontological property of the law does not mean we cannot deepen our understanding of the legal phenomenon, of e.g., how the law develops and evolves.

Such remarks on the complexity of the law, on the emergence of spontaneous orders, on the limits of formalization, do not suggest that the intervention of lawmakers and the use of new technologies cannot impact the tenets of the legal phenomenon. We can even quantify the impact of this intervention through a network analysis. See for example Fig. 4 below, on how a 2001 reform of the Italian legislator affected the number of decisions given by the ICC in Rome.

In addition, the growing use of automated means for legal enforcement is deeply affecting — and even transforming — the very essence of the law. That is what occurs with the use of digital rights management techniques, full packet inspections and filtering systems in the field of copyright law, up to smart contracts through blockchain

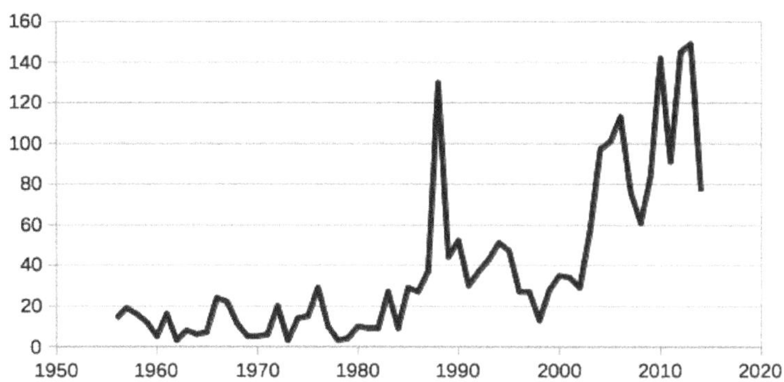

Fig. 4. Distribution in time of the ICC's *in via principale* rulings.

networks and some versions of the principle of privacy by design. Rather than a normative system that hinges on sanctions that should follow terms and conditions of legal accountability, the law is increasingly enforced through outcomes that automatically follow technical instructions, therefore wiping out the interpretation of the law and the exceptions that may be taken into account in each case. Such trends on the re-ontologization of the law have sparked a vibrant debate among scholars, in terms of human dignity and democracy, paternalism and the role of the courts. The more this process intensifies, the less our legal systems will be complex, since no room would be left for the evolution of the law and the understanding of highly context-dependent notions, such as Leibniz's idea of equity, or some shared vision of justice. The result would be the loss of what we dubbed in this chapter as the ontological complexity of the law. Only a weakened version of the algorithmic dimension of the law would be left, namely, the size in bits of the self-enforcing program that automatically predetermines the complexity of the legal output (e.g., a ruling). Chaitin's AIT would still valid in this world and yet, his remarks on the equivalence between the rule of reason and the rule of law, as well as the parallel we proposed — drawing on his insights — between math and the law, warn us against this scenario.

Chaitin once referred to creativity and the beauty of evolution as opposed and superior to any mechanical process in which everything "that ever comes out was already contained in your starting point" (Chaitin 2005, at 142). By considering the troubles of today's legal systems with automation and the use of self-enforcing technologies, Chaitin's remarks hold true in the legal domain as well.

Bibliography

Agnoloni, T. and U. Pagallo (2016) The Case Law of the Italian Constitutional Court between Network Theory and Philosophy of Information, *Informatica e diritto*, XXV, 1, pp. 139–152.

Agnoloni, T. and U. Pagallo (2015) The Power Laws of the Italian Constitutional Court, and Their Relevance for Legal Scholars, in A. Rotolo (eds.), *Legal*

Knowledge and Information Systems - JURIX 2015: The Twenty-Eighth Annual Conference, pp. 1–10. Amsterdam: IOS Press.

Albrecht, A. (2003) Cosmic Inflation and the Arrow of Time, in J.D. Barrow, P.C.W. Davies and C.L. Harper (eds.), *Science and Ultimate Reality: From Quantum to Cosmos*, Cambridge University Press.

Artosi, A., Pieri, B. and G. Sartor, eds. (2013) *Leibniz: Logico-Philosophical Puzzles in the Law: Philosophical Questions and Perplexing Cases in the Law*, Dordrecht: Springer.

Artosi, A. and G. Sartor (2016) Leibniz as Jurist, in M.R. Antognazza (ed.), *The Oxford Handbook of Leibniz*, Oxford University Press.

Barbour, J. (1994) On the Origin of Structure in the Universe, in E. Rudolph and I.-O. Stamatescu (eds.), *Philosophy, Mathematics, and Modern Physics*, Berlin: Springer.

Barbour, J. (2009) The Deep and Suggestive Principles of Leibnizian Philosophy, in S. Phineas Upham (ed.), *All We Need Is a Paradigm: Essays on Science, Economics, and Logic from The Harvard Review of Philosophy*, Chicago and La Salle, Illinois: Open Court.

Boulet, R., Mazzega, P. and Bourcier, D. (2010) Network Analysis of the French Environment Code, in *AI Approaches to the Complexity of Legal Systems. Complex Systems, the Semantic Web, Ontologies, Argumentation, and Dialogue*, edited by P. Casanovas *et al.*, 39–53. Springer, Dordrecht.

Chaitin, G. (2005) *Meta Math!: The Quest for Omega*, New York: Pantheon.

Chaitin, G. (2012) *Proving Darwin: Making Biology Mathematical*, New York: Pantheon.

Chouldechova, A. (2017) Fair Prediction with Disparate Impact: A Study of Bias in Recidivism Prediction Instruments, *Big Data*, 5(2): 153–163.

Fowler, J. H. and S. Jeon (2008) The Authority of Supreme Court Precedent, *Social Networks*, 30: 16–30.

Hayek, F. A. (1982) *Law, Legislation and Liberty. Vol. 1: Rules and Order.* Chicago: University of Chicago Press.

Kleinberg, J., Mullainathan, S. and M. Raghavan (2016) Inherent Trade-offs in the Fair Determination of Risk Scores, *Proceedings of Innovations in Theoretical Computer Science (ITCS)*.

Leibniz, G. W. (2007) *Discourse on Metaphysics*, English trans. and edition by J. Bennett. Available at http://www.earlymoderntexts.com/assets/pdfs/leibniz1686d.pdf (last accessed: 29 March 2017).

Malmgren, S. (2011) *Towards a Theory of Jurisprudential Relevance Ranking: Using Link Analysis on EU Case Law.* Master of Laws degree at Stockholm University under the supervision of C. Magnusson Sjöberg.

Narayanan, A. (2018) 21 Fairness Definitions and their Politics. Technical Report, *Conference on Fairness, Accountability and Transparency*, Tutorial.

Pagallo, U. (2005) *Introduzione alla filosofia digitale: da Leibniz a Chaitin*, Torino, Giappichelli.

Pagallo, U. (2006) *Teoria giuridica della complessità*, Torino, Giappichelli.

Pagallo, U. (2010) As Law Goes By: Topology, Ontology, Evolution, in *AI Approaches to the Complexity of Legal Systems. Complex Systems, the Semantic Web, Ontologies, Argumentation, and Dialogue*, edited by P. Casanovas *et al.*, 12–26. Springer, Dordrecht.

Pagallo, U. (2016) *Leibniz: una breve biografia intellettuale*, Padova: Kluwer.

Prigogine, I. (1980) *From Being to Becoming*, San Francisco: Freeman.

Russell, B. (1900) *A Critical Exposition of the Philosophy of Leibniz*, London: Allen & Unwin.

Ruhl, J. B. and Katz, D. M. (2015), Measuring, Monitoring, and Managing Legal Complexity, *Iowa Law Review*, 101, 191–244.

Simon, H. A. (1996) *The Sciences of the Artificial.* MIT Press, Cambridge, Mass. and London.

Smolin, L. (1991) Space and Time in the Quantum Universe, in A. Ashtekhar and J. Stachel (eds), *Conceptual Problems in Quantum Gravity: Proceedings of the 1988 Osgood Hill Conference, Einstein Studies*, vol. 2, Boston: Birkhäuser.

Weyl, H. (1932) *The Open World*, Yale University Press.

Chapter 9

Learning the Undecidable from Networked Systems

Felipe S. Abrahão[*,§], Ítala M. L. D'Ottaviano[†,¶], Klaus Wehmuth[*,||], Francisco Antonio Doria[‡] and Artur Ziviani[*,]**

This chapter presents a theoretical investigation of computation beyond the Turing barrier from emergent behavior in distributed systems. In particular, we present an algorithmic network that is a mathematical model of a networked population of randomly generated computable systems with a fixed communication protocol. Then, in order to solve an undecidable problem, we study how nodes (i.e., Turing machines or computable systems) can harness the power of the metabiological selection and the power of information sharing (i.e., communication) through the network. Formally, we show that there is a pervasive network topological condition, in particular, the small-diameter phenomenon, which ensures that every node becomes capable of solving the halting problem for every program with a length upper bounded by a loga-

[*]National Laboratory for Scientific Computing (LNCC) 25651-075, Petropolis, RJ, Brazil.
[†]Centre for Logic, Epistemology, and History of Science, University of Campinas (UNICAMP), Brazil.
[‡]Advanced Studies Research Group and Fuzzy Sets Laboratory, PIT, Production Engineering Program, COPPE, UFRJ, P.O. Box 68507, 21945-972, Rio de Janeiro, RJ, Brazil.
[§]fsa@lncc.br.
[¶]itala@cle.unicamp.br.
[||]klaus@lncc.br.
[**]ziviani@lncc.br.

rithmic order of the population size. In addition, we show that this result implies the existence of a central node capable of emergently solving the halting problem in the minimum number of communication rounds. Furthermore, we introduce an algorithmic-informational measure of synergy for networked computable systems, which we call local algorithmic synergy. Then, we show that such algorithmic network can produce an arbitrarily large value of expected local algorithmic synergy.

1. Introduction

We study the general problem of how computable systems may take advantage of an uncomputable environment to solve undecidable problems, in particular, the halting problem. Indeed, although the theory of Turing degrees is a well-established field in computability theory [20, 45] and mathematical logic [41], the possibility of computing beyond the Turing limit, that is, solving problems of Turing degree $\mathbf{0}'$ or above, is one of the major conundrums in the interface of theoretical computer science, mathematics, physics and biology. Such subject has its roots on the incompleteness results in foundational mathematics, mathematical logic, or recursion theory. This posits questions on the computability of the Universe [16, 18, 20–22, 34, 40], living systems [1, 18, 34, 40, 42], or the human mind [17, 18, 43, 50]. On the other hand, besides artificial systems, the computable nature of Life has been supported by current findings and scientific views in computer simulation and evolutionary computation [7, 23, 27, 44] and algorithmic-informational theoretical biology [1, 3, 12–15, 30, 31]. In this way, as a pure mathematical work inspired by important concepts from complex systems science, this chapter has an underlying objective of reconciling an hypothetical computable nature of Life (or artificial systems) with an hypothetical uncomputable Nature (or environment)

We study models of computation for synergistically solving an uncomputable problem in a network of computable systems. We show an abstract model for networked computable systems that can harness the power of random generation of individuals and the power of selection made by an irreducibly more powerful environment. In particular, we investigate the problem of how networked populations of randomly generated programs under metabiological selection of the fit in pervasive topological conditions can solve more and more instances of the halting problem by a fixed global communication protocol. For this purpose, we investigate a class of algorithmic networks based on [4,5] that can solve the halting problem by a synergistic communication protocol of imitation of the fittest neighbor. Thus, our theoretical results show that, within a hypercomputable environment as our assumption, whole populations of computable systems can be hypercomputers. To this end, we present definitions, theoretical models, and theorems based on computability theory, algorithmic information theory, distributed computing, and complex networks theory.

In this sense, this chapter shows how metabiology's early findings in [12, 13] have "opened the gates" not only to unify main concepts of theoretical evolutionary biology, theory of computation, and algorithmic information theory, proving the open-ended Darwin-like evolution [1, 15, 30, 31], but also to initiate in [4, 5] a unifying mathematical theoretical framework for complex systems under a both computational and informational perspective, which combines metabiology with statistical information theory, game theory, multi-agent systems, and network science. In this way, we will briefly discuss in Section 2 how metabiology and algorithmic networks may help bridge abstract formalizations of systemics from mathematical-logical approach in [8] to more realistic models of evolutionary [31] or distributed systems [26, 28, 39].

In this chapter, by adding one more important systemic property from the complex systems' "zoo" to this theoretical framework, we also introduce an algorithmic-informational measure of synergy and we apply it to our theoretical models, proving that the local algorithmic synergy in solving the halting problem can be as asymptotically large as one may want.

2. A Conceptual Background: From Systemics, Information, and Metabiology to Algorithmic Networks

We start by introducing in this section some of the conceptual background of our results. For this purpose, we will briefly cross over more abstract approaches to complex systems theory, intertwining foundational subjects and properties related to mathematical logic [8], information theory [28, 33] and metabiology [1, 12, 15, 31], in order to go toward the theory of algorithmic networks [2, 4, 5].

A system may be initially defined as a unitary entity of a complex and organized nature, made up of a set of active elements [8]. It is characterized as a partial structure with functionality. The general notion of relation used in the definition of system, that of partial relation, is an extension of the usual logical-mathematical concept of relation. It was presented by Mikenberg *et al.* [35] as basic to the introduction of the mathematical concept of pragmatic truth, later called quasi-truth, and has recently received various applications in logic and philosophy of science (see [24]). In this way, it enables one to formally accommodate the incompleteness of information relative to a certain domain of investigation.

Definition 2.1. Let D be a non-empty set. An n-ary partial relation R on D is an ordered term $\langle R_1, R_2, R_3\rangle$, R_1, R_2, and R_3, with no two of them having common elements, and whose union is D, such that:

(1) R_1 is the set of the n-tuples that we know belong to R;
(2) R_2 is the set of the n-tuples that we know do not belong to R;
(3) R_3 is the set of the n-tuples for which we do not know if they belong to R or not.

Definition 2.2. A partial structure is a pair $\langle D, R_i\rangle_{i\in I}$ with D a non-empty set and each R_i, where $i \in I$, a partial relation on D.

Definition 2.3. A system is a partial structure with functionality that can be denoted by

$$\mathbf{S} = \langle D_i, R_{ij}\rangle^{F_k}_{(i,j,k)\in I\times J\times K},$$

D_i being the universe of the partial structure, each R_{ij} a partial relation on D_i, F_k the functionality, with I, J, K being the respective variation indexes.

Note that a system without the possibility of structural or functional alterations has its universe and functionality constant and can be denoted by $\mathbf{S} = \langle D, R_i \rangle_{i \in I}^{F}$. For instance, this is the case for computable systems in which every transition of internal states with external inputs is a partial recursive function and its functionality is the very computation of this function.

We observe that if R_3 does not have any elements ($R_3 = \emptyset$), then R is a usual n-ary relation, identified with R_1, which brings this general definition of a system back to the one in [1]. Hence, as in [1], the immediate definition of a computable system, in which R is a computable (or recursive) relation and it represents a function, becomes well defined. See also [30] for a formalization of this idea in the context of evolutionary systems. In particular, the reader is invited to note that the oracle-sensitiveness [4] property of the population directly assures that the respective algorithmic networks always behave exactly like a total function. The same also holds for early metabiological models in [12] and for subcomputable or hypercomputable versions in [1].

Functionality is a teleological notion, characterized as a certain informational direction. It may be related to the system's goals, targets, or ends, and the potential autonomy of the system's components may lead to processes which are not individually, but instead globally, self-organized. The characteristics of the system may be considered emergences, with systemic synergy, globality and the possibility of novelty being considered among the first properties which appear in the constitution of the very existence of the system. In addition, a system is not completely isolated from its environment, because everything, matter, energy, information, that go into or out of the system comes from, passes through, or goes out to the environment. For example, in this regard, applications of statistical information theory to stochastic dynamical systems have already shown foundational results in defining and measuring such systemic properties [26, 33, 36, 38, 39]. However, apart from evolution as in [1, 12, 15, 31], such study of systemic properties have not been universally applied to deterministic systems. For this purpose, it leaded us to algorithmic networks, as we will explain below.

Creation may be the result of transformations conducted by spontaneous and autonomous activities, or from transformations conducted by constitutive and predetermined activities of elements of the system (and eventually boundary elements). It may be a new product, or be the result of a process of organizational transformation characterized by the formation of new structures or new functions. In both cases, creation may be thought of as the emergence of a system.

The process of evolution is characterized as the sequence of states of equilibrium and disequilibrium, manifested in the succession of distinct organizations which arise through the course of transformation of a system. If every organization that arises is considered a novelty, then one can affirm that evolution is a sequence of organizational innovations that may be rightly referred to as creative evolution. In this way, metabiology [1,12,15,31] gave a way to formalize such abstract notions of system, functionality, emergence, and creation in the context of evolutionary systems, whether subcomputable, computable or hypercomputable. In particular, such models show how sole subcomputable, computable, or hypercomputable systems can become more emergently creative (associated to an irreducible increase in the algorithmic information of the system's behavior) under successive random algorithmic mutations and selection of the fittest, which in turn make the environment define the teleological non-intrinsic functionality of the systems as being the increase of the fitness.

Following this same teleology, we showed that one can indeed formally capture the above notions for non-evolutionary systems, giving rise to a formal theory of algorithmic networks [4]. For example, the model's (non-intrinsic) functionality in [4] is to increase the average fitness of the nodes through the Busy Beaver imitation game (BBIG) at the expense of communication rounds, and the model's (non-intrinsic) functionality in [5] is to increase the average fitness of the nodes through the Busy Beaver imitation game (BBIG) under a Susceptible–Infected–Susceptible contagion scheme at the expense of communication rounds. Moreover, in this direction, whereas the models in [4, 5] do not formally assign any particular intrinsic common goal to the entire algorithmic network, this chapter shows how one can bring the notion of synergistic functionality — translated from statistical-informational measures of synergy (see Section 7) in stochastic dynamical systems to networked deterministic systems — to variations of such models.

3. Preliminary Definitions and Notation

We now restate some main definitions and notation on which the chapter results are based. For a complete introduction to these concepts, see [4, 46].

3.1. *Graphs and networks*

MultiAspect Graphs (MAGs) G are generalized representations for different types of graphs [46, 47]. In particular, a MAG represents dyadic relations between arbitrary n-tuples. Since we aim at a wider range of different network configurations, MAGs allow one to mathematically represent abstract aspects that may appear in complex high-order networks [48]. For example, these may be dynamic (or time-varying) networks, multicolored nodes or edges, multilayer networks, among others. Moreover, this representation facilitates network analysis by showing that their aspects can be isomorphically mapped into a classical directed graph [46]. Thus, the MAG abstraction has proved to be crucial in [4] to establish connections between the characteristics of the network and the properties of the population composed of theoretical machines.

Definition 3.1. Let $G = (\mathscr{A}, \mathscr{E})$ be a MultiAspect Graph (MAG), where $\mathscr{E}$ is the set of existing composite edges of the MAG and $\mathscr{A}$ is a class (or list) of sets, each of which is an *aspect*. Each aspect

$\sigma \in \mathscr{A}$ is a finite set and the number of aspects $p = |\mathscr{A}|$ is called the *order* of G. By an immediate convention, we call a MAG with only one aspect as a *first-order* MAG, a MAG with two aspects as a *second-order* MAG and so on. Each composite edge (or arrow) $e \in \mathscr{E}$ may be denoted by an ordered $2p$-tuple $(a_1, \ldots, a_p, b_1, \ldots, b_p)$, where a_i, b_i are elements of the i-th aspect with $1 \leq i \leq p = |\mathscr{A}|$.

$\mathscr{A}(G)$ denotes the set (or list) of aspects of G and $\mathscr{E}(G)$ denotes the *composite edge set* of G. We denote the i-th aspect of G as $\mathscr{A}(G)[i]$. So, $|\mathscr{A}(G)[i]|$ denotes the number of elements in $\mathscr{A}(G)[i]$. In order to match the classical graph case, we adopt the convention of calling the elements of the first aspect of a MAG as *vertices*. Therefore, we denote the set $\mathscr{A}(G)[1]$ of elements of the first aspect of a MAG G as $\mathrm{V}(G)$. Thus, a vertex should not be confused with a composite vertex.

Note that the terms *vertex* and *node* may be employed interchangeably in this chapter. However, we choose to use the term *node* preferentially in the context of networks, where nodes may realize operations, computations or would have some kind of agency, like in real networks or algorithmic networks. Thus, we choose to use the term *vertex* preferentially in the mathematical context of graph theory.

Dynamic networks represented by $G_t = (\mathrm{V}, \mathscr{E}, \mathrm{T})$ are time-varying graphs (TVGs) as defined in [19, 46]. These are a special case of second-order MAGs which have only one additional aspect relative to variation over time in respect to the set of nodes/vertices.

Definition 3.2. Let $G_t = (\mathrm{V}, \mathscr{E}, \mathrm{T})$ be a *time-varying graph* (TVG), where V is the set of vertices (or nodes), T is the set of time instants, and $\mathscr{E} \subseteq \mathrm{V} \times \mathrm{T} \times \mathrm{V} \times \mathrm{T}$ is the set of edges[a].

For the sake of simplifying our notations in the theorems below, one can take a natural ordering for $\mathrm{T}(G_t)$ such that

$$\forall i \in \mathbb{N} \; (0 \leq i \leq |\mathrm{T}(G_t)| - 1 \implies t_i = i + 1).$$

[a]That is, the set of existent (second-order) composite edges.

Definition 3.3. Let $d_t(G_t, t_i, u, \tau)$ be the minimum number of time intervals (non-spatial steps or, specially in this chapter, node cycles) for a diffusion starting on vertex u at time instant t_i to reach a fraction τ of vertices in the TVG G_t.

Definition 3.4. Let $D(G_t, t)$ denote the *temporal diffusion diameter* of the TVG G_t taking time instant t as the starting time instant of the diffusion process, that is,

$$D(G_t, t) = \begin{cases} \infty & \text{if } \exists u \in \mathrm{V}(G_t) \forall x \in \mathbb{N} \\ & \quad (x \neq d_t(G_t, t, u, 1)) \\ \max\{x \mid x = d_t(G_t, t, u, 1) \\ \quad \land u \in \mathrm{V}(G_t)\} & \text{otherwise} \end{cases}$$

Notation 3.1.1. Let $\lg(x)$ denote the binary logarithm $\log_2(x)$.

And, in the case the TVG G_t is connected, we have the following.

Definition 3.5. Let

$$\mathbb{G}_{sm}(f, t, 1) = \left\{ G_t \left| \begin{array}{l} \quad i = |\mathrm{V}(G_t)| \in \mathbb{N} \\ \land\ f(i, t, 1) = D(G_t, t) = \mathbf{O}(\lg(i)) \\ \land\ \forall i \in \mathbb{N}^* \exists! G_t \in \mathbb{G}(f, t, \tau)(|\mathrm{V}(G_t)| = i) \\ \land\ \forall u \in \mathrm{V}(G_t) \exists x \in \mathbb{N}(x = d_t(G_t, t, u, 1)) \end{array} \right. \right\},$$

where

$$\begin{aligned} f : \mathbb{N}^* \times X \subseteq \mathrm{T}(G_t) \times Y \subseteq]0, 1] &\to \mathbb{N} \\ (x, t, \tau) &\mapsto y, \end{aligned}$$

be a *family* of unique sized time-varying graphs that shares $f(i, t, 1) = D(G_t, t) = \mathbf{O}(\lg(i))$, where i is the number of nodes, as a common property.

3.2. *Formal languages, machines, and algorithmic information theory*

Notation 3.1. Let $(x)_2$ denote the binary representation of the number $x \in \mathbb{N}$. In addition, let $(x)_L$ denote the representation of the number $x \in \mathbb{N}$ in language L. Analogously, let $(w)_{10} \in \mathbb{N}$ denote the decimal representation of the string $w \in L$, where $w = ((w)_{10})_L$.

Notation 3.2. Wherever number $n \in \mathbb{N}$ appears in the domain or in the codomain of a partial (or total) function

$$\begin{aligned} \varphi_{\mathcal{U}} : L &\to L \\ x &\mapsto y = \varphi_{\mathcal{U}}(x), \end{aligned}$$

where $\mathcal{U}$ is a Turing machine, or an oracle Turing machine, running on language L, it actually denotes

$$(n)_L.$$

Definition 3.6. Let

$$\begin{aligned} T : \mathbf{L_U} \times \mathbf{L_U} &\to \mathbb{N} \\ (\mathrm{M}, p) &\mapsto T(\mathrm{M}, p) = n \end{aligned}$$

be the *partial* recursive function that returns the computation time that machine M takes to halt on input p.

As in [3, 12], we have the following definition.

Definition 3.7. Let

$$\begin{aligned} \mathrm{BB} : \{n \mid \min\{|s| \mid s \in \mathbf{L_U} \wedge \exists x(\mathbf{U}(s) = x)\} \leq n\} \subseteq \mathbb{N} &\to \mathbb{N} \\ n &\mapsto \mathrm{BB}(n) = k \end{aligned}$$

be the total[b] function that returns the largest integer that a program $p \in \mathbf{L_U}$ with length $\leq n \in \mathbb{N}$ can output running on machine $\mathbf{U}$. More formally,

$$\mathrm{BB}(n) = \max\{i \mid \mathbf{U}(p) = w \wedge i = (w)_{10} \wedge |p| \leq n\}.$$

[b]Without loss of generality, one can choose a universal self-delimiting programming language $\mathbf{L_U}$ in which there is $x \in \mathbf{L_U}$ such that $\mathbf{U}(w) = x$ and $|w| = \min\{|s| | s \in \mathbf{L_U}\}$, where $w \in \mathbf{L_U}$.

Notation 3.2.1. Let $\langle \cdot, \cdot \rangle$ denote an arbitrary recursive bijective pairing function. This notation can be recursively extended to $\langle \cdot, \langle \cdot, \cdot \rangle \rangle$ and, then, to an ordered tuple $\langle \cdot, \cdot, \cdot \rangle$. This iteration can be recursively applied with the purpose of defining finite-ordered tuples $\langle \cdot, \dots, \cdot \rangle$.

Notation 3.3. Let $\mathbf{L_U}$ denote a recursive binary self-delimiting (or prefix-free) universal programming language for a universal Turing machine $\mathbf{U}$ such that there is a concatenation of strings $w_1, \dots, w_k$ in the language $\mathbf{L_U}$, which preserves[c] the self-delimiting (or prefix-free) property of the resulting string, denoted by

$$w_1 \circ \cdots \circ w_k \in \mathbf{L_U}.$$

In addition, $\mathbf{L_U}$ is a complete binary code with

$$\sum_{p \in \mathbf{L_U}} \frac{1}{2^{|p|}} = 1.$$

The reader may also note that this self-delimiting-preserving concatenation "$\circ$" is just one example of recursive pairing bijective function $\langle \cdot, \cdot \rangle$. In addition, choosing between two distinct recursive pairing bijective functions $\langle \cdot, \cdot \rangle_1$ and $\langle \cdot, \cdot \rangle_2$ can only affect the algorithmic complexity[d] by

$$A(\langle w_1, w_2 \rangle_1) = A(\langle w_1, w_2 \rangle_2) \pm \mathbf{O}(1).$$

Therefore, the reader may equivalently replace[e]

$$w_1 \circ \cdots \circ w_k$$

with

$$\langle w_1, \dots, w_k \rangle$$

---

[c] For example, by adding a prefix to the entire concatenated string $w_1 w_2 \dots w_k$ that encodes the number of concatenations. Note that each string was already self-delimiting. See also [3, 4] for more discussions and properties of the notation "$\circ$".

[d] See Definition 3.10.

[e] Along with the appropriate reinterpretation of what is prefixes or suffixes in language $\mathbf{L_U}$.

in this chapter without affecting the final result.

Definition 3.8. Let $\mathrm{p}_T \in \mathbf{L_U}$ be any program of $\mathbf{U}$ that computes a partial recursive function such that

$$\mathbf{U}(\mathrm{p}_T \circ p) = (T(\mathbf{U}, p))_{\mathbf{L_U}},$$

where function $T(\mathrm{M}, p)$ holds as in Definition 3.6.

Definition 3.9. Let $\mathrm{p}_{+1} \in \mathbf{L_U}$ be any program of $\mathbf{U}$ that computes a partial recursive function such that

$$\mathbf{U}(\mathrm{p}_{+1} \circ p) = ((\mathbf{U}(p))_{10} + 1)_{\mathbf{L_U}}.$$

Definition 3.10. The (unconditional) *prefix algorithmic complexity* (also known as self-delimiting program-size complexity or Solomonoff–Komogorov–Chaitin complexity) of a finite binary string w, denoted by $A(w)$, is the length of the shortest program $w^* \in \mathbf{L_U}$ such that $\mathbf{U}(w^*) = w$.[f] The *conditional* prefix algorithmic complexity of a binary finite string y given a binary finite string x, denoted by $A(y|x)$, is the length of the shortest program $w^* \in \mathbf{L_U}$ such that $\mathbf{U}(\langle x, w^* \rangle) = y$. Note that $A(y) = A(y|\epsilon)$, where ϵ is the empty string. Similarly, we have the *joint* prefix algorithmic complexity of strings x and y defined by $A(x, y) = A(\langle x, y \rangle)$ or $A(x, y) = A(x \circ y)$, the *prefix algorithmic complexity of information* in x about y denoted by $I_K(x : y) = A(y) - A(y|x)$, and the *mutual algorithmic information* of the two strings x and y denoted by $I_A(x; y) = A(y) - A(y|x^*)$.

Now, since we will be dealing with a population of randomly generated arbitrary programs (i.e., Turing machines) in Section 5, we need to define a theoretical machine that can return fitness values for both halting and non-halting programs.

Definition 3.11. Let $\mathbf{L_U}$ be the recursive binary self-delimiting universal programming language $\mathbf{L_U}$ (as in Notation 3.3) for a universal Turing machine $\mathbf{U}$, where there exist a constant $\epsilon \in \mathbb{R}$, with

[f] w^* denotes the lexicographically first $\mathrm{p} \in \mathbf{L_U}$ such that $|\mathrm{p}|$ is minimum and $\mathbf{U}(p) = w$.

$0 < \epsilon \leq 1$, and a constant $C_L \in \mathbb{N}$ such that, for every $N \in \mathbb{N}$,

$$A(N) \leq \lg(N) + (1 + \epsilon)\lg(\lg(N)) + C_L.$$

We define an *oracle*[g] *Turing machine* $\mathbf{U}'$ such that, for every $w \in \mathbf{L_U}$,

$$\mathbf{U}'(w) = \begin{cases} \mathbf{U}(w)\text{“}+1\text{”} & \text{if } \mathbf{U} \text{ halts on } w, \\ \text{“0”} & \text{if } \mathbf{U} \text{ does not halt on } w. \end{cases}$$

Note that, since $\mathbf{L_U}$ is self-delimiting and $\mathbf{U}(w) \in \mathbf{L_U}$, we have that the operator “$+1$” actually means the successor operator in an arbitrary recursive enumeration of language $\mathbf{L_U}$. In the same manner, we have that “0” actually means $(0)_{\mathbf{L_U}}$.

Thus, the oracle Turing machine in Definition 3.11 is basically (except for a trivial bijection) the same as the chosen universal Turing machine. The oracle is only triggered to know whether the program halts or not in first place.

4. Previous Work on Algorithmic Networks

In this section, we remember the previous work [2,4–6] on which this chapter is based.

4.1. *Algorithmic networks*

We remember here the general definition of *algorithmic networks* $\mathfrak{N} = (G, \mathfrak{P}, b)$ in [4]. It is a triple $(G, \mathfrak{P}, b)$ defined upon a population of theoretical machines $\mathfrak{P}$, a generalized graph $G = (\mathscr{A}, \mathscr{E})$, and a function b that makes aspects of G to correspond to properties of $\mathfrak{P}$, so that a node in $\mathrm{V}(G)$ is mapped one-to-one to an element of $\mathfrak{P}$.

Definition 4.1. We define an *algorithmic network* $\mathfrak{N} = (G, \mathfrak{P}, b)$ upon a population of theoretical machines $\mathfrak{P}$, a MultiAspect Graph $G = (\mathscr{A}, \mathscr{E})$, and a function b that causes aspects of G to be mapped[h]

[g] Or any hypercomputer with a respective Turing degree higher than or equal to $\mathbf{0}'$.

[h] See Definition 4.1.1.

into properties of $\mathfrak{P}$, so that a vertex in $\mathrm{V}(G)$ corresponds one-to-one to a theoretical machine in $\mathfrak{P}$ and the communication channels through which nodes can send or receive information from its neighbors are defined precisely by composite edges in G.

We define a *population* $\mathfrak{P}$ as an ordered sequence $(o_1, \ldots, o_i, \ldots, o_{|\mathfrak{P}|})$,[i] where X is the support set of the population and f_o is a labeling surjective function

$$\begin{aligned} f_o : \mathfrak{P} = (o_1, \ldots, o_i, \ldots, o_{|\mathfrak{P}|}) &\to X \subseteq L \\ o_i &\mapsto f_o(o_i) = w, \end{aligned}$$

where L is the language on which the chosen theoretical machine U is running. Each member of this population may receive inputs and return outputs through communication channels. A *communication channel* between a pair of elements from $\mathfrak{P}$ is defined in $\mathscr{E}$ by a composite edge (whether directed or undirected) linking this pair of nodes/programs.

Third, we define function b as follows.

Definition 4.1.1. Let

$$\begin{aligned} b : Y \subseteq \mathscr{A}(G) &\to X \subseteq Pr(\mathfrak{P}) \\ \overline{\mathbf{a}} &\mapsto b(\overline{\mathbf{a}}) = \overline{\mathbf{p_r}} \end{aligned}$$

be a function that *maps* a subspace of aspects Y in $\mathscr{A}$ into a subspace of properties X in the set of properties $Pr(\mathfrak{P})$ of the respective population such that there is a bijective function

$$\begin{aligned} f_{V\mathfrak{P}} \colon \mathrm{V}(G) &\to \mathfrak{P} = \{o_i \mid f_o(o_i) = w \in L\} \\ v \quad &\mapsto \quad f_{V\mathfrak{P}}(v) = o_i, \end{aligned}$$

for every $(v, \overline{\mathbf{x}}) \in Y \subseteq \mathscr{A}(G)$ with $b(v, \overline{\mathbf{x}}) = (o_i, b_{\dim(Y)-1}(\overline{\mathbf{x}})) \in X$, where v is a vertex (or node) and o_i is an element of the sequence/population $\mathfrak{P}$.

---

[i] In which repetitions are allowed.

We say an element $o_i \in \mathfrak{P}$ is *networked* if and only if there exists $\mathfrak{N} = (G, \mathfrak{P}, b)$ such that o_i is running as a node of $\mathfrak{N} = (G, \mathfrak{P}, b)$, where $\mathscr{E}(G)$ is non-empty,[j] otherwise we say o_i is *isolated*, that is, it is only functioning as an element of $\mathfrak{P}$ and not as a node of $\mathfrak{N} = (G, \mathfrak{P}, b)$. We say that an input $w \in L$ is a *network input* if and only if it is the only external source of information every node/program receives and it is given to every node/program before the algorithmic network begins any computation. Note that letter w may also appear across the text as denoting an arbitrary element of a language. It will be specified in the assumptions before it appears or in statement of the respective definition, lemma, theorem or corollary.

Definition 4.2. A *node cycle* in a population $\mathfrak{P}$ is defined as a node/program returning an *output* (which, depending on the language and the theoretical machine the nodes are running on, is equivalent to a node completing a halting computation).

(1) If this node cycle is *not* the last node cycle, then its respective output is called a *partial output*, and this partial output is shared (or not, which depends on whether the population is networked or isolated) with the node's neighbors, accordingly to a specific information-sharing protocol (if any).
(2) If this node cycle is the last one, then this output is called a *final output* such that no more information is shared through the network.
(3) If every node/program in $\mathfrak{P}$ has completed its last node cycle, returning its final outputs, and the population $\mathfrak{P}$ is running networked by an algorithmic network $\mathfrak{N} = (G, \mathfrak{P}, b)$, then we say the algorithmic network $\mathfrak{N}$ as a whole completed an *algorithmic network cycle*.

---

[j]That is, there must be at least one composite edge connecting two elements of the algorithmic network.

In addition, let

$$\mathfrak{C} = \bigcup_{o_i \in \mathfrak{P}} \mathfrak{C}(o_i)$$

be the set of the maximum number of node cycles that any node/program o_i in the population $\mathfrak{P}$ would be able to perform in order to return a final output, where $\mathfrak{C}(o_i) \subseteq \mathbb{N}$ is the set of all node cycles that node/program o_i can perform.

In the particular case a population is defined on the language $\mathbf{L_U}$ and machine $\mathbf{U}'$, we assume the following notation.

Definition 4.3. Let $p_{\text{net}_\mathbf{U}}$ be a program such that $p_{\text{net}_\mathbf{U}} \circ o_i \circ c$ computes on machine $\mathbf{U}'$ cycle-by-cycle what a node/program $o_i \in \mathfrak{P}$ does on machine $\mathbf{U}$ until cycle c when networked by $\mathfrak{N} = (G, \mathfrak{P}, b)$. Let $p_{\text{iso}_\mathbf{U}}$ be a program such that $p_{\text{iso}_\mathbf{U}} \circ o_i \circ c$ computes on machine $\mathbf{U}'$ cycle-by-cycle what a node/program $o_i \in \mathfrak{P}$ does on machine $\mathbf{U}$ until cycle c when isolated. Let $p_{o_i,c}$ be the *partial output* sent by node/program o_i at the end of cycle c. Also, $p_{o_i,\max\{c|c\in\mathfrak{C}(o_i)\}}$ denotes the final output of the node/program o_i.

Definition 4.4. Let $\mathbf{X}_{\text{neighbors}}(o_j, c)$ be the set of incoming neighbors of node o_j that have sent partial outputs to it at the end of the cycle c when running on algorithmic networked $\mathfrak{N} = (G, \mathfrak{P}, b)$. Let $\{p_{o_i,c} \mid o_i \in \mathbf{X}_{\text{neighbors}}(o_j, c) \land i \in \mathbb{N} \land c \in \mathfrak{C}\}$ be the set of partial outputs relative to $\mathbf{X}_{\text{neighbors}}(o_j, c)$.

4.2. *Busy Beaver imitation game*

In [4], we have narrowed our theoretical approach by defining a class of algorithmic networks $\mathfrak{N}_{\text{BBsm}}(N, f, t, 1, j)$ — also denoted by a triplet as $(G_t, \mathfrak{P}_{\text{BB}}(N), b_j)$ — in which their populations $\mathfrak{P}_{\text{BB}}(N)$ and TVGs $G_t \in \mathbb{G}_{\text{sm}}(f, t, 1)$ have determined properties.

As defined in Section 4.1, each element of the population corresponds one-to-one to a node/vertex in G_t and each time instant in

G_t is mapped to a cycle (or communication round). These mappings are defined by the function b_j.

The population $\mathfrak{P}_{\mathrm{BB}}(N)$ is composed of randomly generated Turing machines (or randomly generated self-delimiting programs) which are represented in a self-delimiting universal programming language $\mathbf{L_U}$. This population is synchronous with respect to halting cycles, that is, in the end of a cycle (or communication round, as in distributed computing) every node returns its outputs at the same time. Nodes that do not halt in any cycle always return as final output the lowest fitness, that is, the integer value 0. Here, a straightforward interpretation is that nodes that eventually do not halt in a cycle are "killed", so that their final output has the "worst" fitness. Thus, these nodes are programs that ultimately run on an oracle Turing machine (or a hypercomputable system) $\mathbf{U}'$ — this requirement is also analogous to the one in [3, 14, 30], which deal with a sole program at the time and not with a population of them. However, the oracle is only necessary to deal with the non-halting computations. That is, $\mathbf{U}'$ behaves like an universal Turing machine $\mathbf{U}$ except that it returns zero whenever a non-halting computation occur.

In addition, the networked population $\mathfrak{P}_{\mathrm{BB}}(N)$ follows an *imitation-of-the-fittest protocol* (IFP), diffusing the information of the fittest randomly generated node (i.e., the node that partially outputs the largest integer in cycle 1).[k] Thus, every node in $\mathfrak{N}_{\mathrm{BBsm}}(N, f, t, 1, j)$ obeys the IFP, in which after the first cycle (i.e., after the first round of partial outputs) every node only imitates the neighbor that has partially output the largest integer, repeating this value as its own partial output in the next cycle. Thus, the main idea defining the IFP is a procedure in which each node o_i compares its neighbors' partial output (that is, the integer they have calculated in the respective cycle) and runs the program of the neighbor that have output the largest integer if and only if this integer is larger than the

[k]As in [2, 14, 15, 30], note that we still use the Busy Beaver function as a complexity measure for fitness. Therefore, the largest integer directly represents the fittest final output of a node.

one that the very node o_i has output. Since $\mathfrak{N}_{\text{BBsm}}(N, f, t, 1, j)$ is playing the Busy Beaver game [4] on a network while limited to simple imitation performed by a randomly generated population of programs, we say it is playing a *Busy Beaver Imitation Game* (BBIG). A (network) Busy Beaver game [4] is a game in which each player is trying to calculate the largest integer — as established as our measure of fitness or payoff[l] — it can, using the information shared by its neighbors. Thus, the BBIG is a special case of the Busy Beaver game.

Thus, $\mathfrak{N}_{\text{BBsm}}(N, f, t, 1, j)$ is a synchronous algorithmic network populated by N randomly generated nodes such that, after the first cycle (or arbitrary c_0 cycles), it starts a diffusion process of the biggest partial output (given at the end of the first cycle) determined by network G_t: at the first time instant each node may receive a network input w, which is given to every node in the network, and runs separately (i.e., not networked), returning its respective first partial output; then, the plain diffusion of large integers starts as determined by the IFP through the respective dynamical network G_t. At the last time instant contagion stops and one cycle (or more) is spent in order to make each node to return a final output.

Definition 4.5. Let

$$\mathfrak{N}_{\text{BBsm}}(N, f, t, 1, j) = (G_t, \mathfrak{P}_{\text{BB}}(N), b_j)$$

be an algorithmic network, where f is an arbitrary well-defined function such that

$$f : \mathbb{N}^* \times X \subseteq \mathrm{T}(G_t) \times Y \subseteq]0, 1] \to \mathbb{N}$$
$$(x, t, \tau) \mapsto y$$

and $G_t \in \mathbb{G}_{\text{sm}}(f, t, 1)$, $|\mathrm{V}(G_t)| = N$, $|\mathrm{T}(G_t)| > 0$, and there are arbitrarily chosen[m] c_0, $n \in \mathbb{N}$ with $c_0 + |\mathrm{T}(G_t)| + 1 \leq n \in \mathbb{N}$ such

[l]For more discussions, see [3, 4, 30].

[m]Since they are arbitrarily chosen, one may choose to take them as minimum as possible in order to minimize the number of cycles for example. That is, $c_0 = 0$ and $n = |\mathrm{T}(G_t)| + 1$ for example.

that b_j is an injective function, where

$$\begin{aligned} b_j : \mathrm{V}(G_t) \times \mathrm{T}(G_t) &\to \mathfrak{P}_{\mathrm{BB}}(N) \times \mathbb{N}|_1^n \\ (v, t_{c-1}) &\mapsto b_j(v, t_{c-1}) = (o_i, c_0 + c). \end{aligned}$$

Since the way time instants are mapped into cycles is fixed given values of c_0 and n, we may equivalently denote function b_j as

$$\begin{aligned} b_j : \mathrm{V}(G_t) &\to \mathfrak{P}_{\mathrm{BB}}(N) \\ v &\mapsto b_j(v) = (o_i). \end{aligned}$$

4.3. *Background results*

Following an algorithmic approach to evolutionary open-endedness (EvoOE), we have found in [4] that open-endedness may also emerge as an akin, but different, phenomenon to EvoOE: Instead of achieving an unbounded quantity of algorithmic complexity over time (e.g., after successive mutations), an unbounded quantity of emergent algorithmic complexity is achieved as the population/network size increases indefinitely. The algorithmic complexity of a node/program's final output when networked minus the algorithmic complexity of a node/program's final output when isolated formally defines an irreducible quantity of information that *emerges* with respect to a node/program that belongs to an algorithmic network. We call it as *emergent algorithmic complexity* (*EAC*) of a node/program. The reader may also find more discussions on emergence and open-endedness in [4, 5].

Formally, we have defined *average emergent open-endedness* in the context of general algorithmic networks as follows.

Definition 4.6. We say an algorithmic network $\mathfrak{N}$ with a population of N nodes has the property of *average* (*local*) *emergent open-endedness* (*AEOE*) for a given network input w in c cycles if and only if

$$\lim_{N\to\infty} \mathbf{E}_{\mathfrak{N}} \left(\overset{\text{net}}{\underset{\text{iso}}{\Delta}} A(o_i, c) \right) = \infty,$$

and, in the case of an algorithmic network $\mathfrak{N} = (G, \mathfrak{P}, b)$ with randomly generated nodes, we call this property as *expected (local) emergent open-endedness.* We have

$$\mathbf{E}_{\mathfrak{N}}\left(\underset{\text{iso}}{\overset{\text{net}}{\Delta}} A(o_i, c)\right) = \sum_{b} \frac{\frac{\sum_{o_i \in \mathfrak{P}} \Delta_{\text{iso}}^{\text{net}(b)} A(o_i,c)}{N}}{|\{b\}|},$$

which denotes the *average emergent algorithmic complexity of a node/program (AEAC)* in an algorithmic network $\mathfrak{N} = (G, \mathfrak{P}, b)$ with network input w.

Definition 4.7. The *emergent algorithmic complexity (EAC)* of a node/program o_i in c cycles is given in an algorithmic network that always produces partial and final outputs by

$$\underset{\text{iso}}{\overset{\text{net}(b)}{\Delta}} A(o_i, c) = A(\mathbf{U}(p^b_{\text{net}}(o_i, c)) - A(\mathbf{U}(p_{\text{iso}}(o_i, c)),$$

where $f_o(o_i) \in L$, $p^b_{\text{net}}(o_i, c)$ represents the program that returns the final output of o_i when networked assuming the position v, where $b(v, \bar{\mathbf{x}}) = (o_i, b_{\dim(Y)-1}(\bar{\mathbf{x}}))$, in the MAG G in the specified number of node cycles c with network input w, and $p_{\text{iso}}(o_i, c)$ represents the program that returns the final output of o_i when isolated in the specified number of node cycles c with network input w.

Note that the network input w was omitted in $p^b_{\text{net}}(o_i, c)$ and $p_{\text{iso}}(o_i, c)$, as presented in [4,5]. This is because in the models in [4,5] we were focusing on lower bounds on the expected emergent algorithmic complexity of a node and it was achieved by estimating the occurrence of fittest node/programs that ignores its inputs. However, for the following results in this chapter, we may equivalently denote $p^b_{\text{net}}(o_i, c)$ as $p^b_{\text{net}}(o_i, c, w)$ and $p_{\text{iso}}(o_i, c)$ as $p_{\text{iso}}(o_i, c, w)$.

In [4], we showed that there is a lower bound for the expected emergent algorithmic complexity in algorithmic networks $\mathfrak{N}_{\text{BBsm}}$ such that it depends on how much larger is the average diffusion density (in a given time interval) $\tau_{\mathbf{E}(\max)}(N, f, t, 1)|_t^{t'}$ compared to the cycle-bounded conditional halting probability $\Omega(w, c(x))$.

Theorem 4.1. *Let $w \in \mathbf{L_U}$ be a network input. Let $0 < N \in \mathbb{N}$. Let $\mathfrak{N}_{\mathrm{BBsm}}(N, f, t, 1, j) = (G_t, \mathfrak{P}_{\mathrm{BB}}(N), b_j)$ be well defined. Let $t_0 \leq t \leq t' \leq t_{|\mathrm{T}(G_t)|-1}$. Let*

$$c : \mathbb{N} \to \mathfrak{C}_{\mathfrak{B}\mathfrak{B}}$$
$$x \mapsto c(x) = y$$

be a total computable function where $c(x) \geq c_0 + t' + 1$. Then,

$$\begin{aligned}
&\lim_{N\to\infty} \mathbf{E}_{\mathfrak{N}_{\mathrm{BBsm}}(N,f,t,1)}\left(\overset{\mathrm{net}}{\underset{\mathrm{iso}}{\Delta}} A(o_i, c(x))\right) \\
&\geq \lim_{N\to\infty} (\tau_{\mathbf{E}(\max)}(N, f, t, 1)|_t^{t'} - \Omega(w, c(x))) \lg(N) \\
&- \Omega(w, c(x)) \lg(x) - 2\Omega(w, c(x)) \lg(\lg(x)) - A(w) - C_5.
\end{aligned}$$

This lower bound also depends on the parameter x for which one is calculating the number of node cycles. In fact, we have proved that our results hold even in the case of spending a computably larger number of node cycles compared to x. Furthermore, we have proved that the small-diameter phenomenon is a condition that ensure that there is a central time $t_{\mathrm{cen}_1}(c)$ to trigger expected emergent open-endedness.

Theorem 4.2. *Let $w \in \mathbf{L_U}$ be a network input. Let $0 < N \in \mathbb{N}$. If there exist $0 \leq z_0 \leq |\mathrm{T}(G_t)| - 1$ and ϵ, $\epsilon_2 > 0$ such that*

$$z_0 + f(N, t_{z_0}, 1) + 2 = \mathbf{O}\left(\frac{N^C}{\lg(N)}\right),$$

where

$$0 < C = \frac{\begin{array}{c}\tau_{\mathbf{E}(\max)}(N, f, t_{z_0}, 1)|_{t_{z_0}}^{t_{z_0+f(N,t_{z_0},1)}} \\ -\Omega(w, c_0 + z_0 + f(N, t_{z_0}, 1) + 2) - \epsilon\end{array}}{\Omega(w, c_0 + z_0 + f(N, t_{z_0}, 1) + 2)} \leq \frac{1}{\epsilon_2},$$

then, for every non-decreasing total computable function

$$c : \mathbb{N} \to \mathfrak{C}_{\mathfrak{B}\mathfrak{B}}$$
$$x \mapsto c(x) = y,$$

where t_{z_0}, $t_{z_0+f(N,t_{z_0},1)} \in \mathrm{T}(G_t)$ *and* $c(z_0 + f(N, t_{z_0}, 1) + 2) \geq c_0 + z_0 + f(N, t_{z_0}, 1) + 2$ *and* $\mathfrak{N}_{\mathrm{BBsm}}(N, f, t_{z_0}, 1, j) = (G_t, \mathfrak{P}_{\mathrm{BB}}(N), b_j)$ *is well defined, there exists* $t_{\mathrm{cen}_1}(c)$ *such that* $t_{\mathrm{cen}_1}(c) \leq t_{z_0}$.

Our proofs follow mainly from information theory, computability theory, and graph theory. Therefore, we have shown that there are topological conditions (e.g., the small-diameter phenomenon) that trigger a phase transition in which eventually the algorithmic network $\mathfrak{N}_{\mathrm{BBsm}}$ begins to produce an unlimited amount of bits of average local emergent algorithmic complexity/information. These conditions come from a positive trade-off between the average diffusion density and the number of cycles (i.e., communication rounds). Thus, the diffusion power of a dynamic (or static) network has proved to be paramount with the purpose of optimizing the average fitness/payoff of an algorithmic network that plays the Busy Beaver imitation game in a randomly generated population of Turing machines.

5. A Model of Algorithmic Network for Synergistically Solving a Common Problem

In this section, we present the model of algorithmic networks on which we will prove lemmas and theorems. In this way, we focus on the description and the definition of the new model.

The main idea that defines the algorithmic networks $\mathfrak{N}_{\mathrm{SBB}}$ is to formalize a distinct information-sharing (or communication) protocol that is based on the BBIG, so that nodes can use the largest integer the other nodes are calculating to nourish a global procedure in order to compute a function. So, this is a modification of the algorithmic networks $\mathfrak{N}_{\mathrm{BBsm}}$ in [4]. The latter only follows the IFP with a plain spreading of the largest integer, as described in Section 4.2. In particular, with respect to $\mathfrak{N}_{\mathrm{BBsm}}$ in [4], we will modify the functioning of the IFP in the last node cycle, just in the moment each node is about to return its final output. First, we will describe the properties of $\mathfrak{N}_{\text{SBB}}$ that are common to $\mathfrak{N}_{\text{BBsm}}$. Then, we will describe the functioning of the *synergistic imitation-of-the-fittest protocol* (SIFP) that is different from the IFP in $\mathfrak{N}_{\text{BBsm}}$.

As in [4], pursue overarching mathematical theorems, so we choose to deal with time-varying directed graphs [37, 49]. Note that the static case is covered by a particular case of dynamical networks in which the topology does not change over time. And the undirected case can be seen as a graph in which each undirected edge (or line) represents two opposing arrows. As defined in Section 3.1, $G_t = (\mathrm{V}, \mathscr{E}, \mathrm{T})$ are time-varying graphs (TVGs) as in [19, 46]. These are special cases of MAGs which have only one additional aspect relative to variation over time, besides the set of vertices.

As $\mathfrak{N}_{\text{BBsm}}$, the algorithmic networks $\mathfrak{N}_{\text{SBB}}$, which we will define in Definition 5.4, get their graph topologies from a family of dynamical networks that has a certain diffusion measure as a common feature, in particular, a small diameter compared to the network size (see Definition 3.4). In Definition 3.5, we define $\mathbb{G}_{sm}(f, t, 1)$ as a *family* of unique sized time-varying graphs which shares

$$f(i, t, 1) = D(G_t, t) = \mathbf{O}(\lg(i))$$

as a common property, where i is the number of nodes and $D(G_t, t)$ is the temporal diffusion diameter.

Moreover, as $\mathfrak{N}_{\text{BBsm}}$, the populations $\mathfrak{P}_{\text{SBB}}(N, s)$ of nodes/ programs in Definition 5.3 of $\mathfrak{N}_{\text{SBB}}$ are composed of randomly generated prefix Turing machines (or randomly generated self-delimiting programs) that are represented in a self-delimiting universal programming language $\mathbf{L_U}$. These populations are also synchronous with respect to halting cycles, that is, in the end of a cycle (or communication round, as in distributed computing) every node returns its partial and final outputs at the same time. Nodes that do not halt in any cycle always return as final output the lowest fitness/payoff, that is, the integer value 0. Here, a straightforward interpretation is that nodes that eventually do not halt in a cycle are "killed",[n] so that their final output has the "worst" fitness/payoff.

Now, unlike the networked population $\mathfrak{P}_{\mathrm{BB}}(N)$ in [4], described in Section 4.2, the networked population $\mathfrak{P}_{\mathrm{SBB}}(N, s)$ follows a modified version of the IFP. Instead of just returning as final output the largest integer shared by the neighbors, the SIFP ensures that, at the last cycle, every node employs the network input w together with the largest integer shared by its neighbors to calculate a partial computable function in such a way that every node returns as final output the value of

$$\mathbf{U}(s \circ x \circ w),$$

where x was the latest largest integer shared through the network. Note that this procedure is a *global* information-sharing (or communication) protocol. In summary, the SIFP makes every node obeys a protocol such that, after the first c_0 node cycles (i.e., after the first rounds of isolated partial outputs), every node is obliged to always imitate the neighbor (or itself, if the very node has partially output the largest integer in comparison with its neighbors' partial outputs) that has partially output the largest integer x, repeating this last value as its own partial output in the next node cycle. Since we are dealing only with synchronous algorithmic networks, these global communication protocols apply at the end of each node cycle (or communication round). Finally, the last node cycle is spent in order to cause each node to only return a final output in the form $\mathbf{U}(s \circ x \circ w)$. Thus, the SIFP is formally defined as follows.

Definition 5.1. Let $s \in \mathbf{L_U}$. We say a population $\mathfrak{P}$ follows a (global) *synergistic imitation-of-the-fittest protocol* (SIFP) for program $s \in \mathbf{L_U}$ if and only if every *networked* node/program always obeys the procedure:

for every o_j, $o_i \in \mathfrak{P}$ and c, $c - 1 \in \mathfrak{C}$,

[n]See also [4, 12, 13] for a complete evolutionary formalization of this property. Note that now there is a population of software, while in [12, 13] there is only one single organism at the time.

(a) if $\max\{c \mid c \in \mathfrak{C}\} = 1$, then

$$p_{o_j,c} = \mathbf{U}'(o_j \circ w);$$

(b) if $c = 1$ and $c \neq \max\{c \mid c \in \mathfrak{C}\}$, then

$$p_{o_j,c} = w \circ o_j \circ \mathbf{U}'(o_j \circ w);$$

(c) if $c \neq 1$ and $c \neq \max\{c \mid c \in \mathfrak{C}\}$, then

$$p_{o_j,c} = w \circ o_i \circ k,$$

where

$$k = \max\left\{ x \;\middle|\; \begin{array}{l} p_{o_j,c-1} = w \circ o_i \circ x \\ \lor\, w \circ o_i \circ x \in \left\{ p_{o_i,c-1} \;\middle|\; \begin{array}{l} o_i \in \mathbf{X}_{\text{neighbors}}(o_j, c-1) \\ \land\, i \in \mathbb{N} \\ \land\, c-1 \in \mathfrak{C} \end{array} \right\} \end{array} \right\}$$

(d) if $c = \max\{c \mid c \in \mathfrak{C}\}$ and $p_{o_j,c-1} = w \circ o_i \circ x$, then

$$p_{o_j,c} = \mathbf{U}(s \circ x \circ w).$$

It is important to remark that Definition 5.1 only applies to the networked case. If the population $\mathfrak{P}$ is isolated, no node can communicate with others. Therefore, as we will formalize in Definition 5.2, no protocol applies in the isolated case.

Definition 5.2. Let $\mathbf{L}_{\text{SBB}} \subset \mathbf{L}_{\text{U}}$ be a language of programs in the form $P_{\text{sprot}} \circ p \in \mathbf{L}_{\text{U}}$ where $p \in \mathbf{L}_{\text{U}}$. The prefix P_{sprot} is any program that always ensures that, if the node/program $P_{\text{sprot}} \circ p$ *is networked* and running on $\mathbf{U}'$, then $P_{\text{sprot}} \circ p$ obeys the *synergistic imitation-of-the-fittest protocol* as in Definition 5.1 for some arbitrary program $s \in \mathbf{L}_{\text{U}}$. Otherwise, if the node/program $P_{\text{sprot}} \circ p$ is *isolated* and running on $\mathbf{U}'$, then, for every $w \in \mathbf{L}_{\text{U}}$, $\mathbf{U}'(P_{\text{sprot}} \circ p \circ w) = \mathbf{U}'(p \circ w)$ and every subsequent node cycle works like a reiteration of partial outputs as immediate respective next inputs for the same program p.

Note that the isolated case may be equivalently represented by an algorithmic network built on a population of $p \in \mathbf{L}_{\text{U}}$ that does not follow any information-sharing protocol and the topology of the

MultiAspect Graph (MAG) is composed by one-step self-loops on each node only.

From Definition 5.2, we can now formalize the population $\mathfrak{P}_{\mathrm{SBB}}(N, s)$.

Definition 5.3. Let $\mathfrak{P}_{\mathrm{SBB}}(N, s)$ be the same population $\mathfrak{P}_{\mathrm{BB}}(N)$ in [4], except for using the language $\mathbf{L}_{\mathrm{SBB}}$ as the support set instead of $\mathbf{L}_{\mathrm{BB}}$.

Thus, we can now formally define the studied model of algorithmic networks $\mathfrak{N}_{\mathrm{SBB}}$ as a modification of the algorithmic networks $\mathfrak{N}_{\mathrm{BBsm}}$ in [4]. In summary, $\mathfrak{N}_{\mathrm{SBB}}$ is a synchronous algorithmic network populated by N randomly generated nodes (i.e., programs) such that, after the first cycle (or arbitrary c_0 cycles), it starts a diffusion process of the biggest partial output (given at the end of the first cycle) determined by the network topology of the TVG G_t. More specifically, before the first cycle each node receives a network input w, which is given to every node in the network; then, before the first network time interval, one or c_0 cycles are spent in which each node runs separately, repeating its respective first partial output that will be shared; from then on, as determined by the SIFP in Definition 5.1, the plain diffusion of larger integers starts through the respective dynamical network $G_t \in \mathbb{G}_{\mathrm{sm}}(f, t, 1)$, so that, at each time interval, the SIFP ensures that a fitter node always "infects" its immediate less fit neighbors; finally, at the last time instant, contagion stops and one cycle (or more) is spent in order to make each node return

$$\mathbf{U}(s \circ x \circ w),$$

where x was the latest largest integer shared by the neighbors at the previous node cycle, as final output. This way, we formally define the following.

Definition 5.4. Let $\mathfrak{N}_{\mathrm{SBB}}(N, f, t, 1, j, s)$ denote exactly the same algorithmic network $\mathfrak{N}_{\mathrm{BBsm}}(N, f, t, 1, j)$ in [4] (see Definition 4.5), except for replacing population $\mathfrak{P}_{\mathrm{BB}}(N)$ with $\mathfrak{P}_{\mathrm{SBB}}(N, s)$.

6. Solving the Halting Problem Through the Busy Beaver Imitation Game

In this section, we will prove lemmas, theorems, and corollaries with the purpose of showing that $\mathfrak{N}_{\mathrm{SBB}}$ are algorithmic networks capable of asymptotically solving the halting problem for every network input $w \in \mathbf{L_U}$ with $\lg(N) - C_7 \geq |w|$, where N is the network size (i.e., the number of nodes) and C_7 is a fixed constant that does only depend on the chosen universal programming language. Moreover, since $\mathfrak{N}_{\mathrm{SBB}}$ are dynamic networks whose functioning is based on the diffusion of the fittest node, we will also show that there is a central node for emergently solving the above halting problem in order that the number of necessary communication rounds are minimized. In particular, this node is associated with the highest time-reachability centrality among the nodes.

From [4, 5], it is important to remember that a network Busy Beaver game is a game in which each player is trying to calculate the largest integer it can using the information shared by its neighbors. For the present purposes, as in [4], the population $\mathfrak{P}_{\mathrm{SBB}}(N, s)$ in the studied algorithmic networks $\mathfrak{N}_{\mathrm{SBB}}$, which is playing a particular type of network Busy Beaver game, is in fact limited to simple imitation performed by a randomly generated population of programs. This Busy Beaver imitation game (BBIG) [4] is a particular case of the Busy Beaver game in which every node can only propagate the largest integer. It configures a simple imitation-of-the-fittest procedure. However, unlike [4, 5], the last node cycle is devoted to employ this diffusion of the fittest to solve a problem that can be *partially* computed by program s. Although partial or final outputs are always defined in algorithmic networks $\mathfrak{N}_{\mathrm{SBB}}$ due to the oracle-sensitiveness property [4] of the population $\mathfrak{P}_{\mathrm{SBB}}(N, s)$, these outputs may not match every function value for every input in some cases. For example, our central results in Theorem 6.1 demands a restriction on the domain of possible network inputs in order that every input in this domain generates the correct function value. Nevertheless, in the limit when the population grows indefinitely, one can say that an infinite family of algorithmic networks $\mathfrak{N}_{\text{SBB}}$ make every node asymptotically compute a total function (in the case, the very characteristic function of the halting problem).

While algorithmic networks $\mathfrak{N}_{\text{BB}}$ in [4] can be seen as playing an optimization procedure where the whole pursues the increase of the average fitness/payoff through diffusing on the network the best randomly generated solution, these algorithmic networks $\mathfrak{N}_{\text{SBB}}$ can be seen as playing an optimization procedure where the whole pursues the increase of each node's capability of solving a common problem through diffusing on the network the best randomly generated solution in the smallest number of communication rounds as possible [4]. Thus, the nodes in algorithmic networks $\mathfrak{N}_{\text{BB}}$ may be seen as competing with each other, as in multi-agent systems from a game-theoretical approach [2, 4, 5]; on the other hand, the present model $\mathfrak{N}_{\text{SBB}}$ may be seen as nodes/programs computing using network's shared information to solve a common purpose, as the classical approach in distributed computing. In this sense, this addition of perspective in such models of algorithmic networks is bridging a competition or individualistic-centered view of emergence to a synergistic-centered view of emergence. In Section 7, we will define and explore such synergy in algorithmic networks.

First, we define a total computable function that is capable of deciding whether a program halts or not if the respective large enough computation time is informed as input.

Definition 6.1. Let $\mathrm{p}_{\text{halt}} \in \mathbf{L_U}$ be a program of $\mathbf{U}$ that computes a *total* recursive function such that

$$\mathbf{U}(\mathrm{p}_{\text{halt}} \circ n \circ p) = \begin{array}{l} \text{the output of the algorithm that only lets} \\ \mathbf{U} \text{ runs with } p \in \mathbf{L_U} \text{ as its input} \\ \text{for } t \leq n+1 \text{ computation time until} \\ \text{it halts or not and, then,} \\ \text{returns } \begin{cases} (\text{“non-halting”})_{\mathbf{L_U}} & \text{if } t > n \in \mathbb{N}, \\ (\text{“halting”})_{\mathbf{L_U}} & \text{otherwise.} \end{cases} \end{array}$$

Note that $\mathrm{p}_{\mathrm{halt}}$ in Definition 6.1 always computes a total function for every $n \in \mathbb{N}$ and every $p \in \mathbf{L_U}$ because $\mathbf{U}$ can either halt or not halt on any program p in n computation time. As a consequence, we have the following lemma.

Lemma 6.1. *Let $N \geq |\mathrm{p}_{+1} \circ \mathrm{p}_T \circ p| \in \mathbb{N}$, where $p \in \mathbf{L_U}$. Then, for every $x \geq BB(N)$,*

$$\mathbf{U}(\mathrm{p}_{\mathrm{halt}} \circ x \circ p) = \begin{cases} (\text{"}\textit{non-halting}\text{"})_{\mathbf{L_U}} & \textit{if } \mathbf{U} \textit{ does not halt on } p, \\ (\text{"}\textit{halting}\text{"})_{\mathbf{L_U}} & \textit{if } \mathbf{U} \textit{ halts on } p. \end{cases}$$

Proof. The proof follows directly from Definition 6.1. Since $\mathbf{U}$ can only halt or, exclusively, not halt on p, we divide the proof into the following two cases:

(1) If the universal machine $\mathbf{U}$ halts on p, then the value $T(\mathbf{U}, p)$ will be well defined. Hence, the value $\mathbf{U}(\mathrm{p}_{+1} \circ \mathrm{p}_T \circ p)$ will also be well defined. Therefore, from Definition 3.7, we will have that $\mathrm{BB}(N) \geq (\mathbf{U}(\mathrm{p}_{+1} \circ \mathrm{p}_T \circ p))_{10} > (\mathbf{U}(\mathrm{p}_T \circ p))_{10} = T(\mathbf{U}, p)$. Then,

$$\begin{aligned} x \geq BB(N) &\implies x > T(\mathbf{U}, p) \implies \\ &\implies \mathbf{U}\left(\mathrm{p}_{halt} \circ x \circ p\right) = (\text{"}\textit{halting}\text{"})_{\mathbf{L_U}} \ ; \end{aligned}$$

(2) If the universal machine $\mathbf{U}$ does *not* halt on p, then the value $T(\mathbf{U}, p)$ will not be well defined. Therefore, for every $t \leq x$, we will have $T(\mathbf{U}, p) \neq t$. Then, $\mathbf{U}(\mathrm{p}_{\mathrm{halt}} \circ x \circ p) = (\text{"non-halting"})_{\mathbf{L_U}}$.

□

Now, from [4], we translate its first lemma to the new algorithmic network model $\mathfrak{N}_{\mathrm{SBB}}$, showing how to harness the implications of the law of large numbers in a program-size probability distribution [4].

Lemma 6.2. *Let $\mathfrak{N}_{\mathrm{SBB}}(N, f, t, 1, j, s) = (G_t, \mathfrak{P}_{\mathrm{SBB}}(N, s), b_j)$ be an algorithmic network as in Definition* 5.4. *Then, with probability arbitrarily close to* 1 *as N increases toward infinity, we will have that* *there are constants* C_{BB} *and* C_4 *such that*

$$A(x) \geq \lg(N) - C_4,$$

where

$$x = \max \left\{ y \;\middle|\; \begin{array}{l} p_{o_i,c-1} = w \circ o_k \circ y \;\wedge \\ o_i, o_k \in \mathfrak{P}_{SBB}(N,s) \wedge \\ c = \max\{c \mid c \in \mathfrak{C}\} \end{array} \right\},$$

$\max\{c \mid c \in \mathfrak{C}\} \geq 2$, *and* w *is the network input. In addition, for every* $o_i, o_k \in \mathfrak{P}_{\mathrm{SBB}}(N,s)$ *with* $p_{o_i,c-1} = w \circ o_k \circ x$, $c \geq c_0 + t + f(N,t,1) + 1$, *and* $c = \max\{c \mid c \in \mathfrak{C}\}$, *we will have*

$$x \geq \mathrm{BB}(\lg(N) - C_{\mathrm{BB}}).$$

Proof. This proof of $A(x) \geq \lg(N) - C_4$ is totally analogous to the proof of [4, Lemma 5.1]. Just note that, from Definition 5.4 and [4], we also have that $\mathfrak{N}_{\mathrm{BBsm}}(N,f,t,1,j)$ denotes the same algorithmic network $\mathfrak{N}_{\mathrm{BB}}(N,f,t,\tau,j)$ in [4], except for replacing family $\mathbb{G}(f,t,\tau)$ with family $\mathbb{G}_{\mathrm{sm}}(f,t,1)$. In addition, from Definition 5.1 and the definition of $A_{\max}$ in [4], we have that $A(x) = A_{\max}$, which follows from the fact that the IFP in [4] is only different from the SIFP in just ensuring that nodes returns x instead of $\mathbf{U}(s \circ x \circ w)$ in the last cycle as the final output. To show the second part of the Lemma 6.2 that $x \geq \mathrm{BB}(\lg(N) - C_{\mathrm{BB}})$, it suffices to note that, from Definition 5.1, every networked node only imitates the fittest neighbor after the first node cycle. Thus, since $c \geq c_0 + t + f(N,t,1) + 1$, we have from Definition 3.5 that the number of node cycles will cover the temporal diffusion diameter and it will be enough to make any fittest first partial output, which is at least as fit as $\mathrm{BB}(\lg(N) - C_{\mathrm{BB}})$, propagate to every other node. □

Thus, we can now combine the previous results to build a new theorem. Theorem 6.1 basically asymptotically assures that, if enough communication rounds are expended, matching the network temporal diffusion diameter (which is small compared to the population/network size), then every node is expected to solve the halting problem for program with length dominated by a logarithmic order of the population/network size. Therefore, increasing the population/network size can make such algorithmic networks emergently solve a increasing number of instances of the halting problem such that, in the limit when the population grows indefinitely, all instances of the halting problem will be statistically covered.

Theorem 6.1. *Let* $\mathfrak{N}_{\mathrm{SBB}}(N, f, t, 1, j, s) = (G_t, \mathfrak{P}_{\mathrm{SBB}}(N, s), b_j)$ *be an algorithmic network as in Definition* 5.4 *such that* $\mathfrak{N}_{\mathrm{SBB}}(N, f, t_{z_0}, 1, j, \mathrm{p}_{\mathrm{halt}})$ *is well defined. Let*

$$c : \mathbb{N} \to \mathfrak{C}_{\mathfrak{B}\mathfrak{B}}$$
$$x \mapsto c(x) = y$$

be a non-decreasing total computable function such that

$$c(z_0 + f(N, t_{z_0}, 1) + 2) \geq c_0 + z_0 + f(N, t_{z_0}, 1) + 2,$$

where $t_{z_0} \in \mathrm{T}(G_t)$. *Then, there is a constant* C_7 *such that, for large enough* N *and for every network input* $w \in \mathbf{L}_{\mathbf{U}}$ *with* $\lg(N) - C_7 \geq |w|$, *we will have that every node in* $\mathfrak{N}_{\mathrm{SBB}}(N, f, t_{z_0}, 1, j, \mathrm{p}_{\mathrm{halt}})$ *decides whether* $\mathbf{U}$ *halts or not on* w *in* $c(z_0 + f(N, t_{z_0}, 1) + 2)$ *node cycles with probability arbitrarily close to* 1.

Proof. This proof follows from combining Lemma 6.2 with Lemma 6.1. We have from Clause 5.1(d) in Definition 5.1 that, for every $o_i \in \mathfrak{P}_{\mathrm{SBB}}(N, \mathrm{p}_{\mathrm{halt}})$,

$$p_{o_i, c'} = \mathbf{U}(\mathrm{p}_{\mathrm{halt}} \circ x \circ w),$$

where $c' = c(z_0 + f(N, t_{z_0}, 1) + 2)$. In addition, we have from Lemma 6.2 that $x \geq \mathrm{BB}(\lg(N) - C_{BB})$ holds with probability arbitrarily close to 1 as the population size N tends to infinity. Let C_7 be a constant such that

$$|w| + C_7 \geq |\mathrm{p}_{+1} \circ \mathrm{p}_T \circ w| + C_{\mathrm{BB}}.$$

Therefore, for every network input $w \in \mathbf{L}_{\mathbf{U}}$ with $\lg(N) - C_7 \geq |w|$, we will have that $\lg(N) \geq |\mathrm{p}_{+1} \circ \mathrm{p}_T \circ w| + C_{\mathrm{BB}}$ and, hence, from

Lemma 6.1, that

$$p_{o_i,c'} = \begin{cases} (\text{“non-halting”})_{\mathbf{L_U}} & \text{if } \mathbf{U} \text{ does not halt on } w, \\ (\text{“halting”})_{\mathbf{L_U}} & \text{if } \mathbf{U} \text{ halts on } w. \end{cases}$$

□

Theorem 6.1 looks at all nodes in the algorithmic network. However, one may extract from this result the presence of privileged nodes in solving the respective halting problem. To this end, we first define a general node centrality for distributed processing.

Definition 6.2. Let $\mathfrak{N} = (G, \mathfrak{P}, b)$ be a well-defined algorithmic network in the language $\mathbf{L_U}$ and machine $\mathbf{U}'$. We define a *central node* o_{cen} that can compute a function

$$f : X \subseteq \mathbf{L_U} \to \mathbf{L_U}$$
$$w \mapsto f(w)$$

in the minimum number $c_{\min}$ of node cycles when networked and that does not compute this function when isolated in $c_{\min}$ node cycles if and only if for every $w \in X$, the final output $p_{o_{\text{cen}},\max\{c|c\in\mathfrak{C}(o_{\text{cen}})\}}$ of the networked node o_{cen} holds as

$$p_{o_{\text{cen}},c_{\min}=\max\{c|c\in\mathfrak{C}(o_{\text{cen}})\}} = f(w)$$

such that, for every $c' < c_{\min}$,

$$\mathbf{U}(p_{\text{iso}}(o_{\text{cen}}, c', w)) \neq f(w)$$

and, for every $o \in \mathfrak{P}$, if

$$\mathbf{U}(p_{\text{iso}}(o, c', w)) \neq f(w)$$

and

$$p_{o,c=\max\{c|c\in\mathfrak{C}(o)\}} = f(w),$$

then $c_{\min} \leq c$.

Now, we come back to node centralities in network science in order to rank the node that can be quickly accessible by an arbitrary diffusion from an arbitrary fraction of the nodes. From [19], we have the following definition.

Definition 6.3. Let $\overline{d_t}(G_t, t_i, u, \tau)$ be the minimum number of time intervals (non-spatial steps or, especially in this chapter, node cycles) for a diffusion starting on any vertex $v \in X$ of a fraction $\tau = |X|$ of vertices in the TVG G_t at time instant t_i to reach vertex u, where $X \in \mathcal{P}(\mathrm{V}(G_t))$ is arbitrary.

In this sense, if we consider any possible fraction of nodes at the same time, we can define a node centrality based on the temporal diffusion diameter [4].

Definition 6.4. Let G_t be a TVG with $D(G_t, t) \neq \infty$ and $|\mathrm{V}(G_t)| \geq 2$. We define the *time-reachability centrality* of a vertex u in the TVG G_t from time instant $t \in \mathrm{T}(G_t)$ as

$$\frac{1}{\overline{d_t}(G_t, t, u, 1)}.$$

Definition 6.4.1. We define the set of the vertices with time-reachability centrality $x \leq D(G_t, t) \neq \infty$ in a TVG G_t from time instant $t \in \mathrm{T}(G_t)$ as

$$\mathbf{X}_{\text{treach}}(G_t, t, x) = \left\{ u \,\middle|\, x = \frac{1}{\overline{d_t}(G_t, t, u, 1)} \wedge u \in \mathrm{V}(G_t) \right\}.$$

Note that the condition $D(G_t, t) \neq \infty$ immediately assures that Definitions 6.4 and 6.4.1 are well defined.

Finally, Theorem 6.1 implies that the node centrality for distributed processing and the node centrality for complex network's dynamics can be combined to find a node that can only solve the halting problem when networked in the least amount of communication rounds (i.e., node cycles).

Corollary 6.1.1. *Let* $\mathfrak{N}_{\mathrm{SBB}}(N, f, t, 1, j, s) = (G_t, \mathfrak{P}_{\mathrm{SBB}}(N, s), b_j)$ *be an algorithmic network as in Definition* 5.4 *such that* $\mathfrak{N}_{\mathrm{SBB}}(N, f, t_{z_0}, 1, j, \mathrm{p}_{\mathrm{halt}})$ *is well defined. Let*

$$c : \mathbb{N} \to \mathfrak{C}_{\mathfrak{B}\mathfrak{B}}$$
$$x \mapsto c(x) = y$$

be a non-decreasing total computable function with $t_{z_0} \in \mathrm{T}(G_t)$ *and* $c(z_0 + f(N, t_{z_0}, 1) + 2) \geq c_0 + z_0 + f(N, t_{z_0}, 1) + 2$. *Then, there is a constant* C_7 *such that, for large enough* N *and for every network input* $w \in \mathbf{L_U}$ *with* $\lg(N) - C_7 \geq |w|$, *there is at least one central node (as in Definition* 6.2*) with the respective highest time-reachability centrality (as in Definition* 6.4*) in the algorithmic network* $\mathfrak{N}_{\text{SBB}}(N, f, t_{z_0}, 1, j, \mathrm{p}_{\text{halt}})$ *that decides whether* $\mathbf{U}$ *halts or not on* w *in* $\mathbf{O}(\lg(N))$ *node cycles with probability arbitrarily close to* 1.

Proof. This proof follows from Theorem 6.1. Since $G_t \in \mathbb{G}_{\text{sm}}(f, t_{z_0}, 1)$, we will have that Definition 6.4 is well defined for every vertex in the TVG G_t. Now, we take a vertex $v \in \mathrm{V}(G_t)$ with the highest time-reachability centrality as in Definition 6.4 such that, for every $c'' < c'$,

$$\mathbf{U}(p_{\text{iso}}(o_{\text{cen}}, c'', w)) \neq f(w),$$

where $c' = c(z_0 + f(N, t_{z_0}, 1) + 2)$ and $b_j(v, t_{c-1}) = (o_{\text{cen}}, c_0 + c)$. Thus, we trim the necessary latest time instants in the set of time instant $\mathrm{T}(G_t)$ in order that one can define another TVG $G'_t = (\mathrm{V}(G_t), \mathscr{E}', \mathrm{T}')$ such that

$$\mathrm{T}'(G'_t) = \mathrm{T}(G_t) \setminus \{t | t > t_{z_0} + \overline{d_t}(G_t, t_{z_0}, v, 1)\}$$

and, for every $e \in \mathscr{E}'(G'_t)$, one has $e \in \mathscr{E}(G_t)$. Then, we replace G_t in $(G_t, \mathfrak{P}_{\text{SBB}}(N, s), b_j)$ with G'_t. Note that, from Definition 3.5, we have that

$$\overline{d_t}(G_t, t_{z_0}, v, 1) \leq D(G_t, t_{z_0}) = f(N, t_{z_0}, 1) = \mathbf{O}(\lg(N)).$$

Now, we take any non-decreasing total computable function

$$c : \mathbb{N} \to \mathfrak{C}_{\mathfrak{B}\mathfrak{B}}$$
$$x \mapsto c(x) = y$$

such that $t_{z_0} \in \mathrm{T}(G_t)$ and

$$c_0 + \mathbf{O}(\lg(N)) + 1 = c(z_0 + f(N, t_{z_0}, 1) + 2) \geq c_0 + z_0 + f(N, t_{z_0}, 1) + 2.$$

Therefore, from Definition 5.4, there is a correspondent node $o_{\text{cen}} \in \mathfrak{P}_{\text{SBB}}(N, \mathrm{p}_{\text{halt}})$ that assumes position of the vertex $v \in \mathrm{V}(G_t) =$

$\mathrm{V}(G'_t)$ such that, from Theorem 6.1,

$$p_{o_{\mathrm{cen}},c_0+\mathbf{O}(\lg(N))+1} = \begin{cases} (\text{“non-halting”})_{\mathbf{L_U}} & \text{if } \mathbf{U} \text{ does not halt on } w, \\ (\text{“halting”})_{\mathbf{L_U}} & \text{if } \mathbf{U} \text{ halts on } w. \end{cases}$$

□

7. Algorithmic Synergy

Following the same spirit from the emergent algorithmic complexity of a node introduced in [4, 5], another interesting topic is whether algorithmic networks and algorithmic information theory are sufficient to deal with the problem of measuring synergistic information [28,33] or not. This problem is usually stated within the context of multivariate information theory for stochastic dynamical systems. Generally speaking, it concerns measuring the amount of information in an arbitrary collection of random variables $X_1, \ldots, X_n$ that predicts another random variable Y, but that it is not contained in (or does not derive from) any individual random variable X_i, where $1 \leq i \leq n$, or from combinations of proper subsets of the set $\{X_1, \ldots, X_n\}$, which is given by the partial information diagrams (i.e., PI-diagrams) [28].

On the other hand, from [4, 5], note that emergent algorithmic complexity directly gives a formal measure of irreducible information [9–11, 29, 32] that emerges when comparing the networked case with the isolated case. Thus, if one assumes the definition of synergy as the general phenomenon in which the whole system is irreducibly better in solving a common problem than the "sum" (or the "union") of its parts taken separately, as the problem described in the previous paragraph, then there should be an immediate extension of emergent algorithmic complexity to algorithmic synergistic information.

To tackle this problem, we introduce in this section a formalization of one type of algorithmic synergistic information in the context of algorithmic networks. Thus, instead of studying synergy in stochastic processes, we will be studying synergy in deterministic systems, in particular, in networks of computable systems. That is, we are focusing on the general problem of measuring the amount of algorithmic information in an arbitrary collection of nodes necessary to calculate a function, but that could not be performed by any individual isolated node or by any combination of proper subnetworks of the entire network. In particular, we start by formalizing a measure of average algorithmic synergistic information for individual nodes when comparing the fully networked case with the isolated case. Thus, we leave the joint cases and subnetwork cases for future research.

Definition 7.1. Let $\mathfrak{N} = (G, \mathfrak{P}, b)$ be a well-defined algorithmic network. We define the *local algorithmic synergy* of a node $o_i \in \mathfrak{P}$ toward function

$$f : X \subseteq L \to L$$
$$x \mapsto f(x)$$

in c node cycles with network input w as

$$\overset{\mathrm{net}(b)}{\underset{\mathrm{iso}}{\Delta}} I_K((o_i, c, w) : f(w))$$
$$= I_K(\mathbf{U}(p^b_{\mathrm{net}}(o_i, c, w)) : f(w))$$
$$- I_K(\mathbf{U}(p_{\mathrm{iso}}(o_i, c, w)) : f(w)) \pm \mathbf{O}(1),$$

where $f_o(o_i) \in L$, $p^b_{\mathrm{net}}(o_i, c, w)$ represents the program that returns the final output of o_i when networked assuming the position v, where $b(v, \bar{\mathbf{x}}) = (o_i, b_{\dim(Y)-1}(\bar{\mathbf{x}}))$, in the MAG G in the specified number of node cycles c with network input w, and $p_{\mathrm{iso}}(o_i, c, w)$ represents the program that returns the final output of o_i when isolated in the specified number of node cycles c with network input w.

The reader may find tempting to employ $I_A(x; y)$ instead of $I_K(x : y)$ in Definition 7.1 due to the fact that $I_A(x; y)$ is invertible and $I_K(x : y)$ is not (see Definition 3.10 and [11, 25, 32]). In this sense, note that, since $I_K(x : y) = A(y) - A(y|x)$, we will have that

$$\overset{\text{net}(b)}{\underset{\text{iso}}{\Delta}} I_K((o_i, c, w) : f(w))$$
$$= A(f(w)|\mathbf{U}(p_{\text{iso}}(o_i, c, w))) - A(f(w)|\mathbf{U}(p^b_{\text{net}}(o_i, c, w))) \pm \mathbf{O}(1). \tag{1}$$

However, besides the outputs processed by the algorithmic network be in the form $\mathbf{U}(p)$ and not $\mathbf{U}(p)^*$, the non-invertibility of $I_K(x : y)$ actually captures the notion of towardness in computing the function f. For example, the algorithmic network may be emergently generating the necessary information to compute function f at the same time that function f does not give the necessary information to determine the emergent behavior of the algorithmic network. This way, the non-invertibility would be a sound property. In fact, investigating the cases in which the measure in Definition 7.1 is invertible is an interesting future research.

Moreover, a constant represented by $\mathbf{O}(1)$ in the Definition 7.1 is employed in order to deal with some non-intuitively correct cases that may appear depending on the chosen universal programming language. For example, when

$$A(f(w) \mid \mathbf{U}(p_{\text{iso}}(o_i, c, w))) < A(f(w) \mid \mathbf{U}(p^b_{\text{net}}(o_i, c, w)))$$
$$= A(f(w) \mid f(w))$$

with $\mathbf{U}(p_{\text{iso}}(o_i, c, w)) \neq f(w)$ or when

$$A(f(w) \mid \mathbf{U}(p_{\text{iso}}(o_i, c, w))) = A(f(w) \mid f(w))$$
$$> A(f(w) \mid \mathbf{U}(p^b_{\text{net}}(o_i, c, w)))$$

with $\mathbf{U}(p^b_{\text{net}}(o_i, c, w)) \neq f(w)$. Thus, there may be "blurred intervals" with respect to algorithmic synergy, so that one cannot decide whether there is a positive value of local algorithmic synergy or not. In fact, as odd as it may seem, it is in consonance with the equalities and inequalities in algorithmic information theory that hold, except for a constant that only depends on the chosen universal programming language. Note that these complexity/information oscillations are expected to happen in algorithmic information theory.

In order to specify the type of algorithmic network from which one is calculating the local algorithmic synergy, we may also denote $\Delta_{\text{iso}}^{\text{net}(b)} I_K((o_i, c, w) : f(w))$ by

$$\underset{\text{iso}}{\overset{\mathfrak{N}}{\Delta}} I_K((o_i, c, w) : f(w))$$

or

$$\underset{\text{iso}}{\overset{(G,\mathfrak{P},b)}{\Delta}} I_K((o_i, c, w) : f(w)).$$

Thus, for the current studied model, we have the following definition.

Definition 7.1.1. We denote the *local algorithmic synergy* of a node $o_i \in \mathfrak{P}_{\text{SBB}}(N, s)$ in an algorithmic network $\mathfrak{N}_{\text{SBB}}(N, f, t, 1, j, s)$ toward
function

$$\begin{aligned} f : X \subseteq \mathbf{L_U} &\to \mathbf{L_U} \\ x &\mapsto f(x) \end{aligned}$$

in c node cycles with network input w as

$$\begin{aligned} &\underset{\text{iso}}{\overset{\mathfrak{N}_{\text{SBB}}(N,f,t,1,j,s)}{\Delta}} I_K((o_i, c, w) : f(w)) \\ &= I_K(\mathbf{U}(p_{\text{net}}^{b_j}(o_i, c, w)) : f(w)) \\ &\quad - I_K(\mathbf{U}(p_{\text{iso}}(o_i, c, w)) : f(w)) \pm \mathbf{O}(1) \end{aligned}$$

where $f_o(o_i) = P_{\text{sprot}} \circ p_i \in \mathbf{L}_{\text{SBB}}$, $p_{\text{net}}^{b_j}(o_i, c, w)$ represents the program that returns the final output of o_i *when networked* assuming the position v, where $b_j(v) = (o_i)$, in the TVG G_t in c node cycles with network input w, and $p_{\text{iso}}(p_i, c, w)$ represents the program that returns the final output of p_i *when isolated* in c node cycles with network input w.

Furthermore, for a fixed function b, one can define the average value of local algorithmic synergy.

Definition 7.2. Let $\mathfrak{N} = (G, \mathfrak{P}, b)$ be a well-defined algorithmic network. We define the *average local algorithmic synergy* of a node $o_i \in \mathfrak{P}$ toward function

$$\begin{aligned} f : X \subseteq L &\to L \\ x &\mapsto f(x) \end{aligned}$$

in c node cycles with network input w as

$$\frac{\sum_{o_i \in \mathfrak{P}} \Delta_{\text{iso}}^{\mathfrak{N}} I_K((o_i, c, w) : f(w))}{|\mathfrak{P}|}.$$

Then, since population $\mathfrak{P}_{\text{SBB}}(N, s)$ is randomly generated, one can define the expected local value of the algorithmic synergy of a node for a fixed function b_j in the current studied case.

Definition 7.2.1. We define the *expected local algorithmic synergy* of a node $o_i \in \mathfrak{P}_{SBB}(N, s)$ toward function

$$\begin{aligned} f : X \subseteq L &\to L \\ x &\mapsto f(x) \end{aligned}$$

in c node cycles (or communication rounds) with network input w as

$$\begin{aligned} &\mathbf{E}_{\mathfrak{N}_{\text{SBB}}} \left(\underset{\text{iso}}{\overset{\mathfrak{N}_{\text{SBB}}(N,f,t,1,j,s)}{\Delta}} I_K((o_i, c, w) : f(w)) \right) \\ &= \frac{\sum_{o_i \in \mathfrak{P}_{\text{SBB}}(N,s)} \Delta_{\text{iso}}^{\mathfrak{N}_{\text{SBB}}(N,f,t,1,j,s)} I_K((o_i, c, w) : f(w))}{N}. \end{aligned}$$

Now, we can combine the results from Section 6 with the definition of expected local algorithmic synergy in order to make it as large as one may want:

Theorem 7.1. *Let*

$$f_h : X \subseteq \mathbf{L_U} \to \{h, \overline{h}\} \subset \mathbf{L_U}$$
$$x \mapsto f_h(x) = \begin{cases} \overline{h} = (\text{“}\textit{non-halting}\text{”})_{\mathbf{L_U}} & \textit{if } \mathbf{U} \textit{ does not halt on } x \\ h = (\text{“}\textit{halting}\text{”})_{\mathbf{L_U}} & \textit{if } \mathbf{U} \textit{ halts on } x \end{cases}$$

be a function defined on arbitrary $h \in \mathbf{L_U}$. *Let* $\mathfrak{N}_{\text{SBB}}(N, f, t, 1, j, s) = (G_t, \mathfrak{P}_{\text{SBB}}(N, s), b_j)$ *be an algorithmic network as in Definition* 5.4. *Let*

$$c : \mathbb{N} \to \mathfrak{C}_{\mathfrak{B}\mathfrak{B}}$$
$$x \mapsto c(x) = y$$

be a non-decreasing total computable function such that

$$c(z_0 + f(N, t_{z_0}, 1) + 2) \geq c_0 + z_0 + f(N, t_{z_0}, 1) + 2,$$

where $t_{z_0} \in \mathrm{T}(G_t)$. *Let* $x \in \mathbb{N}$ *be an arbitrary number. Then, there are constants* C_7 *and* $\mathrm{p}_{\text{halt}} \in \mathbf{L_U}$ *such that, for large enough* N *and for every network input* $w \in \mathbf{L_U}$ *with* $\lg(N) - C_7 \geq |w|$, *we will have that the expected local algorithmic synergy of a node* $o_i \in \mathfrak{P}_{\text{SBB}}(N, \mathrm{p}_{\text{halt}})$ *in algorithmic network* $\mathfrak{N}_{\text{SBB}}(N, f, t_{z_0}, 1, j, \mathrm{p}_{\text{halt}})$ *toward solving*[o] *the halting problem with domain*

$$X = \{w \mid \lg(N) - C_7 \geq |w|\} \subset \mathbf{L_U}$$

[o]That is, toward

$$f_h : \{x \mid \lg(N) - C_7 \geq |x|\} \subset \mathbf{L_U} \to \{h, \overline{h}\} \subset \mathbf{L_U}$$
$$x \mapsto f_h(x) = \begin{cases} \overline{h} & \text{if } \mathbf{U} \text{ does not halt on } x, \\ h & \text{if } \mathbf{U} \text{ halts on } x. \end{cases}$$

in $c' = c(z_0 + f(N, t_{z_0}, 1) + 2)$ *node cycles is larger than* x, *i.e.*,

$$\lim_{N\to\infty} \mathbf{E}_{\mathfrak{N}_{\text{SBB}}} \left(\mathop{\Delta}_{\text{iso}}^{\mathfrak{N}_{\text{SBB}}(N,f,t_{z_0},1,j,\mathrm{p}_{\text{halt}})} I_K((o_i, c', w) : f_h(w)) \right) \geq x,$$

with probability arbitrarily close to 1.

Proof. This proof follows from a combination of Theorem 6.1 with Definition 7.2.1 for a sufficiently complex $h \in \mathbf{L_U}$. First, we know from Definition 7.2.1 and Eq. (1) that

$$\begin{aligned}
&\mathbf{E}_{\mathfrak{N}_{\text{SBB}}} \left(\mathop{\Delta}_{\text{iso}}^{\mathfrak{N}_{\text{SBB}}(N,f,t_{z_0},1,j,s)} I_K((o_i, c', w) : f_h(w)) \right) \\
&\quad = \frac{\sum_{o_i \in \mathfrak{P}_{\text{SBB}}(N,s)} \Delta_{\text{iso}}^{\mathfrak{N}_{\text{SBB}}(N,f,t,1,j,s)} I_K((o_i, c', w) : f_h(w))}{N} \\
&\quad = \frac{\begin{array}{c}\sum_{o_i \in \mathfrak{P}_{\text{SBB}}(N,s)} A(f_h(w) \,|\, \mathbf{U}(p_{\text{iso}}(o_i, c', w))) \\ -A(f_h(w) \,|\, \mathbf{U}(p^b_{\text{net}}(o_i, c', w))) \pm \mathbf{O}(1)\end{array}}{N}. \qquad (2)
\end{aligned}$$

Let $w_{\min}$ denote the element of language $\mathbf{L_U}$ in which its length is the minimum length larger than zero. From Definition 5.3, we know every node belongs to $\mathbf{L_U}$. Therefore, since there always are randomly generated nodes that ignore any input and keep returning $w_{\min}$ as output in any node cycle when running isolated, then, from Definitions 3.3, 3.10, and 5.3 and the law of large numbers, there is $\epsilon > 0$ such that

$$\begin{aligned}
&\frac{\sum_{o_i \in \mathfrak{P}_{\text{SBB}}(N,s)} A(f_h(w) \,|\, \mathbf{U}(p_{\text{iso}}(o_i, c', w)))}{N} \\
&\quad \geq \frac{\sum_{o_i \in \{o | \ \mathbf{U}(p_{\text{iso}}(o, c', w)) = w_{\min}\}} A(f_h(w) \,|\, w_{\min})}{N} \\
&\quad + \frac{\sum_{o_i \in \{o | \ \mathbf{U}(p_{\text{iso}}(o, c', w)) \neq w_{\min}\}} A(f_h(w) \,|\, \mathbf{U}(p_{\text{iso}}(o, c', w)))}{N} \\
&\quad \geq \epsilon A(f_h(w) \,|\, w_{\min}) + 0. \qquad (3)
\end{aligned}$$

Now, choose p_{halt} with $\overline{h} = (\text{“non-halting”})_{\mathbf{L_U}}$ and $h = (\text{“halting”})_{\mathbf{L_U}}$ as in Definition 6.1, where[p]

$$\begin{aligned} A(h \,|\, w_{\min}) &> \epsilon^{-1}(x + 2\mathbf{O}(1)), \\ A(\overline{h} \,|\, w_{\min}) &> \epsilon^{-1}(x + 2\mathbf{O}(1)), \\ h &\neq \overline{h}. \end{aligned} \tag{4}$$

Furthermore, from Theorem 6.1, there is a constant C_7 such that, for large enough N and for every network input $w \in \mathbf{L_U}$ with $\lg(N) - C_7 \geq |w|$, we have that, for every node $o_i \in \mathfrak{P}_{\text{SBB}}(N, \mathrm{p}_{\text{halt}})$,

$$A(f_h(w) \,|\, \mathbf{U}(p^b_{\text{net}}(o_i, c', w))) = A(f_h(w) \,|\, f_h(w)) = \mathbf{O}(1) \tag{5}$$

in $c(z_0 + f(N, t_{z_0}, 1) + 2)$ node cycles with probability arbitrarily close to 1. Therefore, from Eqs. (2)–(4), we will have that there is a constant C_7 such that, for every network input $w \in \mathbf{L_U}$ with $\lg(N) - C_7 \geq |w|$,

$$\begin{aligned} &\lim_{N\to\infty} \mathbf{E}_{\mathfrak{N}_{\text{SBB}}} \left(\overset{\mathfrak{N}_{\text{SBB}}(N,f,t_{z_0},1,j,\mathrm{p}_{\text{halt}})}{\underset{\text{iso}}{\Delta}} I_K((o_i, c', w) : f_h(w)) \right) \\ &\geq \lim_{N\to\infty} \frac{\sum_{o_i \in \mathfrak{P}_{\text{SBB}}(N,\mathrm{p}_{\text{halt}})} A(f_h(w) \,|\, \mathbf{U}(p_{\text{iso}}(o_i, c', w))) - \mathbf{O}(1) \pm \mathbf{O}(1)}{N} \\ &\geq \epsilon\epsilon^{-1}(x + 2\,\mathbf{O}(1)) - 2\mathbf{O}(1) \end{aligned}$$

holds with probability arbitrarily close to 1. □

Indeed, for some universal programming languages and classical labelings on halting computation and non-halting computation, e.g., 1 and 0, respectively, the expected local algorithmic synergy of a node may not be positive. What Theorem 7.1 assures is that, for any chosen universal self-delimited programming language and any arbitrarily chosen x, there are h and $\overline{h}$ that can univocally represent the halting case and the non-halting case, respectively, and that the expected local algorithmic synergy of a node becomes larger than x.

[p]The number 2 is not really necessary here. We chose to employ it in order to avoid minor ambiguities in the asymptotic dominance.

8. Conclusion and Future Work

We have studied a particular model of algorithmic networks $\mathfrak{N}_{\mathrm{SBB}}$. These are composed of randomly generated self-delimiting programs as nodes, which share information accordingly to the synergistic imitation-of-the-fittest protocol (SIFP). From this model, we studied how to make the nodes asymptotically solve the halting problem as the population grows indefinitely. In this way, we have shown how a fixed global information-sharing (or communication) protocol can exploit the power of random generation of individuals and the power of selection made by an irreducibly more powerful environment in order to solve an uncomputable problem.

To this end, we have modified the model introduced in [4] to enable each node to calculate a partial recursive function for the network input and the latest largest integer shared by the neighbors. Specifically, this modification was made in the imitation-of-the-fittest protocol (IFP) in [4].

First, we proved that, if the population/network size is large enough, the network diameter is small compared to the population/network size, and enough communication rounds (i.e., node cycles) are expended (in particular, matching the network diameter), then every node is expected to solve the halting problem for any program with length dominated by a logarithmic order of the population/network size. In other words, nodes can emergently solve an increasing number of instances of the halting problem as the population grows indefinitely. This way, for algorithmic networks $\mathfrak{N}_{\mathrm{SBB}}$, all instances of the halting problem are statistically covered in the limit when the population grows indefinitely. This result shows that there is at least one fixed algorithm that can be distributedly run on networked randomly generated universal Turing machines, so that the entire algorithmic network $\mathfrak{N}_{\mathrm{SBB}}$ can compute a function in the Turing degree $\mathbf{0}'$, if the population/network size is large enough. Therefore, besides computation time and memory, networked randomly generated environment-evaluable nodes (which we may call **o**-nodes) can be regarded as a third type of computational resource. Thus, for algorithmic networks $\mathfrak{N}_{\mathrm{SBB}}$, any set with Turing degree $\mathbf{0}'$ is indeed decidable if enough (but still finite) time, memory, and **o**-nodes are given. As already studied for time hierarchies and space hierarchies, we propose as future research the investigation of **o**-node hierarchies. Furthermore, we also propose the investigation of resource-bounded versions of our present results, for example, in the case nodes belong to a time complexity class and the environment (i.e., the machine in which each node is being simulated) belongs to sufficiently higher time complexity class.

Secondly, we introduced two types of node centralities, in particular, one for distributed processing and one for network diffusion. From these and from the previous results, we proved that these two centralities can be intrinsically combined to show that, in the previously described conditions, there is one central node that can solve the halting problem in the minimum amount of communication rounds and only if networked. This result may help understand how node centralities in network science may be related to emergently privileged nodes in distributed processing.

Third, we introduced one type of algorithmic-informational measure of synergy, bridging previously studied concepts in multivariate information theory for stochastic processes to algorithmic information theory and algorithmic networks. With this respect, the general problem of synergy in networked computable systems can be translated as the problem of measuring the amount of algorithmic information in an arbitrary collection of nodes strictly necessary to calculate a function, but that could not be obtained by any individual isolated node or by any combination of proper subnetworks of the entire network. Then, narrowing our approach, we defined a measure of average algorithmic synergistic information for individual nodes in the specific case there is a comparison of the totally networked case with the totally isolated case. We call it local algorithmic synergy. Further, we showed that, for any chosen universal self-delimited programming language, one can make the algorithmic networks $\mathfrak{N}_{\mathrm{SBB}}$ produce as much expected local algorithmic synergy of a node as one may want. In this way, we related the emergent algorithmic complexity in [4] to a new type of emergent property, in the case, synergy. Thus, showing how systemic properties commonly studied in complex systems science, such as synergy, can be formalized in the context of networked deterministic systems. Moreover, with respect to synergy and networked computable systems, our results may help unlocking a formalism to find new mathematical phenomena in future work, such as new types of algorithmic measures of synergy for a comparison of the fully networked case with proper subnetworks.

This chapter follows a general pursuit of an abstract mathematical theory for systemic properties in complex systems (especially, living systems), such as evolution [12, 15, 30, 31], emergence of complexity [4, 5, 30], and emergence of creativity [4, 6, 15]. Thus, the theory of algorithmic networks goes toward the direction of establishing formal theories for other common systemic properties usually attributed to complex systems, like synergy and centralities presented in this chapter. In this direction, if one assumes the hypothesis that hypercomputation is possible in Nature, our results in this chapter have shown how Life might have found a way to synergistically harness the power of selection of individuals in sufficiently random population of individuals, even if every living being remains as a computable system. This way, the present work may also "open the gate" for the study of other systemic properties in future work, such as self-organization and autopoiesis, within the context of distributed deterministic systems.

Acknowledgments

Authors acknowledge the partial support from CNPq through their individual grants: F. S. Abrahão (313.043/2016-7), K. Wehmuth (312599/2016-1), and A. Ziviani (308.729/2015-3). Authors acknowledge the INCT in Data Science, INCT-CiD (CNPq 465.560/2014-8). Authors also acknowledge the partial support from CAPES/STIC-AmSud (18-STIC-07), FAPESP (2015/24493-1), and FAPERJ (E-26/203.046/2017). We also thank the comments and critiques from Hector Zenil and Mikhail Prokopenko.

Bibliography

[1] F. S. Abrahão, Metabiologia, Subcomputação e Hipercomputação: em direção a uma teoria geral de evolução de sistemas, Ph.d. thesis (2015).

[2] F. S. Abrahão, Emergent algorithmic creativity on networked Turing machines, in *The 8th International Workshop on Guided Self-Organization at the Fifteenth International Conference on the Synthesis and Simulation of Living Systems (ALIFE)* (2016), Extended Abstract available at http://guided-self.org/gso8/program/index.html.

[3] F. S. Abrahão, The "paradox" of computability and a recursive relative version of the Busy Beaver function, in *Information and Complexity* (World Scientific, 2016) pp. 3–15. Available at https://doi.org/10.1142/9789813109032_0001.

[4] F. S. Abrahão, K. Wehmuth and A. Ziviani, Algorithmic Networks: central time to trigger expected emergent open-endedness, *Theoretical Computer Science* (2019), Available at https://doi.org/10.1016/j.tcs.2019.03.008.

[5] F. S. Abrahão, K. Wehmuth and A. Ziviani, Emergent open-endedness from contagion of the fittest. *Complex Systems* **27**(4), (2018), Available at https://www.complex-systems.com/abstracts/v27_i04_a03/.

[6] F. S. Abrahão, K. Wehmuth and A. Ziviani, Expected emergent algorithmic creativity and integration in dynamic complex networks, in *Meeting on Theory of Computation (etc), Congress of the Brazilian Computer Society (sbc)* (2018). Available at https://doi.org/10.5281/zenodo.1241236.

[7] M. A. Bedau, Four puzzles about life, *Artificial Life* **4**(2) (1998) 125–140. Available at http://doi.org/10.1162/106454698568486.

[8] E. Bresciani and Í. M. Loffredo D'Ottaviano, Basic concepts of systemics, in in *Systems, Self-Organisation and Information: An Interdisciplinary Perspective* (Routledge, 2019).

[9] C. S. Calude, *Information and Randomness: An Algorithmic Perspective*, 2nd edn. (Springer, 2002).

[10] C. S. Calude, Information: the algorithmic paradigm, in *Formal Theories of Information* (Springer, 2009), pp. 79–94.

[11] G. Chaitin, *Algorithmic Information Theory*, 3rd edn. (Cambridge University Press, 2004).

[12] G. Chaitin, Life as evolving software, in *A Computable Universe* (World Scientific, 2012), pp. 277–302.

[13] G. Chaitin, *Proving Darwin: Making Biology Mathematical* (Vintage Books, 2013).

[14] G. Chaitin, V. M. F. G. Chaitin and F. S. Abrahão, Metabiología: los orígenes de la creatividad biológica, *Investigación y Ciencia* **448** (2014), 74–80 (Spanis).

[15] V. M. F. G. Chaitin and G. J. Chaitin, A philosophical perspective on a metatheory of biological evolution, in *The Map and the Territory: Exploring the Foundations of Science, Thought and Reality* (Springer, 2018), pp. 513–532.

[16] S. Barry Cooper, Emergence as a computability-theoretic phenomenon, *Applied Mathematics and Computation* **215**(4) (2009), 1351–1360.
[17] B. Jack Copeland, Turing's O-machines, Searle, Penrose and the brain, *Analysis* **58**(2) (1998) 128–138.
[18] B. Jack Copeland, Hypercomputation, *Minds and Machines* **12**(4) (2002), 461–502.
[19] E. C. Costa, A. B. Vieira, K. Wehmuth, A. Ziviani and A. P. Couto da Silva, Time centrality in dynamic complex networks, *Advances in Complex Systems* **18** (2015). Available at http://arxiv.org/abs/1504.00241http://dx.doi.org/10.1142/S021952591550023X.
[20] G. Dodig Crnkovic and M. Burgin, Unconventional algorithms: complementarity of axiomatics and construction, *Entropy* **14**(11) (2012), 2066–2080.
[21] N. C. A. da Costa and F. A. Doria, Some thoughts on hypercomputation, *Applied Mathematics and Computation* **178**(1) (2006), 83–92.
[22] N. C. A. da Costa and F. A. Doria, How to build a hypercomputer, *Applied Mathematics and Computation* **215**(4) (2009), 1361–1367.
[23] K. Dingle, C. Q. Camargo and A. A. Louis, Input–output maps are strongly biased towards simple outputs, *Nature Communications* **9**(1) (2018), 761.
[24] Í. M. Loffredo D'Ottaviano, *On the Theory of Quasi-truth*, Series special issues of epistemology: Relations between natural sciences and human sciences (2010), pp. 325–340.
[25] R. G. Downey and D. R. Hirschfeldt, *Algorithmic Randomness and Complexity*, Theory and Applications of Computability (Springer New York, 2010).
[26] N. Fernández, C. Maldonado and C. Gershenson, Information measures of complexity, emergence, self-organization, homeostasis, and autopoiesis, in *Guided Self-Organization: Inception* (Springer, 2014), pp. 19–51.
[27] J. D. Fouks, Towards an algorithmic theory of adaptation, *Theoretical Computer Science* **223**(1–2) (1999), 121–142.
[28] V. Griffith and C. Koch, Quantifying synergistic mutual information, in *Guided Self-Organization: Inception* (Springer, 2014), pp. 159–190.
[29] P. D. Grünwald and P. M. B. Vitányi, Algorithmic information theory, in *Philosophy of Information* (Elsevier, 2008), pp. 281–317.
[30] S. Hernández-Orozco, F. Hernández-Quiroz and H. Zenil, Undecidability and irreducibility conditions for open-ended evolution and emergence, *Artificial Life* **24**(1) (2018), 56–70. Available at http://doi.org/10.1162/ARTL_a_00254.
[31] S. Hernández-Orozco, N. A. Kiani and H. Zenil, Algorithmically probable mutations reproduce aspects of evolution, such as convergence rate, genetic memory and modularity, *Royal Society Open Science* **5**(8) (2018), 180399. Available at 1709.00268.
[32] M. Li and P. Vitányi, *An Introduction to Kolmogorov Complexity and Its Applications*, 2nd edn. (Springer Science & Business Media, New York, 1997).

[33] J. Lizier, N. Bertschinger, J. Jost and M. Wibral, Information decomposition of target effects from multi-source interactions: perspectives on previous, current and future work, *Entropy* **20**(4) (2018), 307. Available at http://doi.org/10.3390/e20040307.

[34] G. Longo, Incomputability in physics and biology, *Mathematical Structures in Computer Science* **22**(5) (2012), 880–900.

[35] I. Mikenberg, N. C. A. da Costa and R. Chuaqui, Pragmatic truth and approximation to truth, *Journal of Symbolic Logic* **51**(1) (1986), 201–221.

[36] M. Oizumi, L. Albantakis and G. Tononi, From the phenomenology to the mechanisms of consciousness: integrated information theory 3.0, *PLoS Computational Biology* **10**(5) (2014), e1003588. Available at http://doi.org/10.1371/journal.pcbi.1003588.

[37] R. Kumar Pan and J. Saramäki, Path lengths, correlations, and centrality in temporal networks, *Physical Review E* **84**(1) (2011), 016105. Available at http://doi.org/10.1103/PhysRevE.84.016105.

[38] M. Prokopenko (ed.), *Guided Self-Organization: Inception*, Emergence, Complexity and Computation, Vol. 9 (Springer, Berlin, 2014).

[39] M. Prokopenko, F. Boschetti and A. J. Ryan, An information-theoretic primer on complexity, self-organization, and emergence, *Complexity* **15**(1) (2009), 11–28. Available at http://doi.wiley.com/10.1002/cplx.20249.

[40] M. Prokopenko, M. Harré, J. Lizier, F. Boschetti, P. Peppas and S. Kauffman, Self-referential basis of undecidable dynamics: From the Liar paradox and the halting problem to the edge of chaos, *Physics of Life Reviews* (2019). Available at http://doi.org/10.1016/j.plrev.2018.12.003.

[41] H. Rogers Jr., *Theory of Recursive Functions and Effective Computability* (MIT Press, Cambridge, MA, 1987).

[42] Y. P. Shimansky, Trans-algorithmic nature of learning in biological systems, *Biological Cybernetics* **112**(4) (2018), 357–368.

[43] H. T. Siegelmann and E. D. Sontag, On the computational power of neural nets, *Journal of Computer and System Sciences* **50**(1) (1995), 132–150.

[44] R. K. Standish, Open-ended artificial evolution, *International Journal of Computational Intelligence and Applications* **03**(02) (2003), 167–175.

[45] A. Syropoulos, *Hypercomputation: Computing Beyond the Church-Turing Barrier* (Springer, 2008).

[46] K. Wehmuth, É. Fleury and A. Ziviani, On MultiAspect graphs, *Theoretical Computer Science* **651** (2016), 50–61. Available at http://doi.org/10.1016/j.tcs.2016.08.017.

[47] K. Wehmuth, É. Fleury and A. Ziviani, MultiAspect graphs: algebraic representation and algorithms, *Algorithms* **10**(1) (2017), 1–36. Available at http://doi.org/10.3390/a10010001.

[48] K. Wehmuth and A. Ziviani, Centralities in high order networks, in *Meeting on Theory of Computation (ETC), Congress of the Brazilian Computer Society (SBC)* (2018) (English).

[49] K. Wehmuth, A. Ziviani and E. Fleury, A unifying model for representing time-varying graphs, *Proc. IEEE Int. Conf. Data Science and Advanced Analytics* (*DSAA*) (2015), pp. 1–10.
[50] H. Zenil and F. Hernandez-Quiroz, On the possible computational power of the human mind, in *Worldviews, Science and Us* (World Scientific, 2006), pp. 315–337.

Chapter 10

Limits of Computation and Physical Reality

Apostolos Syropoulos*

The physical laws of the universe dictate what can and what cannot be computed. The Church-Turing Thesis (CTT) is based on the assumption that the Turing machine is the ultimate computing device and according to the CTT something is computable only if it can be computed by a Turing machine. However, the CTT has not been proved and it can be definitely proved in a universe with specific properties. Whether our universe is such a universe is an open problem as it is an open problem whether the universe behaves like a machine. A broader approach that rejects the CTT and shows that the mechanistic universe and the CTT are incompatible is proposed and discussed thoroughly.

1. Introduction

In 2003, three collaborators of mine and I attended a workshop that took place in Spain. In this workshop, people discussed and presented papers about a new model of computation. During the workshop, someone talked about an alternative model of computation that was closely related to the theme of the workshop. He talked about many things but at the end of his talk he *dared* to say that his model of computation is more expressive than the Turing machine. I was shocked and I immediately asked him: "How can anything do things no Turing machine can?" It seems that he was not really sure about

*Xanthi, Greece; asyropoulos@yahoo.com.

his ideas (after all he was a doctoral student...) so his answer was not so convincing. Others asked similar questions and soon people were just shouting.

A few months later and during a walk out in the woods, I "reprocessed" this exchange of questions and answers. After a while my conclusion was that possibly I was wrong and he was right! Why? Mainly because I finally realized that we all silently assume that the *Church-Turing Thesis* (CTT), which is an unproven hypothesis, is actually valid. Yes, I knew that no one had proved the CTT when I asked the student at the workshop, but since no one had questioned its validity to the best of my knowledge at that time, I assumed that it is indeed valid as most people do. At that moment, I started thinking about the possibility that the CTT is not valid and what this would mean in general. Later, I searched the literature and I discovered that there are some people who actually do not believe in the validity of the CTT. Then, I tried to write down all the different attacks against the validity of the CTT. This work culminated in the form of a published book. In 2008, when this book on *hypercomputation* [24] was finally published, I faced a crowd similar to the one I described above at a conference of the Greek Mathematical Society. At that conference, I dared to say that quite possibly the CTT is false. Then, someone from the audience got really angry and told me unspeakable things. Unfortunately, in this case history repeated itself as a farce...

In general we do not know if the CTT is true or false, nevertheless, the CTT can be proved if we are working in a mathematical universe that has specific properties. Such a mathematical universe is the *effective topos* **Eff**. Without going into the details of the proof of the validity of the CTT (see [15] for details), it suffices to say that the proof follows from the fact that the effective topos is a mathematical universe whose *internal logic*[a] allows one to prove that every

[a]Very roughly, when we have a mathematical structure and we find a way to interpret its constituents as logical propositions and the various operations as logical connectives, then these form the internal logic of the structure. For example, the internal logic of the power set of some set is the classical logic.

total function from natural numbers to natural numbers is recursive. Hence, the proof of the CTT is almost trivial. Consequently, one can say that a proof or a disproof of the CTT depends on the logic of the universe. Going one step further we may ask: What is the logic of the universe?

The Turing machine (TM) is a conceptual (or idealized) device and not a physical device, nevertheless, there are many computer simulations of a TM. The functionality of a TM can be described by a *transition function* that fully specifies the behavior of this particular TM. The simplicity and elegance of the TM is chiefly "responsible" for its use in models of the universe and the human brain. Thus, the CTT is trivially valid if the universe is a TM. Of course, the question is whether the universe is a TM. Here, it is rather important to exactly specify what the word "universe" means. Obviously, we should consider only scientifically proven facts and not (wild) speculations about parallel universes that are created each time someone performs some sort of measurement. In other words, we should not pay attention to "scientific" ideas related to the multiverse and related theories. Whether the universe is a mathematical entity is an idea closely related to the idea that the universe is a computational machine and so these two ideas should be examined together.

If the CTT is not valid, then is there something else that dictates what can and what cannot be computed? *Hypercomputation* [24,28] is primary about proposals that transcend the capabilities of the Turing machine and secondary about the *limits* of computation in general. In different words, hypercomputation is about what can be computed mechanically (i.e., by a computing device real or conceptual).

Plan of the Chapter. In what follows, I first try to see whether there is a logic of the Universe compatible with the CTT by examining various foundational approaches to the problem of defining formal systems capable of describing Nature. Since each such system should have an internal logic, I discuss these logics and their possible use in the "proof" of the CTT. Next, I examine whether the universe is a machine and/or a mathematical universe, an idea akin to the

Pythagorean *musica universalis*. Assuming that the CTT is invalid I discuss what can be said about the ultimate limits of computation. The final section summarizes the ideas presented in the main text.

2. On the "Logic" of the Universe

Typically, physical laws are described using mathematical equations. Of course, one can express a physical law using Portuguese, Greek, or any other natural language for that matter. Of course, it is far less ambiguous to use mathematical formulas instead of natural languages. And this is exactly the reason why algorithms are defined using *pseudo-code* (i.e., a precisely defined notation) instead of a natural language. More generally, it is a trend to see disciplines becoming sciences once they adopt a mathematical formalism that is used to describe phenomena and make "predictions." At the same time, mathematics is a branch of knowledge by itself so it is developing independently from its use in sciences.

Occasionally, scientists need more tools to describe and express new ideas and phenomena. This leads to the introduction of new mathematics or the development of an existing field of mathematics. Spontaneously, for example, computer science is the reason why mathematical logic flourished. But not all scientists have a common way of thinking or a philosophical basis that guides them in their work. In fact, we need not to confuse the logic or philosophical basis used to develop a theory with the actual tools and their internal logic. For example, it seems that most people that have contributed to the development of topology are employing some sort of trial and error approach. The logic of this methodology is the classical one (see [30]), while the internal logic of topological spaces is the intuitionistic logic.

The pillars of modern physics are the theory of relativity and quantum mechanics. It is not totally wrong to say that relativity theory is about planets, stars, and galaxies and quantum mechanics is about atoms and subatomic particles. Also, objects in between and their interactions can still be described by classical mechanics with extremely good accuracy, provided the objects do not move with speeds close to the speed of light. Certainly, a complete theory would be one that would make relativity theory compatible with quantum mechanics (i.e., one that would at least quantize gravity, provided this is actually possible). Unfortunately, there is no such theory and this explains why each theory is examined individually. On the other hand, classical mechanics is only approximately correct and only under specific conditions, so it has no place in the discussion that follows.

The axiomatic method is a way to formulate a scientific theory. According to this method, one first specifies a number of primitive notions (e.g., how one can prove something). Then, one specifies a number of postulates that are called *axioms* and possibly a set of *inference rules*, which are used to derive new truths (i.e., theorems) from existing ones. Clearly, any proposition of the theory is obtained from the axioms using the inference rules. In many cases, there are only axioms and the inference rules are just the "classical" ones (e.g., modus ponens, modus tollens, etc.). In a sense, Euclid developed plane geometry using the axiomatic method. He formulated five postulates and used them to prove the various propositions in his elements. His inference rules were the ones used by people to reason about things in general. As an alternative example of an axiomatic system, think that if one assumes that the negation of Euclid's fifth postulate (i.e., the one about parallel lines) is true, then one can develop a non-Euclidean geometry.

It should not surprise anyone the fact that there are many works that propose an axiomatic system for relativity theory. In a way, Robb's *A Theory of Time and Space*, which was published later as [21], was the first effort to axiomatize the special theory of relativity. In this work, Robb presented 21 postulates, which have been used to derive the various propositions of his system. In addition, he has made heavy use of the *after* relation that he introduced in the first postulate of his system:

Postulate I. If an element [of time] B is after an element A, then the element [of time] A is not after the element B.

The relation B *after* A means that B lies inside or on the future light cone based at A. This relation is a partial order on the set of elements of time. Robb's work has been overviewed by Suppes [22] and Goldblatt [10]. However, Goldblatt [10] used the primitive notions of *orthogonality* and *betweenness* to develop a similar system. Walker [32] used *particles* and *instants* of particles in his axiomatization. Also, Andréka, Madarász, and Németi [3] developed axioms for both special and general relativity and applied the resulting systems to black holes, wormholes, and timewarps. Fortunately or unfortunately, these approaches cannot be used to readily prove or disprove the CTT. The reason of course is they employ classical logic.

All the aforementioned axiomatizations have one thing in common — they use set theory. Although sets are considered by many as the building blocks of mathematics, still there are others who advocate that mathematics can be best explained by *categories*. Category theory is a very general formalism with many applications. Instead of giving a *informal* description of category theory, I will quote Tom Leinster's [16] picturesque description of category theory:

> Category theory takes a bird's eye view of mathematics. From high in the sky, details become invisible, but we can spot patterns that were impossible to detect from ground level. How is the lowest common multiple of two numbers like the direct sum of two vector spaces? What do discrete topological spaces, free groups, and fields of fractions have in common?

Mathematical structures (e.g., Hilbert spaces) are used to describe physical processes so category theory may give a bird's eye view of physics. Here is a formal definition of categories, which is based on the one given in [11].

Definition 2.1. A category $\mathcal{C}$ is made up of

(1) a collection of things that are called $\mathcal{C}$-objects;
(2) a collection of "bridges" between $\mathcal{C}$-objects that are called $\mathcal{C}$-arrows;

(3) each arrow f has as *domain* the object $\operatorname{dom} f$ and as *codomain* the object $\operatorname{cod} f$. If $A = \operatorname{dom} f$ and $B = \operatorname{cod} f$, then we write $f : A \to B$;
(4) an operation that assigns to each pair (g, f) of $\mathcal{C}$-arrows, such that $\operatorname{dom} g = \operatorname{cod} f$, a $\mathcal{C}$-arrow $g \circ f$, such that $\operatorname{dom}(g \circ f) = \operatorname{dom} f$ and $\operatorname{cod}(g \circ f) = \operatorname{cod} g$, that is, $g \circ f : \operatorname{dom} f \to \operatorname{cod} g$. In addition, given the arrows $f : A \to B$, $g : B \to C$, and $h : C \to D$, then $h \circ (g \circ f) = (h \circ g) \circ f$;
(5) for each $\mathcal{C}$-object A there is a $\mathcal{C}$-arrow $\mathrm{id}_A : A \to A$ called the *identity* arrow, such that for any $f : A \to B$ and $g : B \to C$, $\mathrm{id}_B \circ f = f$ and $g \circ \mathrm{id}_B = g$.

Category theory is more about arrows and their properties than about objects and their properties (roughly, on could say that set theory is more about object than about arrows). Typically, arrows become interesting when they have additional properties. For example, an arrow $f : A \to B$ is called monic if there are two arrows $g, h : C \to A$ such that the equality $g \circ f = h \circ f$ implies $g = h$. Monic arrows are used to define subobjects: Given a $\mathcal{C}$-object D, a subobject of D is a monic $\mathcal{C}$-arrow $f : A \to D$. Clearly, when a category has special arrows and/or special objects, their arrows have more properties that the standard ones. For example, when there is a category $\mathcal{C}$ the collection of arrows from object A to object B is itself an object of $\mathcal{C}$, then this category is *closed*. However, the best examples of such categories are the Cartesian closed categories (CCCs) and topoi. CCCs are interesting because their internal language is the λ-calculus. On the other hand, the internal logic of a topos is the typed higher-order intuitionistic logic.

Guts [12] proposed the use of elementary *topoi* (i.e., a category with a terminal object, binary products and equalizers, exponentials, and a subobject classifier) for the description of space–time. More specifically, Guts considered an elementary topos that has a continuous real number object. This is a special kind of object that represents the set of real numbers. The “standard” definition of this object can be found in [18]. Guts used the real number object $\mathbf{R}_T$

of some elementary topos $\mathcal{C}$ and two morphisms ϕ and ψ of the same elementary topos to define a $\mathcal{C}$-object a, which is called an *affine* object. Also, he notes that not every topos with a real number object has an affine object. In the end he concludes that "the language of topos theory allows us to axiomatize in a unified way both special and general theories of relativity, the axioms being the same in both cases." Interestingly, the **Eff** seems to be an elementary topos, which means that in Gut's axiomatization the CTT is true.

The novelty introduced by quantum mechanics is that there are limitations to what can be measured in the physical world. So when we try to measure the speed of a subatomic particle, we affect its position by introducing uncertainty in the measurement of its position. Obviously, the same happens when we try to measure the position of a moving particle. Many call this an *inherent uncertainty* but personally I view this as a manifestation of vagueness in Nature. Also, when we actually measure something, we measure one of the possible states that coexist before the measurement. Since quantum mechanics is so "weird" many researchers have tried to put forth a logic that would allow for describing the events that happen at the subatomic level. Quantum logic, which was proposed by Garrett Birkhoff and John Von Neumann [5], was an attempt to provide logical system capable of describing quantum mechanics. While one can provide a semantic description of classical logic in terms of Boolean algebras, a semantic description of quantum logic is using the lattice of closed subspaces of some Hilbert space. Unfortunately, many, including the famous French logician Jean-Yves Girard [8], do not consider quantum logic a real logic.

Coherent spaces have been used by Girard [7] to give a semantic description of linear logic. This is a logic that is aware of resources. Thus (atomic) propositions are viewed as resources and cannot be used freely. Assume that $\mathcal{X}$ is a set. Then, two subsets $a, b \subset \mathcal{X}$ are polar when their intersection is at most a singleton set:

$$a \mathrel{\overset{\frown}{\smile}} b \quad \Leftrightarrow \quad \mathrm{card}(a \cap b) \leq 1.$$

The polar $\sim A$ of a set $A \subset \wp(X)$ of subsets of $\mathcal{X}$ has the following property:

$$b \in \sim A \quad \Leftrightarrow \quad \forall a \in A, a \curlywedgedownarrow b.$$

A coherent space with carrier $\mathcal{X}$ is a subset $X \subset \wp(X)$ equal to its bipolar, that is, $X = \sim\sim X$. Assume that X is a coherent space with carrier $\mathcal{X}$. Then,

(1) X contains the empty set and all singletons;
(2) if $a' \subset a \in X$, then $a' \in X$;
(3) if $a \subset \mathcal{X}$ and $a \notin X$, there are $x, y \in X$, $x \neq y$ such that $\{x, y\} \notin X$;
(4) if $x, y \in X$ are distinct, then $\{x, y\} \notin X$ if and only if $\{x, y\} \in \sim X$.

In addition, we define a binary relation on $\mathcal{X}$

$$x \frown_X y \quad \Leftrightarrow \quad \{x, y\} \in X.$$

This relation is called coherence. Girard has used coherent spaces to give a semantic description of linear logic. Each coherent space represents a proposition and the logical connectives become operations between coherent spaces. For example, the operation $X \otimes Y$ is defined as follows:

$$(x, y) \frown_{X \otimes Y} (x', y') \quad \Leftrightarrow \quad x \frown_X x' \text{ and } y \frown_Y y'.$$

In order to "define" the logic of quantum mechanics, Girard proposed quantum coherent spaces. He started with a finite-dimensional (complex) Hilbert space $\mathcal{X}$. Then, he formed the set $\mathcal{E} = \mathcal{H}(\mathcal{X})$ of Hermitian (self-adjoint) operators on $\mathcal{X}$. It turns out that $\mathcal{E}$ is a real vector space that is equipped with the scalar product

$$\langle u \,|\, v \rangle = \operatorname{tr}(uv),$$

where $\operatorname{tr}(uv)$ is the trace of the product uv. Two Hermitian operators are said to be polar when $0 \leq \langle u \,|\, v \rangle \leq 1$. A quantum coherent space with carrier $\mathcal{X}$ is a subset $X \subset \mathcal{H}(\mathcal{X})$ equal to its bipolar.

Quantum coherent spaces can be used to give an alternative semantic description of linear logic in terms of Hilbert spaces. In a sense, Girard provided "quantum" semantics for linear logic but his work did not provide new insight to quantum phenomena.

There are fuzzy models of linear logic [4, 23] and this may imply that linear logic can be used to reason in a vague environment. This is particularly interesting, despite of Girard's "allergy" to fuzzy "logic". After all it makes sense to reinterpret quantum mechanics using vagueness and possibilities instead of indeterminacy and probabilities. This quite justified because possibility theory is about events and the degree to which they surprise us. On the other hand, probability theory is roughly about events and how likely they are. Thus there is some superficial similarity that can be used to adopt possibility theory and abandon probability theory. Thus if we accept the view that vagueness is a fundamental property of our cosmos, then certain quantum phenomena can be reinterpreted accordingly. For example, the standard interpretation of $\alpha|0\rangle + \beta|1\rangle$, where $|0\rangle$ and $|1\rangle$ are the *basic* states of a very simple quantum system, is that a particle is in states $|0\rangle$ or $|1\rangle$ with probability that depends on α and β. According to a layman's interpretation of probability theory, these two numbers express the change that a particle is in one of these states. A fuzzy theoretic interpretation of this state is that the particle is in both states but with some degree (see [29]). The "trick" that allows us to replace probabilities with possibilities is the remark that all probabilities of an event sum to 1, whereas all possibilities should be less than or equal to 1. Thus, the possibilistic approach is more general than the probabilistic one and by using possibilities instead of probabilities one gets an interpretation that is completely compatible with "old" theory. In addition, this interpretation leaves room for new things which might be excluded by the "old" theory.

Categorical quantum mechanics [1] is a scientific program to understand the structural and operational features of quantum theory using the language of dagger symmetric monoidal categories (see [14] for "usage" examples). These are monoidal categories, which means that they are equipped with a functor (i.e., an arrow between categories) $\mathcal{C} \otimes \mathcal{C} \to \mathcal{C}$ that has a number of properties. In addition, they have an object I called the unit of the tensor product $\otimes$. Also, they are equipped with a functor $\dagger : \mathcal{C}^{\mathrm{op}} \to \mathcal{C}$, where $\mathcal{C}^{\mathrm{op}}$ is the opposite of $\mathcal{C}$ (i.e., it has the same objects but the arrows are inverted, that is, the domain becomes the codomain and vice versa). This functor maps each object to itself and each arrow $f : A \to B$ to an arrow $f^\dagger : B \to A$ such that for all $f : A \to B$ and $g : B \to C$

$$\mathrm{id}_A^\dagger = \mathrm{id}_A, \quad (g \circ f)^\dagger = f^\dagger \circ g^\dagger, \quad \text{and} \quad f^{\dagger\dagger} = f.$$

Furthermore, the dagger structure should preserve the symmetric monoidal structure, which means that, among others, for all $f : A \to B$ and $g : C \to D$ the following condition holds:

$$(f \otimes g)^\dagger = f^\dagger \otimes g^\dagger.$$

The category **fRel** of finite sets and relations between them as arrows is an example of a dagger symmetric monoidal category [9]. In this category, the Cartesian product $\times$ of sets/relations is the tensor, the set $1 = \{*\}$ is the tensor unit, and the dagger is defined by

$$R^\dagger \stackrel{\text{def}}{=} \{(y, x) \mid (x, y) \in R\}.$$

Abramsky and Duncan [2] describe the internal logic of any dagger symmetric monoidal category $\mathcal{A}$. In particular, the formulas of this logic are built using the following grammar:

$$F ::= \mathbf{0} \mid \mathrm{I} \mid A \mid A^* \mid F \otimes F \mid F \oplus F;$$

here A is any object of $\mathcal{A}$. The operation $(F)^*$ on any formula F is defined by the following equations:

$$X^{**} = X,$$

$$(X \otimes Y)^* = X^* \otimes Y^*,$$

$$(X \oplus Y)^* = X^* \oplus Y^*,$$

$$\mathrm{I}^* = \mathrm{I},$$

$$\mathbf{0}^* = \mathbf{0}.$$

Although this approach can be used to mathematically describe the quantum world, still it seems that a version of category theory that incorporates vagueness would be a much better tool. Clearly, such a tool would provide a natural way to handle vagueness, while this approach cannot directly handle any kind of uncertainty. A fuzzy version of category theory [27] was introduced by this author a few years ago.

The discussion so far has revealed that there are attempts to axiomatically define relativity theory or, more generally, to define mathematical universes capable of describing the physical world. However, all approaches have different internal logics and certainly in most systems one cannot trivially prove or disprove the CTT. So we cannot settle the CTT by relying on the internal logic of mathematical systems that (currently best) describe the cosmos we know and understand it. So, we definitely need to investigate whether there are other ways to prove or disprove the CTT.

3. The Universe as a Machine

Recently, I read an article in the Aeon digital magazine[b] that was entitled "The idea of creating a new universe in the lab is no joke" by Zeeya Merali. In this article, she discussed the idea of creating a universe in a laboratory and other closely related ideas. Merali mentions the work of Andrei Linde [17] who proposed that one can create a universe in a laboratory at a very high temperature. However, Linde is not the first person who has advocated the idea that a universe can be created by an intelligent being. For example, the motif that the universe (or cosmos) was created by a (divine) creature is a common theme in most religions and mythologies that I happen to be aware of. Today most scholars assume that this theme is actually an allegorical description of how the cosmos was created. But Linde's ideas are very different from such religious and mythological beliefs

[b]See https://goo.gl/dn4YrQ.

and stories, respectively, but they are similar in the sense that they describe the idea that someone creates a universe.

Let us assume that somehow it is *possible* to create a universe in a laboratory. Then, it should be possible to design a computer program that *implements* a universe. Indeed, Hayes [13] has concluded that it is possible to implement a universe in a conventional programming language (e.g., Java or Perl), nevertheless, he recognized that the implementation would be a particularly difficult one. But if a universe can be realized as a computer program, then it does make sense to conclude that the universe is a computer.[c] Indeed, this is an idea that was introduced by Zuse [36]. Later on, others have elaborated on this idea that culminated to the mathematical universe hypothesis [31], according to which our physical world is an abstract mathematical structure. One basic implication of this hypothesis is that only computable and decidable (in Gödel's sense) structures exist. Although I have discussed why these ideas make no sense (see [24–26, 28]) still it is necessary to present again the core of these counter-arguments here.

Assume that it is possible to write a computer program that implements or realizes a universe (i.e., not necessarily our universe). According to the accepted scientific theory, the universe was created by an enormous explosion, which is known as the big bang. Astrophysicists theorize that there was a singularity, something like an enormous black hole, which exploded and created the universe. To the best of my knowledge, no astronomer has reported anything about exploding black holes. Also, we know very few things about what actually happens inside a black hole. This simply means that our knowledge is quite limited in this respect. So the natural question is — If we do not know enough about black holes and about the origins of the universe, then how can we write a computer program

[c]In order to understand the logical leap, think of the Java Virtual Machine that is, the first *in silico* or real machine that could run Java bytecodes. Later on, the JVM was realized in the form of real processors that can execute Java bytecode in hardware. Thus a computer program realizing a universe can be the *in silico* counterpart of a real universe.

that implements astronomical objects for which we do know very few things? Moreover, physicists have not managed to devise a theory that unifies gravity with the other three fundamental forces. This simply means that we do not know if there is some kind of connection between the fundamental forces and so we lack basic knowledge that prohibits us from creating or realizing a universe in a computer system.

One may argue that we have no knowledge of these things today but with the advancement of technology and new knowledge that will obtain in the course of time, we will get a deep understanding of all these things and so will be able to realize things. However, there are more obstacles that do prohibit one to create universes. Wharton [33–35] claimed that Heisenberg's uncertainty principle prohibit us from knowing the initial state of the universe with enough precision to compute the future. Thus it is impossible to create a replica of our universe. Also, according to Wharton, Schrödinger's equation computes something that does not correspond to what we actually observe. Its output allows us to compute the probabilities of different possible outcomes. And if these remarks are not enough to convince someone about the implausibility of the universe-as-a-computer thesis, as a recent discovery made by Ringel and Kovrizhin [20] made it clear that we are not living in a computer simulation. According to the *simulation argument* [6] "we are almost certainly living in a computer simulation". If we live in a computer simulation, then our universe is either a computer program or just a computer realizing our universe. The work by Ringel and Kovrizhin shows that our universe is not a simulation because there is a noncomputable phenomenon. This phenomenon is the Quantum Hall Effect, which briefly described in the next paragraph.

Consider a thin plate of a conductive material (e.g., a material that allows electric current to flow, such as copper). Electric current is supplied by a battery and flows through this plate. Also, we position a pair of probes connected to a voltmeter opposite each other along the sides of the plate. Initially, the voltmeter shows no voltage difference. However, when a magnetic field perpendicular to the current flow is applied to the plate, the voltmeter shows a small voltage difference. By reversing the direction (polarity) of the magnetic field, the polarity of the voltage will also reverse. This phenomenon is called the Hall Effect, named after Edwin Hall. In 1980, Klaus von Klitzing discovered the Quantum Hall Effect, which is a very important achievement in condensed matter physics. Von Klitzing found that at low temperatures and in strong magnetic fields, the Hall resistance of a two-dimensional electron system assumes quantized values that depend only on fundamental constants and integer numbers.

Ringel, one of the authors of the aforementioned paper, was not really sure if his work disproved the simulation argument and said: "Who knows what are the computing capabilities of whatever simulates us".[d] In different words, he proposes the idea that the there are computing capabilities that are not known to us or, going one step further, that we do not know if the CTT is true! Thus it is possible to have a universe which is a computer or to live in a computer simulation but in both cases we need models of computation that are more powerful than the Turing machine (also known as *hypercomputational* models of computation).

The idea that the universe is a computer is based on the assumption that computation is a natural phenomenon. This simply means that computation occurs spontaneously in Nature. However, computation does not occur in Nature. Intelligent beings set up systems (e.g., electronic computers) that can perform certain actions that are interpreted by these beings as computation. In addition, only these beings can understand the result of some computation. To think otherwise is some kind of mysticism.

Thesis 3.1. *Computation is not a physical phenomenon, nevertheless, intelligent beings can set up physical system to perform some computation whereas the result can be interpreted only by them.*

---

[d]See https://bit.ly/2WFVucx.

4. Hypercomputation and the Limits of Computation

There is a problem that the *universal* TM (UTM)[e] cannot solve. In modern parlance, this problem is about our inability to construct a computer program that given a program that got stalled can determine whether the second program has entered an infinite loop or performs some extremely demanding computations. The problem is known as the *halting problem* (HP). Any machine transcending the capabilities of the TM should be able to solve the HP. Thus machines that simulate or realize universes should be able to solve the HP. Nevertheless, each kind of machine has its own HP that cannot be solved by machines of its kind (see [19]). So regardless of the computational power of a machine, there is always a problem that it cannot solve. Eventually, there must be some ultimate computing device that will dictate what can be and cannot be computed, nevertheless, we currently cannot say what this machine is. And this is exactly the essence of hypercomputation.

All the machines that are supposed to simulate or realize a universe are "crisp" machines. Of course we have no idea what kind of machines these are, but all the theories assume that these machines operate in exact ways just like the TM. However, our universe is vague, which means that these theories are flawed because they do not take under consideration vagueness.[f] Thus supporters of the universe-as-a-machine theories need a broad theory of vague hypercomputation, which will definitely include as a special case recursion theory, to better support their views.

5. Conclusions

I have tried to see whether it is possible to *prove* the CTT. It is a fact that currently we cannot rely on mathematical formalisms, which are

[e]A UTM is a TM that has an input a number that represents another TM and the input of the second TM and computes exactly what the second TM is supposed to compute.
[f]After all, one could argue that this is exactly the reason why Ringel and Kovrizhin [20] reached their conclusion.

used to describe our cosmos, in order to prove or disprove the CTT. However, there are physical systems that cannot be computed by a TM, which means that CTT is probably wrong. However, there is definitely some generalized CTT but we do not know anything about it yet.

Acknowledgement

I would like to thank George Papadopoulos for this comments and suggestions that helped me to improve the final text.

Bibliography

[1] Abramsky, S., and Coecke, B. Categorical quantum mechanics. In *Handbook of Quantum Logic and Quantum Structures*, K. Engesser, D. M. Gabbay, and D. Lehmann, Eds. Elsevier, Amsterdam, 2009, pp. 261–323.

[2] Abramsky, S., and Duncan, R. A categorical quantum logic. *Mathematical Structures in Computer Science 16*, 3 (2006), 469–489.

[3] Andréka, H., Madarász, J. X., and Németi, I. Logic of space-time relatibity theory. In *Handbook of Spactial Logics*, M. Aiello, I. Pratt-Hartmann, and J. van Benthem, Eds. Springer, Dordrecht, The Netherlands, 2007, pp. 607–711.

[4] Barr, M. Fuzzy models of linear logic. *Mathematical Structures in Computer Science 6*, 3 (1996), 301–312.

[5] Birkhoff, G., and Neumann, J. V. The logic of quantum mechanics. *Annals of Mathematics 37*, 4 (1936), 823–843.

[6] Bostrom, N. Are you living in a computer simulation? *Philosophical Quarterly 53*, 211 (2003), 243–255.

[7] Girard, J.-Y. Between logic and quantic: a tract. In *Linear Logic in Computer Science*, T. Ehrhard, J.-Y. Girard, P. Ruet, and P. Scott, Eds., London Mathematical Society Lecture Note Series. Cambridge University Press, Cambridge, UK, 2004, pp. 346–381.

[8] Girard, J.-Y. *The Blind Spot: Lectures on Logic.* European Mathematical Society, Zürich, 2011.

[9] Gogioso, S. A bestiary of sets and relations. arXiv:1506.05025v2 [quant-ph], 2015.

[10] Goldblatt, R. *Orthogonality and Spacetime Geometry.* Universitext. Springer-Verlag, New York, 1987.

[11] Goldblatt, R. *Topoi: The Categorial Analysis of Logic.* Dover Publications, Mineola, New York, 2006.

[12] Guts, A. K. A topos-theoretic approach to the foundations of relativity theory. *Doklady Akademii Nauk SSSR 318*, 6 (1991), 1294–1297. English translation in *Soviet Mathematics Doklady*, 43(3), 904–907, 1991.

[13] Hayes, B. Debugging the universe. *International Journal of Theoretical Physics 42*, 2 (2003), 277–295.

[14] Heunen, C. *Categorical Quantum Models and Logics.* Amsterdam University Press, Amsterdam, 2009.

[15] Jacobs, B. *Categorical Logic and Type Theory.* Elsevier Science B.V., Amsterdam, 1999.

[16] Leinster, T. *Basic Category Theory.* Cambridge Studies in Advanced Mathematics, Vol. 143. Cambridge University Press, Cambridge, 2014.

[17] Linde, A. Stochastic approach to tunneling and baby universe formation. *Nuclear Physics B 372*, 1–2 (2006), 421–442.

[18] MacLane, S., and Moerdijk, I. *Sheaves in Geometry and Logic.* Universitext. Springer-Verlag, New York, 1992.

[19] Ord, T., and Kieu, T. D. The Diagonal Method and Hypercomputation. *The British Journal for the Philosophy of Science 56*, 1 (2005), 147–156.

[20] Ringel, Z., and Kovrizhin, D. L. Quantized gravitational responses, the sign problem, and quantum complexity. *Science Advances 3*, 9 (2017).

[21] Robb, A. A. *Geometry of Time and Space.* Cambridge University Press, Cambridge, 1936. A paperback edition of the original hardback was published in 2014.

[22] Suppes, P. Some open problems in the philosophy of space and time. In *Space, Time and Geometry*, P. Suppes, Ed. D. Reidel Publishing Company, Dordrecht, The Netherlands, 1973, pp. 383–401.

[23] Syropoulos, A. Yet another fuzzy model for linear logic. *International Journal of Uncertainty, Fuzziness and Knowledge-Based Systems 14*, 1 (2006), 131–136.

[24] Syropoulos, A. *Hypercomputation: Computing Beyond the Church-Turing Barrier.* Springer, New York, 2008.

[25] Syropoulos, A. Can we debug the Universe? arXiv:0910.2859v2 [cs.OH], 2009.

[26] Syropoulos, A. Hypercomputation: fantasy or reality? a position paper. *Parallel Processing Letters 23*, 1 (2013), 1350005.

[27] Syropoulos, A. Fuzzy categories. arXiv:1410.1478v1 [cs.LO], 2014.

[28] Syropoulos, A. *Demystifying Computation: A Hands-On Introduction.* World Scientific, London, 2017.

[29] Syropoulos, A. On vague computers. In *Emergent Computation: A Festschrift for Selim G. Akl*, A. Adamatzky, Ed. Springer International Publishing, Cham, 2017, pp. 393–402.

[30] Syropoulos, A. On TAE machines and their computational power. *Logica Universalis 13*, 2 (2019), 165–170.

[31] Tegmark, M. *Our Mathematical Universe: My Quest for the Ultimate Nature of Reality.* Alfred A. Knopf, New York, 2014.

[32] Walker, A. G. Foundations of relativity: Parts I and II. *Proceedings of the Royal Society of Edinburgh Section A: Mathematics 62*, 3 (1948), 319–335.

[33] Wharton, K. The universe is not a computer. arXiv:1211.7081v2 [quant-ph], 2015.

[34] Wharton, K. The universe is not a computer. In *Questioning the Foundations of Physics: Which of Our Fundamental Assumptions Are Wrong?* A. Aguirre, B. Foster, and Z. Merali, Eds. Springer International Publishing, Cham, 2015, pp. 177–189.

[35] Wharton, K. Against the cosmic computer. *New Scientist 217*, 2903 (9 February 2013), 30–31.

[36] Zuse, K. *Rechnender Raum.* Friedrich Vieweg & Sohn, Braunschweig, Germany, 1969. English translation available at ftp://ftp.idsia.ch/pub/juergen/zuserechnenderraum.pdf (last accessed on 2019/01/24).

Chapter 11

Compression is Comprehension and the Unreasonable Effectiveness of Digital Computation in the Natural World*

Hector Zenil†

Chaitin's work, in its depth and breadth, encompasses many areas of scientific and philosophical interest. It helped establish the accepted mathematical concept of randomness, which in turn is the basis of tools that I have developed to justify and quantify what I think is clear evidence of the algorithmic nature of the world. To illustrate the concept, I will establish novel upper bounds of algorithmic randomness for Elementary Cellular Automata. I will discuss how the practice of science consists in conceiving a model that starts from certain initial values, running a computable instantiation, and awaiting a result in order to determine where the system may be in a future state — in a shorter time than the time taken by the actual unfolding of the phenomenon in question. If a model does not comply with all or some of these requirements, it is traditionally considered useless or even unscientific, so the more precise and faster, the better. A model is thus better if it can explain more with less, which is at the core of Chaitin's "compression is comprehension". I will pursue these questions related to the random versus possibly algorithmic

*This chapter is based on an invited talk delivered at the Workshop on *Patterns in the World*, Department of Philosophy, University of Barcelona on December 14, 2018.

†Oxford Immune Algorithmics, Oxford University Innovation, Oxford, UK; Algorithmic Dynamics Lab Unit of Computational Medicine, Center for Molecular Medicine, Karolinska Institute, Stockholm, Sweden; Algorithmic Nature Group, LABORES for the Natural and Digital Sciences, Paris, France; hector.zenil@algorithmicnaturelab.org.

nature of the world in two directions, drawing heavily on the work of Chaitin. I will also discuss how the algorithmic approach is related to the success of science at producing models of the world, allowing computer simulations to better understand it and make more accurate predictions and interventions.

1. Introduction

The work of Greg Chaitin has been at the center of my intellectual interests. He was not only one of the examiners for my PhD in computer science in 2011, but my work would have not been possible without his seminal ideas. I met Greg in 2006, having been invited to his home and to his office at IBM's legendary Thomas J. Watson Research Center (see Fig. 1), both in Yorktown Heights, NY, USA. His house was typical of an intellectual, full of objects and books piled one on top of the other. But Greg also had exotic pieces of sculpture scattered around his house, and an over-representation of Leibniz- and Gödel-related books, the two major sources of influence on his work and thinking.

In 2008, I organized a two-part panel discussion (Fig. 2) during the Wolfram Science Conference at the University of Vermont in Burlington, an event that, together with a second meeting I organized (with Adrian German) in 2008 at Indiana University Bloomington 2, will, I believe, come to be regarded as a historic. The first part of the panel discussion addressed the question "Is the Universe random?". The participants were Cris Calude, John Casti, Greg Chaitin, Paul Davies, Karl Svozil and Stephen Wolfram. The second part addressed the question "What is computation and how does nature compute?" The participants were Ed Fredkin, Greg Chaitin, Tonny Legett, Cris Calude, Rob de Ruyter, Tom Toffoli, and Stephen Wolfram, with George Johnson, Gerardo Ortiz and myself serving as moderators. Transcriptions of both parts were published in my volume [21, 22].

In 2007, I also helped Stephen Wolfram design a commemorative medal to celebrate the 60th birthday of Gregory Chaitin, which also involved minting for the first time a 300-year-old medal originally designed by Gottfried Leibniz to celebrate the invention or discovery

Fig. 1. (Top) A picture of Greg Chaitin I took outside his longtime office at IBM Research headquarters, the Thomas J. Watson Research Center. (Bottom) The two sides of the medal I helped Wolfram design, featuring material relating to Chaitin's life achievements on the obverse and Leibniz' own medal celebrating binary arithmetic on the reverse.

of binary arithmetic (see Fig. 1). I published a blog post soon after we came up with the idea for the medal, explaining the pre- and post-minting story of the medal [26].

At the center of my research is Greg Chaitin's work on what is known as algorithmic probability, which in turn is related to Chaitin's omega (Ω) number, also called the halting probability. My very first research paper [24] was published in Chaitin's 60th birthday festchrift [25]. The paper advanced the claim and provided the first numerical evidence of a bias present in nature that I later advanced in a contribution that won an FQXi prize [36] on the digital or analog nature of the universe. Such claims have more recently been noticed

Part I Panel: Is the universe random?

Part II Panel: What is computation and how does nature compute?

Fig. 2. Part I and Part II panel discussion pictures. (Top) From left to right: Hector Zenil, Stephen Wolfram, Paul Davies, Ugo Pagallo, Greg Chaitin, Cris Calude, Karl Svozil, Gordana Dodig Crnkovic, and John Casti. (Bottom) From left to right: Hector Zenil, Greg Chaitin, Ed Fredkin, Rob de Ruyter, Tonny Legett, Cris Calude, Tom Toffoli, and Stephen Wolfram. Transcriptions in [21] and [22].

and expanded [27], a further confirmation in which my own methods played an important role again. In a later follow-up piece, I connected Turing patterns with Turing machines, explaining how symmetry-breaking creates structure — as opposed to randomness — out of nothing, with only computability being assumed [23].

2. Complexity from Computation

Not only did Chaitin help found one of the most exciting areas of modern science (computer science), but it may turn out that his contribution, together with that of Alan Turing, may have had a more profound effect on our understanding of our physical reality than we had hitherto supposed.

At the beginning of the 20th century and through the end of the Second World War, computers were human, not electronic, mainly women. The work of a computer consisted precisely in solving tedious arithmetical operations with paper and pencil. This was looked upon as work of an inferior order. At an international mathematics conference in 1928, David Hilbert and Wilhelm Ackermann suggested the possibility that a mechanical process could be devised that was capable of proving all mathematical assertions. This notion is referred to as *Entscheidungsproblem* (in German), or "the decision problem". If a human computer did no more than execute a mechanical process, it was not difficult to imagine that arithmetic would be amenable to a similar sort of mechanization. The origin of *Entscheidungsproblem* dates back to Gottfried Leibniz, who having (around 1672) succeeded in building a machine based on the ideas of Blaise Pascal that was capable of performing arithmetical operations (named Staffelwalze or the Step Reckoner), imagined a machine of the same kind that would be capable of manipulating symbols to determine the truth value of mathematical principles. To this end, Leibniz devoted himself to conceiving a formal universal language, which he designated characteristica universalis, a language that would encompass, among other things, binary language and the definition of binary arithmetic.

In 1931, Kurt Gödel [9] arrived at the conclusion that Hilbert's intention (also referred to as "Hilbert's program") of proving all theorems by mechanizing mathematics was not possible under certain reasonable assumptions. Gödel advanced a formula that codified an arithmetical truth in arithmetical terms and that could not be proved without arriving at a contradiction. Even worse, it implied that there was no set of axioms that contained arithmetic free of true formulas that could not be proved.

In 1944, Emil Post, another key figure in the development of the concepts of computation and computability (focusing especially on the limits of computation) found [10] that this problem was intimately related to one of the 23 problems (the tenth) that Hilbert, speaking at the Sorbonne in Paris, had declared the most important challenges for 20th century mathematics.

Usually, Hilbert's program is considered a failure, though in fact it is anything but. Even though it is true that Gödel debunked [9] the notion that what was true could be proved, presenting a negative solution to the "problem of decision", and Martin Davis [11] (and independently, Julia Robinson [12]) used Gödel's negative result to provide a negative solution to Hilbert's tenth problem (the argument for which was completed by Yuri Matiyasevich [13]), Hilbert's supposedly failed program originated what we now know as Computer Science, the field that wouldn't have been possible without Alan M. Turing's concept of the universal machine.

2.1. *One machine for everything*

Not long after Gödel, Alan M. Turing made his appearance. Turing contemplated the problem of decision in much cruder terms. If the act of performing arithmetical operations is mechanical, why not substitute a mechanical device for the human computer? Turing's work represented the first abstract description of the digital general-purpose computer as we know it today. Turing defined what in his article he termed an "a" computer (for "automatic"), now known as a Turing machine.

A Turing machine is an abstract device which reads or writes symbols on a tape one at a time, and can change its operation according to what it reads, and move forwards or backwards through the tape. The machine stops when it reaches a certain configuration (a combination of what it reads and its internal state). It is said that a Turing machine produces an output if the Turing machine halts, while the locations on the tape the machine has visited represent the output produced.

The most remarkable idea advanced by Turing is his demonstration that there is an "a" machine that is able to read other "a" machines and behave as they would for an input "s". In other words, Turing proved that it was not necessary to build a new machine for each different task; a single machine that could be reprogrammed sufficed. This erases the distinction between program and data, as well as between software and hardware, as one can always codify data as a program to be executed by another Turing machine and vice versa, just as one can always build a *universal machine* to execute any program and vice versa.

Turing also proved that there are Turing machines that never halt, and if a Turing machine is to be *universal*, and hence able to simulate any other Turing machine or computer program, it is actually expected that it will never halt for an infinite number of inputs of a certain type (while halting for an infinite number of inputs of some other type). This is what Turing would have expected, given Gödel's results and what he wanted to demonstrate: that Hilbert's mechanization of mathematics was impossible. This result is known as the undecidability of the halting problem. In his seminal article, Turing defined not only the basis of what we today know as digital general-purpose computers, but also software, programming and subroutines. Thus without a doubt it represents the best answer to date that we have to the question "What is an algorithm?"

In fact in Alan Turing's work on his universal machine, he even introduced the concept of a subroutine that helped him in his machine construction. These notions are today the cornerstone of the field that Turing, more than anyone else, helped establish, namely Computer Science.

Once we approach the problem of defining what an algorithm is and arrive at the concept of *universality* that Turing advanced, the question to be considered in greater detail concerns the nature of algorithms. Given that one now has a working definition of the algorithm, one can begin to think about classifying problems, algorithms and computer programs by, for example, the time they take or the storage memory they may require to be executed. One may assume that the time required for an algorithm to run would depend on the type of machine, given that running a computer program on a Pentium PC is very different from executing it on a state-of-the-art supercomputer. This is why the concept of the Turing machine was so important — because any answers to questions about problem and algorithm resources will only make sense if the computing device is always the same, and that device is none other than the universal Turing machine. So, for example, every step that a Turing machine performs while reading its tape is counted as a time step. Many algorithms can arrive at the same conclusion taking different paths, but some may be faster than others, but this is now a carefully considered matter when fixing the framework for Turing's model of computation: one asks whether there is an algorithm that surpasses all others in economy as regards the resources required when using exactly the same computing device. These are the questions that opened up an entire new field in the wake of Turing's work, the development of which Turing would certainly have been delighted to witness. This field is today referred to as the theory of Computational Complexity, which would not have been possible without a concept such as that of the universal Turing machine. The theory of Computational Complexity focuses on classifying problems and algorithms according to the time they take to compute when larger inputs are considered, and on how size of input and execution time are related to each other. This is all connected to two basic resources needed in any computation: space (storage) and time. For example, one obvious observation relating to this theory is that no algorithm will need more space than time to perform a computation. One can then quickly proceed to ask more difficult but more interesting questions, such as whether a machine can execute a program faster if it is allowed to behave probabilistically instead of deterministically. What happens if one adds more than one tape to a universal Turing machine operation? Would that amount to implementing an algorithm to solve a problem much faster? Or one may even ask whether there is always an efficient algorithm for every inefficient algorithm, a question that may lead us to a fascinating topic connecting computers and physics.

2.2. *The world of simple programs*

If algorithms can come in a variety of types, some slow, others faster, what is it that allows nature to produce, with no apparent effort, what seem to us to be complex objects? These range from the laws of physics to the formation of matter and galaxies to the beginning of life on Earth (and possibly in other parts of the universe). In the end, one can see all these natural phenomena as a kind of computation, regardless of whether it is of exactly the same type as that performed by a Turing machine. This latter possibility cannot be completely disregarded. Thanks to Turing we know that even simple devices such as universal Turing machines possess incredible power.

One of the natural world's most fascinating characteristics is that it presents a wide range of physical and biological systems that behave in different ways, just like algorithms, most of them having some regular features while nonetheless being hard to predict. Climate is a case in point. Even though it is cyclical, it is impossible to predict its details more than a week in advance. Where does nature's complexity come from? Throughout human history we have encountered objects, in particular mathematical ones, that seem complex to us. One set of such objects comprises numbers that can be expressed as the division p/q, with p and q being integers. Numbers 5, 0.5 or even infinite numbers such as $0.333\ldots$ can be written as $5/1$, $1/2$ and $1/3$, respectively. But as far back as the ancient Greeks numbers have been known, such as π and the square root of 2, which cannot be expressed in this way. One could think of arithmetical division as an algorithm that can be performed by a Turing machine, the result being provided in the output tape. Multiplication, for example, is an algorithm to shorten the number of steps needed to perform the same operation using only addition. In the case of numbers that admit a rational representation p/q, the algorithm of the division of integers consists of the common procedure of finding quotients and remainders. In the case of numbers such as π and the square root of 2, the algorithm produces an infinite non-periodic expansion, so that the only way to represent them is symbolically (i.e., π and $\sqrt{2}$). The Pythagoreans found that those numbers with ostensible infinite complexity could be produced from very simple operations, for example, when seeking the value of the hypotenuse of a right triangle with sides of length 1. Since Euclid, it has also been known that such numbers are not the exception among real numbers that are found, for example, in the continuous interval (0, 1).

In algorithmic terms, rational and irrational numbers are different in nature. When one starts a Turing machine that implements the algorithm for division, there is no algorithm that allows for the production of an irrational number followed by halting, whereas the division of rational numbers can halt (when the remainder is zero) or enter an infinite cycle that will produce a repetitive decimal expansion.

In engineering, including systems programming, the intuition of what is complex (in comparison to an irrational number in mathematics, for example) has been radically different.

The usual assumption has been that to produce something that seems complex, a process that is just as complex has to be devised. This issue, however, is closely connected to Turing's concept of universality, given that a universal Turing machine that is programmable is, in principle, capable of producing any degree of "complexity", for example, the type of complexity (or randomness) that one can see in the decimal expansion of π.

If Euclid's algorithm for division or π can produce such apparent complexity, how usual is it to run a random computer program that produces the same complexity? If computer programs that produce complexity need a very complex description, the probability of finding one small enough would be very low. For example, even though Turing's 1936 article contains all the main elements of the traditional description of a universal Turing machine capable of reproducing the type of complexity to be found in the digits of π, the construction of his universal machine requires at least 18 states, and at least 23 instructions (the exact number cannot be calculated on the basis of Turing's article due to the fact that he uses subroutines that can be implemented on machines of different sizes).

Whatever the actual threshold for reaching Turing universality, it had typically been thought to be high (a case in point: von Neumann's universal builder, a system that was the anticipation of the modern concept of cellular automata, requires 29 states), and it was thought that a universal machine would require a certain minimum complexity (at least as to the number of states and symbols required to describe it). In an experiment with extremely small and simple computer programs, Stephen Wolfram found that this threshold of complexity and universality was likely to be extremely low [32], and that very little was required to find a machine that produced high complexity or that was capable of being Turing universal. Indeed, not only did Wolfram find a very small Turing machine with only two symbols and five states that was capable of carrying out universal computation under some very simple conditions, that is, a computer that was powerful enough to emulate any standard computer, there was another Turing machine that Wolfram suspected was Turing-universal, and to ascertain whether this was so, in 2007 Todd Rowland and I organized a prize competition (https://www.wolframscience.com/prizes/tm23/) with a view to determining how simple the rules for a universal Turing machine could in fact be.

Wolfram's other favorite computer programs are called Elementary Cellular Automata [32] and it is on one of these that the 2,5 original smallest universal Turing machines are based.

Elementary Cellular Automata (ECA) are another superb example of a computing model capable of very rich behavior. ECA are minimalistic computer programs that are very visual, as they print out their space–time evolution in two dimensions and can be subjected to easy visual inspection. Recently, my colleagues and I found another extremely small Turing-universal computer by combining two ECA, which are themselves extremely simple [8].

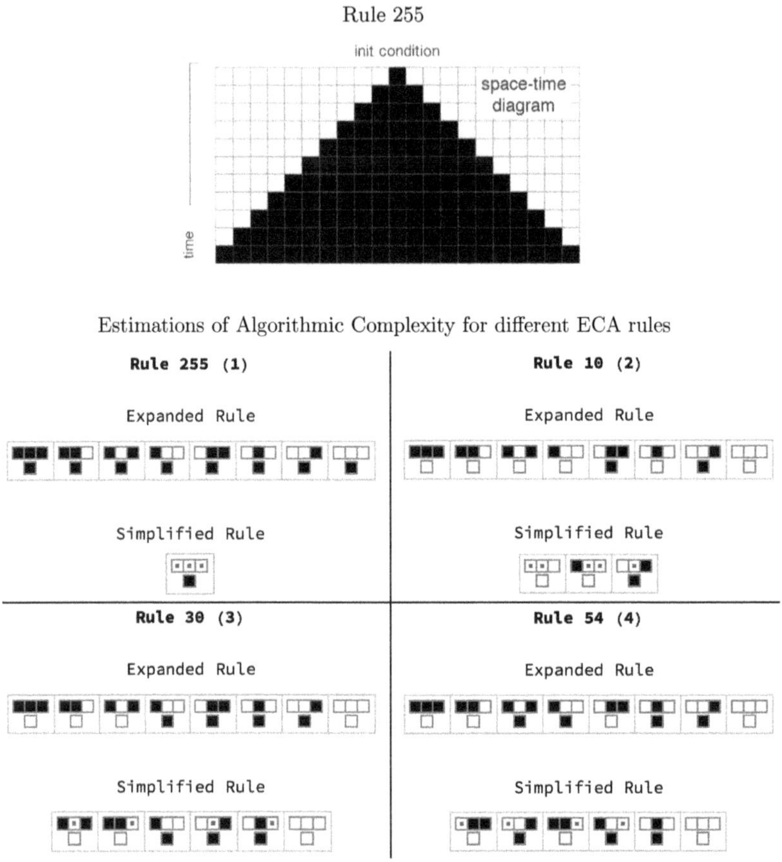

Fig. 3. (Top) An example of a cellular automaton, in this case Elementary Cellular Automaton (ECA) rule 255 which consists in traversing the cells 3 × 3 along each row and converting any three cells starting from the original initial condition (a single black cell) and printing another black cell below. (Middle) ECA cases showing the process of wildcard rule simplification. For rule 255 the simplified description of the ECA tells us that any black or white combination of three cells always leads to the same black cell.

In Fig. 3, I showed how the ideas of Chaitin (and Kolmogorov and Solomonoff) can help understand the complexity of ECA by looking at how difficult it is to describe them succinctly from their generating model. The simplified rule shown in Fig. 3 is an upper bound on their Kolmogorov–Chaitin complexity. Such simple estimations already provide a much better characterization than other simplifications, such as that of the so-called Langton's λ parameter, as they correspond better to the literature on the complexity of ECA. And even better estimations and tighter bounds can be found using more powerful approaches based on Chaitin's (and Levin's) work, notably with the help of two methods that I and my team put together called CTM and BDM (as shown in Fig. 3), rooted in a beautiful concept called algorithmic probability, which is deeply related to Chaitin's own Ω number.

2.3. *Algorithmic probabilities*

Just as the formulas for the production of the digits of π are compressed versions of π, the laws of physics can be seen as systems that compress natural phenomena. These laws are valuable because it is thanks to them that the result of a natural phenomenon can be predicted without having to wait for it to unfold in real time, e.g., one can solve the equations that describe planetary movement instead of waiting two years to know the future positions of a planet. For all practical purposes, the laws of physics are like computer programs and the scientific models we have of them: executable on digital computers and susceptible of numerical solutions.

At the center of Chaitin's work related to his Chaitin Ω [28], and also the variation introduced by Solomonoff [31] and Levin [30], is the notion of algorithmic probability, which can be defined as,

$$\mathrm{AP}(s) = \sum_{p:U(p)=s} 1/2^{|p|},$$

that is, the sum over all the programs p for which a universal Turing machine U outputs s and halts.

The notion behind AP is very intuitive. If one wished to produce the digits of π randomly, one would have to try time after time until one managed to hit upon the first numbers corresponding to an initial segment of the decimal expansion of π. The probability of success is extremely small: 1/10 digits multiplied by the desired quantity of digits. For example, $1/10^{2400}$ for a segment of 2400 digits of π. But if instead of shooting out random numbers one were to shoot out computer programs to be run on a digital computer, the result would be very different.

A program that produces the digits of π would have a higher probability of being produced by a computer. Concise and known formulas for π could be implemented as short computer programs that would generate any arbitrary number of digits of π.

2.4. *Algorithmic Chaitin complexity*

The length of the shortest program that produces a string is today the mathematical definition of randomness, as introduced by Kolmogorov [29], Chaitin [28], and Solomonoff [31], and later expanded by Levin [30], and also called Kolmogorov–Chaitin complexity.

The idea is relatively simple. If a string s of length $|s|$ cannot be produced by a program $|p|$ shorter than $|s|$ in bits, then the string s is random because it cannot be effectively described in a shorter way than by s itself, there being no program p that generates s whose length is shorter than s. Formally,

$$C_U(s) = \min\{|p| : U(p) = s\},$$

where U is a universal Turing machine. The invariance theorem [28, 29, 31] guarantees that the value of C, whether calculated with one particular universal Turing machine or any other universal Turing machine, is the same at the limit. Formally, if U_1 and U_2 are two universal Turing machines and $C_{U_1}(s)$ and $C_{U_2}(s)$ are the values for the algorithmic complexity of s for U_1 and U_2, respectively, there exists a constant c such that

$$|C_{U_1}(s) - C_{U_2}(s)| < c.$$

Thus, the longer the string, the less important c is, and the more stable, the algorithmic complexity value C is. One of the disadvantages of C is that, given the halting problem for Turing machines, C is not computable, which is to say that given a string, there is no algorithm that returns the length of the shortest computer program that produces it.

2.5. *Complexity is in inverse relation to probability*

Algorithmic probability and algorithmic complexity K are inversely proportional, as established by the so-called algorithmic Coding theorem [16, 25]:

$$|-\log_2 \mathrm{AP}(s) - C(s)| < c,$$

where c is a constant independent of s. The Coding theorem implies that the algorithmic complexity can be estimated from the frequency of a string.

To illustrate the above condition let us consider π. Under the assumption of Borel's absolute normality of π, whose digits appear randomly distributed, and with no knowledge of the deterministic source and nature of π as produced by short mathematical formulas, we ask how an entropy versus an algorithmic metric performs. First, the Shannon entropy rate (thus assuming the uniform distribution along all integer sequences of N digits) of the N first digits of π, in any base, would suggest maximum randomness at the limit. However, without access to or without making assumptions as regards the probability distribution, approximations to algorithmic probability would assign π high probability, and thus the lowest complexity by the Coding theorem, as has been done in [17, 18, 33, 34].

Just as with π but in application to graphs and networks, it has been proven how certain graphs can be artificially constructed to target any level of Shannon entropy [19, 20], preserving low algorithmic randomness.

But how relevant is the algorithmic Coding theorem in explaining, e.g., natural phenomena, if it only applies to Turing-universal systems? We know that the natural world allows and can carry out Turing-universal computation because we have been able to build computers that take elements from nature and make them perform as general-purpose computers. However, we don't know how much of the natural world is Turing-computable or how physical laws may naturally implement any form of computation. So, in [15] we showed that up to 60% of the bias found in the output of systems that are not Turing universal may be explained by the algorithmic Coding theorem. This means that this theorem is far more relevant than expected, as it not only explains the way data and patterns distribute for a very particular computational model, but does so, to a considerable extent, for basically any computing model. It is not difficult to imagine that nature operates at some of these levels, perhaps not even on a fixed one but at multiple scales, with the algorithmic Coding theorem being relevant to all of them.

2.6. *The Coding Theorem Method*

The algorithmic *Coding Theorem Method* (CTM) [33, 34] provides the means for approximation via the frequency of a string. Now, why is this so? The underlying mathematics originates from the relation specified by algorithmic probability between frequency of production of a string from a random program and its algorithmic complexity. It is also therefore denoted as the algorithmic *Coding theorem*, in contrast to another well-known coding theorem in classical information theory. Essentially, the numerical approximation hinges on the fact that the more frequently a string (or object) occurs, the lower its algorithmic complexity. Conversely, strings with a lower frequency have higher algorithmic complexity. Otherwise stated,

$$\mathrm{CTM}(s) = -\log_2 \mathrm{AP}(s) + c.$$

The way to implement a compression algorithm at the level of Turing machines, unlike popular compression algorithms, which are heavily based on Shannon entropy, is to go through all possible compression schemes. This is equivalent to traversing all possible programs that are a compressed version of a piece of data, which is exactly what the CTM algorithm does.

2.7. *The Block Decomposition Method*

Our approach to Chaitin's halting probability and Solomonoff–Levin's algorithmic probability consists in asking after the probability of a matrix being generated by a random Turing machine on a two-dimensional array, also called a *termite* or *Langton's ant* [6]. Hence an accounting procedure is performed using Turing machines that aims to approximate the algorithmic complexity of the identified structures. This technique is referred to as the *Block Decomposition Method* (BDM), as introduced in [18,37]. The BDM technique requires a partition of the adjacency matrix corresponding to the graph into smaller matrices. With these building blocks at hand, we numerically calculate the corresponding algorithmic probability by running a large set of small two-dimensional deterministic Turing machines, and then — by applying the algorithmic Coding theorem, as discussed above — its algorithmic complexity.

Following such a divide-and-conquer scheme we can then approximate the overall complexity of the original adjacency matrix by the sum of the complexity of its parts. Note that we have to take into account a logarithmic penalization for repetition, given that n repetitions of the same object only add $\log n$ to its overall complexity, as one can simply describe a repetition in terms of the multiplicity of the first occurrence. Technically, this translates into the algorithmic complexity of a labeled graph G by means of BDM which is defined as follows:

$$\mathrm{BDM}(G,d) = \sum_{(r_u,n_u)\in A(G)_{d\times d}} \log_2(n_u) + K_m(r_u), \tag{1}$$

where $K_m(r_u)$ is the approximation of the algorithmic complexity of the sub-arrays r_u arrived at by using the algorithmic Coding theorem, while $A(G)_{d\times d}$ represents the set with elements (r_u, n_u), obtained by decomposing the adjacency matrix of G into non-overlapping squares, i.e., the block matrix, of size $d \times d$. In each (r_u, n_u) pair, r_u is one such square and n_u its multiplicity (number of occurrences). From now on $\mathrm{BDM}(G, d = 4)$ will be denoted only by $\mathrm{BDM}(G)$, but it should be taken as an approximation to $C(G)$ unless otherwise stated (e.g., when taking the theoretical true $C(G)$ value). Once CTM is calculated, BDM can be implemented as a look-up table, and hence runs efficiently in linear time for non-overlapping fixed size submatrices.

2.8. *Algorithmic information dynamics*

Unlike most complexity measures that are designed for static objects, except those related to dynamical systems (e.g., Lyapunov exponents), I have led the introduction of a measure of algorithmic complexity adapted for dynamical systems and designed to characterize the change of algorithmic complexity of an object evolving over time. The measure is universal in the sense that it can deal with any computable feature that the system may display over time, either spontaneously or as a result of an external perturbation/intervention/interaction.

At the core of Algorithmic Information Dynamics [3,35], the algorithmic causal calculus that we introduced is the quantification of the change of complexity of a system under natural or induced perturbations, particularly the direction (sign) and magnitude of the difference of algorithmic information approximations denoted by C between an object G, such as a cellular automaton or a graph and its mutated version G', e.g., the flip of a cell bit (or a set of bits) or the removal of an edge e from G (denoted by $G\backslash e = G'$). The difference $|C(G) - C(G\backslash e)|$ is an estimation of the shared algorithmic mutual information of G and $G\backslash e$. If e does not contribute to the description of G, then $|C(G) - C(G\backslash e)| \leq \log_2 |G|$, where $|G|$ is the uncompressed size of G, i.e., the difference will be very small and at most a function of the graph size, and thus $C(G)$ and $C(G\backslash e)$ have almost the same complexity. If, however, $|C(G) - C(G\backslash e)| \leq \log_2 |G|$ bits, then G and $G\backslash e$ share at least n bits of algorithmic information in element e, and the removal of e results in a loss of information. In contrast, if $C(G) - C(G\backslash e) > n$, then e cannot be explained by G alone, nor is it algorithmically not contained/derived from G, and it is therefore a fundamental part of the description of G, with e as a generative causal mechanism in G, or else it is not part of G but has to be explained independently, e.g., as noise. Whether it is noise or part of the generating mechanism of G depends on the relative magnitude of n with respect to $C(G)$ and to the original causal content of G itself. If G is random, then the effect of e will be small in either case, but if G is richly causal and has a very small generating program, then e as noise will have a greater impact on G than would removing e from an already short description of G. However, if $|C(G) - C(G\backslash e)| \leq \log_2 |G|$, where $|G|$ is, e.g., the vertex count of a graph, or the runtime of a cellular automaton, G, then e is contained in the algorithmic description of G and can be recovered from G itself (e.g., by running the program from a previous step until it produces G with e from $G\backslash e$).

We have shown how we can infer and reconstruct space–time evolutions by quantification of the disruptiveness of a perturbation. We can then extract the generating mechanism from the ordered time indices, from least to most disruptive, and produce candidate generating models. Simpler rules have simpler hypotheses, with an almost perfect correspondence in row order. Some systems may look more disordered than others, but locally the relationship between single rows is for the most part preserved (indicating local reversibility).

We have shown that the later in time a perturbation is injected into a dynamical system the less it contributes to the algorithmic information content of the overall space–time evolution. We then move from discrete two-dimensional systems to the reconstruction of phase spaces and space–time evolutions of N-dimensional, continuous, chaotic, incomplete and even noisy dynamical systems.

3. Computability in Nature

Chaitin's work has allowed me to frame and attempt to answer questions related to the random or algorithmic nature of the world and universe: whether it is in some sense mechanistic and can be reproduced by an artificial mechanism such as an electronic computer. A first point to note is that we as human observers access the world through what we call science, at least when we do so formally, and the aim of science is to perform causal inference from empirical observations about the world or the universe, and the driver is twofold: to understand the world and to formulate predictions regarding natural phenomena. The result of this approach over the last millennia has been quite successful; some may describe it as incredible, even unreasonable.

3.1. *Cellular automata as inexhaustible pattern generators*

The cellular automata (CA) model can be seen as a pattern-generating computational system, and because the model is Turing-complete it is in a formal sense an inexhaustible source of complexity generation. CA have an intrinsic parallel nature that make them ideal for visually representing a computation. We recognize certain CA performing computation as having specific meanings, for instance, colliders, counters, and majority deciders.

To illustrate the concept of algorithmic complexity, we can represent ECA rules in a simplified fashion as shown in Fig. 3, featuring the original rule and the simplified one using wildcards.

The use of "wildcards" allows some sort of trivial compression allowing a shorter description, and thus tighter upper bounds to Kolmogrov–Chaitin complexity as compared to their original, longer 8-bit descriptions. In Fig. 3 four ECA are shown, with their rules represented as sets of local rules or icons followed by their wildcard simplified version, which consists in reducing the description length of the global rule by looking at ways to compress the local rules. An implementation of this type of wildcard simplification can be found in the Wolfram Language under the RulePlot[ ] function with Appearance → "Simplified" indicated. Clearly, the more random-looking and complex, the larger the wildcard description. Each rule allows a greater simplification (compared to its original 8-bit description) when the rule produces less algorithmically complex patterns (the number followed by each rule is its Wolfram class).

In Fig. 4, it is clear how an a priori parameter such as Langton's λ that measures the density of non-zeros in the description of a rule as a measure of complexity is outperformed by methods based on Chaitin's algorithmic randomness, one of which (BDM) is numerically based on a variation of Chaitin's Ω.

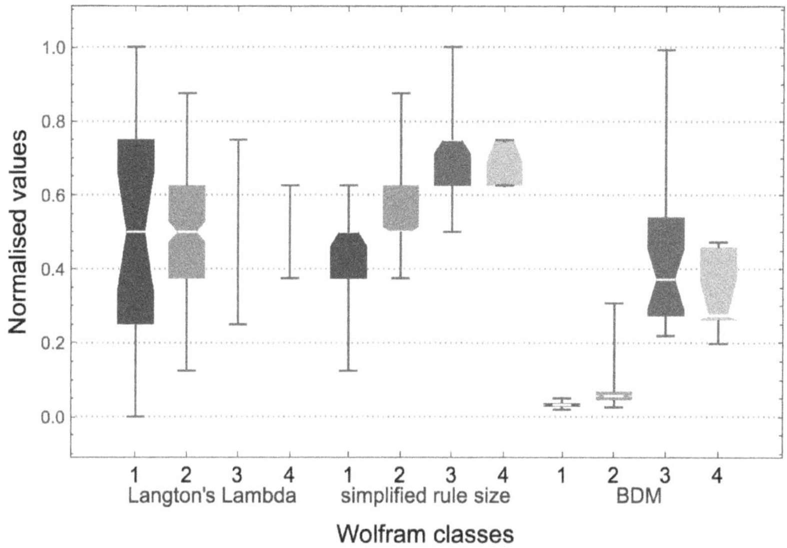

Fig. 4. An a priori measure based on non-zero density (called Langton's lambda or λ) of cellular automata versus tighter bounds of algorithmic randomness according to the rule simplification scheme in Fig. 3. The simplified rule size is an a priori algorithmic randomness measure based on a simplification of the CA global rule, and BDM is an a posteriori observer-oriented measure based Chaitin's randomness. Introduced by Wolfram [32], Wolfram's classes correspond to the typical behavior of each ECA based on how often a rule behaves in a certain way for a random initial condition. Classes 1 and 2 are the most simple, while class 3 displays statistically-random behavior and class 4 displays random behavior together with sophisticated structures with which computation can be implemented [1, 32]. Values normalized by the maximum value in each test set.

3.2. *Science: mapping data to models*

While the practice of science, and mechanistic models in particular, points towards an algorithmic universe governed by ordered laws and computable principles, it is only with the theory of algorithmic randomness that we are able to mathematically understand and quantify this content.

The methods that I have developed allow a partial mapping between the space of all possible computer programs and data, making them natural candidate mechanistic models of the data observations, as illustrated in Fig. 5. To do this, data is chunked in smaller pieces, with each piece having a corresponding candidate

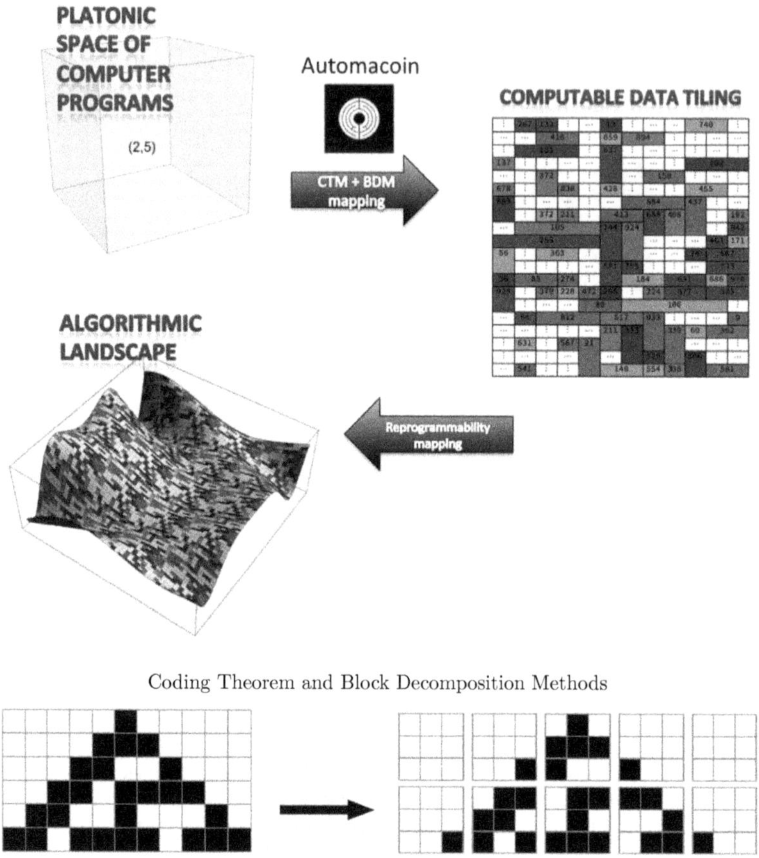

Fig. 5. Algorithmic Information Dynamics (AID) guides the exploration and matching. (Top) Together CTM and BDM are the methods at the heart of AID [33,34,39], helping to mine and guide the search for computable candidate models reproducing a piece of data (or the components of a smaller piece of data) from the space of all computer programs. Automacoin is a cryptocurrency currently being designed to be more helpful in such a task. (Bottom) The process of block decomposition of a ECA (or any other system) consists in dividing the CA time-space evolution into smaller blocks for which estimations of algorithmic randomness are precomputed using the CTM. This image is licensed under a Creative Commons Attribution 3.0 License.

computable model from the space of computer programs, with the help of both CTM and BDM. Because the computer power needed to traverse the space is huge (as difficult as the most difficult computable function, the so-called Busy Beaver problem [2]), we have used supercomputers in the past and have now proposed a new kind of cryptocurrency that will be used to compute the space, rather than computing useless one-way functions as is the typical approach of other, non-ecologically friendly cryptocurrencies. The resulting algorithmic landscape in Fig. 5 is not flat because different pieces of data have associated computable models (computer programs) of different lengths, so some parts of the data are more difficult to produce than others, i.e., require more information to be specified, either because they are disconnected from the rest or are parts in contact with other systems injecting new information.

Randomness may sometimes seem to be dominating science, because if we separate noise from signal it is what appears as noise that bothers science and is the subject of current investigation. Once science tames noise, it becomes signal, and science moves to the next apparent source of noise.

Science has provided us with all sorts of models, such as statistical and mechanistic models. Mechanistic models are very important because they suggest underlying mechanisms as candidates for what may occur with actual phenomena. A mechanistic/algorithmic approach provides a causal model from first principles. Unlike a statistical description, it is prescriptive, as it provides guidance for manipulating causes.

For example, in Fig. 7, a mechanistic model of the solar system is shown. In a mechanistic model such as this one, if we were forced to place the sun and the moon at a certain distance from each other in order to reproduce solar and lunar eclipses, then such a model would suggest a distance that can be verified or falsified by other means, that in turn may feed back to our model to make it better. The traditional approach is to consider a model that starts from certain values, and after some time provides some clue as to where the system may be at a future time, with some precision and in less time than the time taken by the actual unfolding of the phenomenon being studied. If a model does not comply with these requirements then it is traditionally seen as less valuable, so in general, the more precise and the faster it can be, the better. But more importantly, a model is better if it can explain more with less.

Fig. 6. Patterns all over the surface of the planet made by living organisms other than humans and far removed from chance, such as fairy rings, basaltic hexagonal columns, termite mounds. *Left*: The Giant's Causeway in Northern Ireland. Hexagonal basalt columns formed by sea water cooling incandescent lava relatively quickly, around 60 million years ago, as a result of the interplay of a small number of forces. *Right*: A "fairy ring" on the ground is formed in a natural manner by the differential growth of Micelios mushrooms, i.e., an extremely simple mechanical process. This image is licensed under a Creative Commons Attribution 3.0 License.

The movement of planets initially appeared to us as ungoverned, and some of them even seemed to be randomly moving in the sky, and were thus called *wanderers* in Greek. It was not until Galileo, Copernicus, Kepler and later Newton that it was found that planets followed rules. But there were some movements of the internal planets such as Venus, as well as other celestial phenomena, which could not be fully explained by Newtonian mechanics, and thus General Relativity was needed — another mechanistic model to explain what again was apparently random but was found not to be so under the new model (Fig. 6).

It is true that moving from one theory to another reveals that we cannot ever be sure that a given theory definitively explains the data, nor can we say that these rules or differential equations are followed by nature or physics, yet it is also undeniable that each iteration of a model is more encompassing, able to explain not only previously puzzling phenomena but many others as well. In this sense, the compass of what new theories explain, generally speaking, exceeds their own lengths. In other words, we gain much more explanatory and even predictive power with an increasingly compact number of theories over time. This is not only a clear indication that our world and universe is removed from randomness, with a handful of models being able to explain almost every aspect of physical experience, if not human experience, it is also strong evidence of the algorithmic nature of the world. Indeed, the universe that can be explained with theories and models of apparently ever-decreasing length means that the data they describe is formally of low algorithmic randomness, and that finding ever smaller models among all possible computable models is algorithmically very likely, which in turn, by the coding theorem, means that the world is algorithmically simple.

3.3. *Pervasiveness of Turing universality*

While it was known that Turing universality could be achieved with extreme simplicity (in terms of resources, i.e., state + symbols), it was not known how pervasive it was. Once a computer program is Turing universal it can simulate any other computer program. Our work is the first to shed some light on quantifying how pervasive Turing universality may be. It turns out that the number of Turing-universal machines is of density one, meaning that basically almost all computer programs are capable of universal computation, given the appropriate initial conditions [7]. We have also shown how computer programs previously thought to be essentially different can actually simulate other computer programs of unbounded complexity. In this sense, we have shown a complete collapse of the so-called Wolfram classes in a fundamental way [7]. But we have also shown that there is an asymmetry in the number of possible compilers accessible during a given exhaustive compiler exploration time for performing these simulations, which reintroduces a hierarchy and re-establishes the classes from an epistemological perspective. The fundamental collapse, however, strengthens Wolfram's so-called Principle of Computational Equivalence. By way of the aforementioned asymmetry we also defined a topological measure based upon simulation networks, establishing the degree of difficulty involved in finding the right compilers to make a program simulate other programs. All this constitutes a completely novel Bayesian approach to Turing universality and a powerful framework within which to reprogram a system's behavior. These ideas are also consistent with my many first views on programmability, based on Turing's "imitation game", for testing black box capabilities from input perturbation (interrogation) and output assessment. We propose to use the tools of algorithmic complexity to study such mappings as an ultimate behavioral evaluation framework.

We have conceived these tools based on the work of Chaitin and the other founders of AIT as alternatives to compression, tools which allow us to find underlying models. The methods represent a way to navigate through the space of all possible (small) computer programs in order to match computable models with data observations from the real world.

My tools based on Chaitin's work are helping us understand some aspects of dynamic systems useful in the context of living systems, and moreover able to steer and manipulate their behavior.

Another question of relevance concerns computability versus uncomputability. Here again, even assuming that there are non-computable processes, our best tools consist in simulating everything in a digital computer by representing systems and their respective models with computable algorithms, even in those cases where we are dealing with apparently continuous differential equations, which, thanks to CPU's finite precision capabilities, in the end become discrete and symbolic approximations of these apparent and convenient continuous representations (see Fig. 7).

We have seen how we have contributed to matching data with computable models [35], but how successful and relevant are computable models? The distinction between models as equations and models as computer programs represents a gap between the discrete and the continuous that does not seem necessarily fundamental (see Fig. 7). While mechanistic models can also be wrong, they have the advantage of conforming to the laws of physics on the basis of which they are constructed, depending on the nature of the model, whether

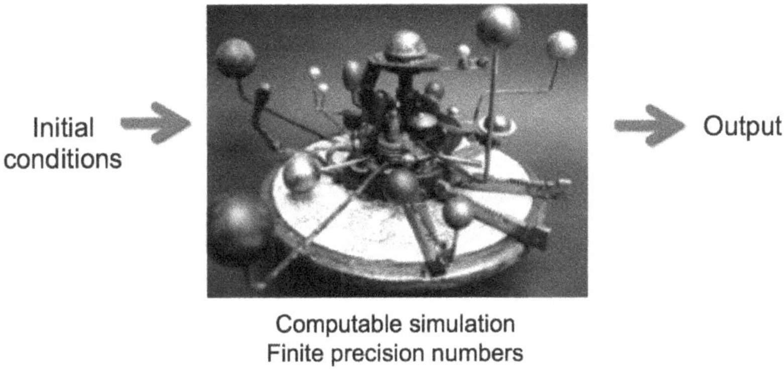

Fig. 7. A mechanistic model is a model that can be run and simulated by mechanical means by setting it in an initial condition, following a set of instructions, and determining an output, e.g., the state of the model. In this case, for example, the placement of the planets. Image from Wikipedia under Creative Commons license.

classical or not. While the model may look analog because it moves in apparently continuous space and time, measurements are of finite accuracy. Most models in science today are not only mechanistic but algorithmic, and they run on electronic digital computers. They are thus computable models of the physical phenomena they are meant to describe. In other words, modern science has become a computational science, both in practice and in principle in many cases, and its power has not diminished. On the contrary, it has never been more powerful, in both its explanatory and its predictive capacity.

4. Complexity in Nature

4.1. *Algorithmic probability in biology*

I aim to understand the behavior of artificial and biological systems from a computational perspective, and to develop methods to reprogram biological cells at a molecular/genetic level as we do computers. I do this by combining the power of the conceptual and methodological advances represented by the formal theories of information (Shannon, Kolmogorov, Chaitin, Solomonoff, Levin, *et al.*) and computation (as advanced by Turing, Church, Kleene, *et al.*), and by exploiting the versatility of principles from areas such as dynamical systems and complex networks, to develop a novel, elegant, and very sophisticated framework and field, namely algorithmic information dynamics, to help deal with questions of causation and produce model-driven, generative and mechanistic explanations for natural and artificial processes.

Biology is mostly discrete, without much room for apparent continuity. Not only is the genetic code discrete, finite, and composed of four letters, genes produce an integral number of viable proteins that either dock or do not, and genes can be simulated closely by modeling them as being either on or off. Proteins are very finite objects. They converge in great numbers, and when they don't they most likely never dock or fulfill their purpose. Chemical concentrations that appear continuous are only a useful representation. Living systems, however, are, in a manner of speaking, assembled out of chemical LEGO pieces. Even signals among cells are proteins that can be counted and followed one by one. Living systems themselves depend on the very binary choice that enables cells to differentiate between themselves and everything else by way of their perfectly limited membranes. Communication among cells is through the physical interchange of proteins — the same kind of LEGO pieces are involved here as everywhere else. And just as with everything else, in most regular cases the piece either docks and fulfills its purpose or it does not. Even Brownian motion that plays a role in, e.g., protein folding, is produced by finite and well-defined particles or molecules interacting with other molecules.

We have explored how non-trivial deterministic systems can display unbounded changes in complexity. We have proved that undecidability is essential for legitimate forms of OEE definitions and that this is also deeply connected to open and closed systems, with self-referential state-dependent systems displaying a greater degree of versatility. For OEE to be possible (under reasonable definitions), then open and undecidable dynamic systems are necessary. But even more important are the results that we have reported regarding the substitution of uniformly distributed mutations for mutations occurring according to the so-called Universal Distribution that predicts the way in which rule-based systems behave according to the computational power of their source (see paper J49). It turns out that this simple and sound substitution (because neither the world nor physical laws are truly random) can significantly speed up evolutionary convergence, may justify the need of genetic memory and modularity (e.g., gene organization), and even explain phenomena such as diversity explosions and mass extinctions that have no extrinsic (e.g., climate) explanations.

Equipped with the measures that we have developed, my team and I aim to tackle a fundamental challenge in science: that of developing tools for causal discovery [14, 35]. This in order to unveil the design principles and generating mechanisms of arbitrary dynamic systems. In particular, the development of an interventional calculus based upon the theory of computability and algorithmic probability that identifies the key markers in genetic networks with which to steer and manipulate cell function and cell fate. The range of application of this work is very general, and aims to generate effective intervention tools to steer the causal content and thus the fate of artificial and natural complex systems. I am engaged in devising strategies to understand, test and reprogram artificial and natural systems as models of computation. I have also devised two measures for testing the "algorithmicity" and programmability of a system. We also introduced a perturbation-based calculus to study the information landscape of complex systems and networks — a calculus capable of unveiling information on the energy landscape — using a measure of algorithmic (re)programmability that connects the theory of dynamic systems with (algorithmic) information theory to reveal the dynamic landscape at the heart of the problem of causality. I use computer programs as models of the world. Determinism from classical mechanics implies that everything can be seen as a computer program and that the complexity of systems led purely by causal elements included in their description are dominated by their evolving time. This allows us to find clues to move systems between different functions, thereby effectively reprogramming them [35].

4.2. *Algorithmic intelligence*

The human mind is algorithmically biased to suit specific purposes. Some of these algorithmic mechanisms can replace previously considered biases based on prior experience (e.g., the so-called system 1). We have known that random generation tasks, in which people have to generate random-looking sequences, are linked to some important aspects of mental and cognitive capacities. We have introduced a new approach to the study of the behavior of animals and humans based on algorithmic information theory. We have shown that humans best outsmart computers when they are tested on randomness generation at the age of 25 [4]. When competing against all possible algorithms, we have found that human behavior is at its most algorithmically random at 25 years of age, thereby introducing a new measure of cognitive complexity possibly related to other biological and cultural phenomena, such as creativity. Indeed, humans produce more randomness than most computer programs when they are 25 years old. In parallel, we are also working on measures of integrated information with which to profile brain networks.

In Fig. 8, ECA can be seen to be interacting with each other and generating structures according to a meta CA rule. Often these interactions appear random, and to an observer they may be difficult to deconvolve or disentangle, but new methods introduced by my team and based on computability theory and the work of Chaitin and Levin (together with the work of Kolmogorov and Solomonoff) help new machine learning algorithms to distinguish the sources by their most likely causal generating mechanisms (in this case, each side is generated by either rules 54 and 50 or 82 and 60) [14].

Most systems in nature are like black boxes that we cannot open and fully understand, and we are therefore left with approaches not very different from the black box approach of Alan Turing — his imitation game — for dealing with the challenge and question of machine intelligence. What we conceived was a test for systems where questions would be inputs and answers would be compressed. One could then discern the variation among the compressed outcomes for

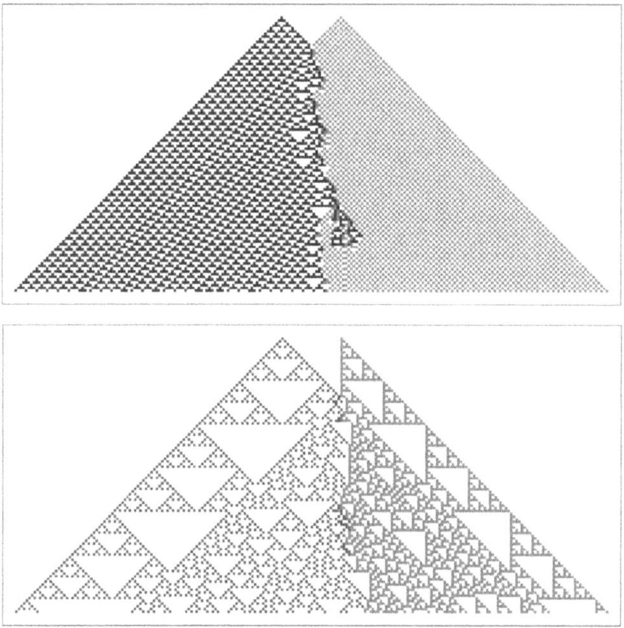

Fig. 8. Examples of simple interacting computer programs (top: ECA rules 54 and 50, bottom: ECA rules 82 and 60) with a simple function determining the interaction from which data emerges and gets convoluted. Different rules come from different generating sources. In this case, different Elementary Cellular Automata rules. Observers will often face this problem and will not be able to disentangle data without the right tools. The world is made of this type of interaction, producing apparent complexity and concealing the underlying generating rules that new methods can help disentangle [14]. They do not even have to be complicated in themselves, but when interacting they produce a highly complex pattern that cannot be quantified with traditional tools, and may be irreducible.

answers of varying complexity and assess the system's capabilities. The idea has evolved into a perturbation algorithmic calculus able to characterize artificial and biological systems [14, 35].

These tools are becoming a new approach to the whole area of machine learning and even of Artificial Intelligence, introducing the inferential power of algorithmic probability to complement the current state of areas such as deep learning, which are weak on the fundamentals for understanding human intelligence, such as symbol manipulation, logic, inference, and the understanding of cause and effect.

4.3. *An algorithmic universe*

Certain parts of the universe seem ordered and structured (see Fig. 6). On Earth, for example, life is an example of organization and structure, contrasting with what can be deemed background noise (comparable to what appears on the screen of an old non-tuned analogue television) left over by the Big Bang and serving as proof of the state in which the universe found itself after its first moments of existence. What Turing would probably never have guessed is that when running random Turing machines, the machines produce highly structured objects. Could this be merely an interesting analogy or could it perhaps be an actual indication that the universe is more algorithmic than initially expected? If so, then Turing machines and computer programs are not just products of the human imagination, they are perhaps responsible for the order in the universe. Alan Turing may have had an intuitive answer to this question, as he was also interested in structure formation and helped found another area in biochemistry called *morphogenesis* [5], the study of pattern formation, starting from a simple shape which would first break its symmetries at random.

We have made great progress at taming apparent randomness since ancient times, with mathematical logic and rationality to begin with — even if we sometimes seem to be regressing to ancient times. A long time ago we left behind explanations based on divinity, magic or paranormal phenomena in exchange for tools such as inference, and statistics. Approaches such as correlation in regression were very useful but they have been overused and abused, and new and better tools for dealing with modern causality are badly needed.

With ever-increasing predictive power, science moves random observations under the explanation of computable models and merges previously computable models into ever more encompassing models whose program-size may be difficult to quantify but whose explanatory power is increasingly greater, hence indicating a clear pattern whereby the universe looks increasingly ordered and of low Kolmogorov–Chaitin complexity (Fig. 10). Science has been mostly motivated by models, has been a model-driven practice. A case can be made regarding the size of new models/theories. For example, taken at face value, the description of the equations for General Relativity (GR) may give the impression of being longer than those for Newtonian classical gravity, hence a possible indication (upper bounds of comparable measure) of their underlying algorithmic complexity. However, the GR equations are strictly shorter than the equations specifying the Newtonian version of gravity plus all the corrections that they require to account for the phenomena that they cannot account for by themselves, such as the movement of the internal planets, in particular the rate of precession of the perihelion of Mercury's orbit. In comparison, General Relativity requires few to no corrections. In other words, Newtonian mechanics would require a stream of regular adjustments to match the computable accounts of General Relativity for such phenomena.

One may also argue that even GR can only predict the precession movement of Mercury for a certain time because of 3-body kinds of phenomena, and this is right, but the fact that we can predict with extremely high accuracy the movements of Mercury for the next 1000 years is different from not been able to account for its movements at all using Newtonian mechanics.

So, in some sense, science can be illustrated as the practice of moving natural phenomena from what we as observers perceive as random, towards non-random phenomena. Science moves empirical data from apparent randomness to algorithmic non-randomness (Fig. 9). Each one of these phenomena under the non-random quadrant was previously on the random phenomena quadrant. One can see that there is also a cascade effect in the right quadrant, with models explaining all or several other previously advanced models. So the overwhelming evidence is that the world has a strong algorithmic component, because science can explain, if not predict, most of what happens in the universe to an increasingly high degree of accuracy and comprehensiveness. The fact that these models compress more observations from natural phenomena in a reduced number of models is an indication that, indeed, compression is comprehension.

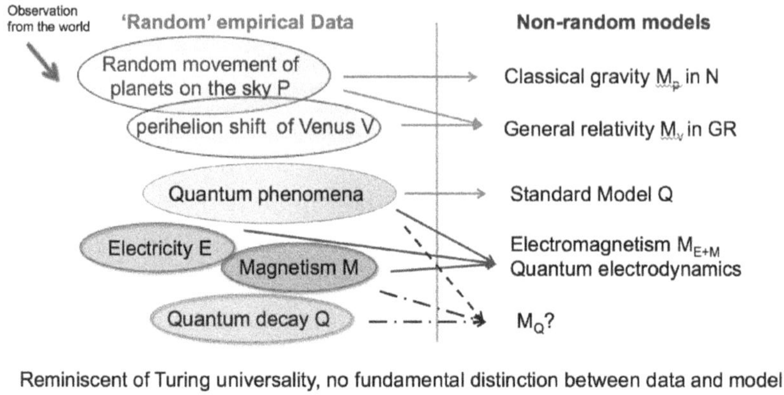

Fig. 9. Science moves apparently random data from observation to model/theory. Indeed, scientific models tend to cover previously random data together with other previously unexplained phenomena. Whether there is a model that can encompass all other models is an open question, but the tendency so far has been clear, and not an artifact of humans as intelligent or conscious observers. This image is licensed under a Creative Commons Attribution 3.0 License.

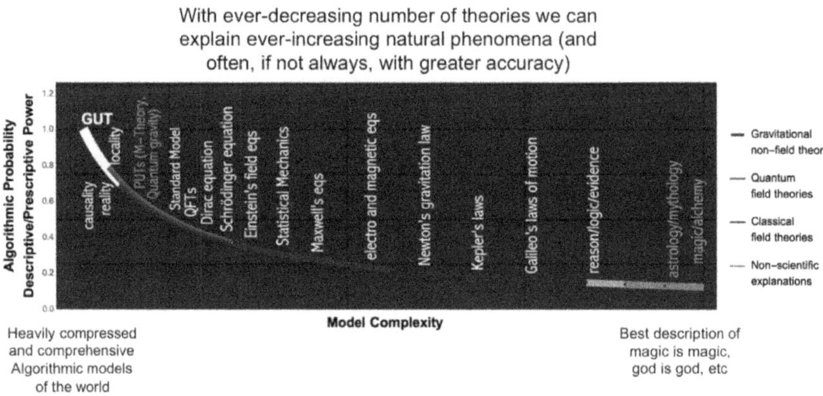

Fig. 10. This diagram shows in what direction science has successfully been operating in previous centuries, and how this has been increasingly sped up in the last century, relying on a handful of fundamental theories that have unified seemingly disparate areas of science, each time explaining more with less. The existence of science is evidence of the algorithmic nature of the world, and the fact that more modern scientific models encompass larger observations strengthen the hypothesis of an algorithmic universe. This image is licensed under a Creative Commons Attribution 3.0 License.

Even in some non-classical interpretations noise may be only apparent, because in fact all deterministic trajectories are explored and we just happen to be in one particular branch of a multiverse. That of course does not solve the problem of the source of global randomness at the level of the multiverse itself, determining why we experience only one random universe and not all others. However, evidence (Figs. 9 and 10) also suggests that what we think is noise is often a signal whose source is unknown or irrelevant to a system of interest.

5. Conclusions

Studying randomness from the computing perspective affords us a framework for studying the nature of the world that discerns the way in which patterns in the universe are distributed. In a world of computable processes in which the laws (like programs) do not have a slantwise distribution, AP(s) would indicate the probability of a natural phenomenon occurring as a result of running a program. Distribution AP(s) has an interesting particularity: it can start from basically anything, and, like a robust distribution, it remains qualitatively unchanged. It is the process that determines the form of the distribution and not the initial distribution of the programs. This is important because one does not make any strong initial assumption about the distribution of the initial conditions or of the laws of physics.

Computer programs can be looked at from a certain vantage point as laws of physics. If one starts with a random initial condition (input) and executes a program chosen at random, there is a very good probability that its final appearance will be regular, and frequently very well organized. By the same token, if one were to shoot out particles at random, the probability of groups forming in the way they do would be so small — in the absence of laws of physics — that nothing whatsoever would happen in the universe. Perhaps Alan Turing has not only helped us understand the world of computing machines and computing programs, founding an entire scientific area, but has also taught us a great deal about the universe in which we live and how it came to be the way it is. Algorithmic complexity and algorithmic probability may explain the unreasonable effectiveness of mathematics in the natural sciences, showing it to have been based on very solid foundations.

Bibliography

[1] M. Cook, Universality in elementary cellular automata, *Complex Sys.* (2004) 1–40.

[2] T. Radó, On non-computable functions, *Bell Sys. Tech. J.* **41**(3) (1962), 877–884.

[3] H. Zenil, N.A. Kiani and J. Tegnér, Algorithmic information dynamics of emergent, persistent, and colliding particles in the game of life, in *From Parallel to Emergent Computing*, ed. A. Adamatzky (Taylor & Francis/CRC Press, 2019), pp. 367–383.

[4] N. Gauvrit, H. Zenil, F. Soler-Toscano, J.-P. Delahaye and P. Brugger, Human behavioral complexity peaks at age 25, *PLoS Comput. Biol.* **13**(4) (2017), e1005408.

[5] A. M. Turing, The chemical basis of morphogenesis, *Philos. Trans. Roy. Soc. London B* **237**(641) (1952), 37–72.

[6] C. Langton, Computation at the edge of chaos: phase transitions and emergent computation, in *Emergent Computation*, ed. S. Forest (The MIT Press, 1991), pp. 12–37.

[7] J. Riedel and H. Zenil, Cross-boundary behavioural reprogrammability reveals evidence of pervasive universality, *Int. J. Unconventional Comput.* **13**(14–15) (2018), 309–357.

[8] J. Riedel and H. Zenil, Rule primality, minimal generating sets and turing-universality in the causal decomposition of elementary cellular automata, *J. Cellular Automata* **13** (2018), 479–497.

[9] K. Gödel, Über formal unentscheidbare Sätze der Principia Mathematica und verwandter Systeme, I. *Monats. Math. Phys.* **38** (1931), 173–98.

[10] E. L. Post, Finite Combinatory Processes — Formulation 1, *J. Symbolic Logic* **1** (1936), 103–105.

[11] M. Davis, Hilbert's tenth problem is unsolvable, *Amer. Math. Monthly* **80** (1973), 233–269.

[12] M. Davis, Y. V. Matiyasevich and J. Robinson, Hilbert's tenth problem: diophantine equations: positive aspects of a negative solution, in *Mathematical Developments Arising from Hilbert Problems*, ed. F. E. Browder, Proceedings of Symposia in Pure Mathematics, Vol. 28, Part 2 (American Mathematical Society, 1976), pp. 323–378.

[13] Y. V. Matiyasevich, Enumerable sets are Diophantine, *Dok. Akad. Nauk SSSR* **191** (1970), 279–282 (in Russian). MR 0258744. English translation in *Soviet Mathematics* **11**(2) (1970), 354–357.

[14] H. Zenil, N. A. Kiani, A. Zea and J. Tegnér, Causal deconvolution by algorithmic generative models, *Nat. Mach. Intell.* **1** (2019), 58–66.
[15] H. Zenil, L. Badillo, S. Hernández-Orozco and F. Hernández-Quiroz, Coding-theorem like behaviour and emergence of the universal distribution from resource-bounded algorithmic probability, *Int. Parallel Emergent Distributed Sys.* (2018).
[16] T. M. Cover and J. A. Thomas, *Elements of Information Theory*, 2nd edn. (Wiley-Blackwell, 2009).
[17] H. Zenil, F. Soler-Toscano, J.-P. Delahaye and N. Gauvrit, Two-dimensional Kolmogorov complexity and validation of the coding theorem method by compressibility, preprint (2012), arXiv:1212.6745.
[18] H. Zenil, F. Soler-Toscano, N. A. Kiani, S. Hernández-Orozco and A. Rueda-Toicen, A decomposition method for global evaluation of Shannon entropy and local estimations of algorithmic complexity. preprint (2016), arXiv:1609.00110 [cs.IT].
[19] H. Zenil, N. A. Kiani and J. Tegnér, Low algorithmic complexity entropy-deceiving graphs, *Phys. Rev. E* **96** (2017), 012308.
[20] M. Morzy, T. Kajdanowicz and P. Kazienko, On measuring the complexity of networks: Kolmogorov complexity versus entropy, *Complexity* **2017** (2017), 3250301.
[21] H. Zenil, *Randomness Through Computation: Some Answers, More Questions* (World Scientific, Singapore, 2011).
[22] H. Zenil, *A Computable Universe: Understanding and Exploring Nature as Computation* (World Scientific, Singapore, 2012).
[23] H. Zenil, Turing patterns with Turing machines: emergence and low-level structure formation, *Nat. Comput.* **12**(4) (2013), 499–515.
[24] J-P. Delahaye and H. Zenil, On the Kolmogorov-Chaitin complexity for short sequences, in *Randomness and Complexity: From Leibniz to Chaitin*, ed. C. Calude (World Scientific, 2007).
[25] C. Calude (ed.), *Randomness and Complexity: From Leibniz to Chaitin* (World Scientific, Singapore, 2007).
[26] H. Zenil, http://www.mathrix.org/liquid/archives/the-history-of-the-chaitin-leibniz-medallion (accessed on February 26, 2019).
[27] K. Dingle, C. Camargo and A. A. Louis, Input-output maps are strongly biased towards simple outputs, *Nat. Commun.* **9**(761) (2018).
[28] G. J. Chaitin, On the length of programs for computing finite binary sequences: statistical considerations, *J. ACM* **16**(1) (1969), 145–159.
[29] A. N. Kolmogorov, Three approaches to the quantitative definition of information, *Int. J. Comput. Math.* **2**(1–4) (1968), 157–168.
[30] L. A. Levin, Laws of information conservation (nongrowth) and aspects of the foundation of probability theory, *Prob. Peredachi Inf.* **10**(3) (1974), 30–35.
[31] R. J. Solomonoff, A formal theory of inductive inference. Parts I and II, *Inform. Control* **7**(1) (1964), 1–22 and 224–254.

[32] S. Wolfram, *A New Kind of Science* (Wolfram Media, Champaign, IL, 2002).

[33] J.-P. Delahaye and H. Zenil, Numerical evaluation of the complexity of short strings: a glance into the innermost structure of algorithmic randomness, *Appl. Math. Comput.* **219** (2012), 63–77.

[34] F. Soler-Toscano, H. Zenil, J.-P. Delahaye and N. Gauvrit, Calculating Kolmogorov complexity from the frequency output distributions of small turing machines, *PLoS ONE* **9**(5) (2014), e96223.

[35] H. Zenil, N. A. Kiani, F. Marabita, Y. Deng, S. Elias, A. Schmidt, G. Ball and J. Tegnér, An algorithmic information calculus for causal discovery and reprogramming systems, bioaRXiv185637; https://doi.org/10.1101/185637.

[36] H. Zenil, The World is Either Algorithmic or Mostly Random, *Third Prize Winning Essay — Foundational Questions Institute (FQXi) Contest "Is Reality Digital or Analog*?" (2011).

[37] H. Zenil, F. Soler-Toscano, K. Dingle and A. Louis, Graph automorphisms and topological characterization of complex networks by algorithmic information content, *Physica A* **404** (2014), 341–358.

[38] W. Kirchherr, M. Li and P. Vitányi, The miraculous universal distribution, *Math. Intelligencer* **19**(4) (1997), 7–15.

[39] H. Zenil, F. Soler-Toscano, N. A. Kiani, S. Hernández-Orozco and A. Rueda-Toicen, A decomposition method for global evaluation of Shannon entropy and local estimations of algorithmic complexity, *Entropy* **20**(8) (2018), 605.

Chapter 12

On Information, Quanta, and Context

J. Acacio de Barros*

Classical information theory relies on the probabilities that signals appear in a source. As such, information depends on Kolmogorov's probability theory. However, it is well known that some quantum sources are context-dependent, and therefore may not conform to classical probability theory. For such sources, information can be defined through von Neumann's entropy. However, there are possible contextual sources that are neither quantum nor classical. In this chapter, we explore a simple example of one such source, and show that, similarly to the quantum case, the presence of contextuality increases the amount of information. We also show that, by extending Shannon's entropy to accommodate negative quasi-probability distributions, we are able to provide a measure of informational content that is neither described by Shannon nor von Neumann's entropies.

1. Introduction

Information is a fundamental concept not only in engineering, but also in physics. It is present in statistical mechanics, where the thermodynamical behavior of a system can be derived from our lack of information about the underlying states of particles in either a classical or quantum statistics [2]. It is an essential idea in certain interpretations of quantum mechanics [24]. And it is even considered

*School of Humanities and Liberal Studies, San Francisco State University, 1600 Holloway Ave., San Francisco, CA 94132, USA; barros@sfsu.edu.

by some physicists as part of the underlying fabric of reality, such as in Wheeler's it-from-bit [47] or D'Arianno's derivation of QFT from information [9].

Information is also key in philosophy of language and philosophy of mind [20]. For example, a very popular view of consciousness is the Integrated Information Theory [38], known as IIT. As its name reflects, IIT relies on the concept of information, and posits that consciousness is a specific type of integration of information where the whole has more information than its parts.

In physics and in engineering, information is usually related to Claude Shannon's definition stemming from his mathematical theory of communications [41]. However, Shannon's theory requires that the information of a set of sources be computed based on the joint probability distribution over the content outcomes of those sources. This creates an issue in physics, where for certain systems no such joint probability exists [23], a phenomenon known as contextuality.[a]

Contextual sources appear in physics as a consequence of the impossibility of measuring all observables simultaneously for a quantum system. For instance, Kochen and Specker (KS) [31] proved that for quantum systems whose Hilbert spaces have dimension three or greater, it is possible to construct a set of dichotomic observables (117 in their original paper) and a collection of contexts where some of those observables could be measured simultaneously but where the totality of them could not. They then showed that attempting to assign a truth-value to such observables would lead to an inconsistency, unless the truth-values changed with the context.

This above result from KS implies that we cannot construct a non-negative joint probability distribution for the observables [11], and therefore cannot use Shannon's definition of information. However, for quantum systems, it is possible to define information by

[a]Some readers may associate the lack of joint probabilities to non-locality, as is the case in Bell's theorem, but the concept of contextuality is more general, with non-locality being a particular case of contextuality where the contexts are selected in a space-like way. See, for example, the discussions about contextuality and non-locality as it relates not only to physics, but also to psychology, in [17].

using von Neumann's entropy, which is shown to be somewhat equivalent to Shannon's [40]. Furthermore, von Neumann's entropy can be shown relevant to extensions of the quantum case, when there is an orthomodular lattice for the set of observables [28].

In this chapter, we investigate the concept of information for contextual sources. In particular, we discuss how Shannon entropy fails to measure the amount of information of contextual sources, and how, for some systems, von Neumann's entropy is not adequate. To do so, we organize this chapter in the following way. In Section 2, we provide a definition of context and context dependency, also known as contextuality. In Section 3, we show that quantum theory, defined by the algebra of observables on a Hilbert space, requires a description of properties that are context dependent, i.e., that quantum observables are contextual. We do this by following a Kochen–Specker-type argument given in [7]. In Section 4, we present Shannon's entropy, considered a measure of information for classical systems, and also von Neumann's entropy, which measures information for quantum systems. Then, in Section 5, we discuss the informational content of several sources that are correlated, some contextual and some noncontextual. We use those examples to show that Shannon's entropy cannot be used in a straightforward way to describe the expected amount of information for contextual systems, a failure that cannot be resolved by resorting to von Neumann's entropy. We end with some final remarks and comments in Section 6.

2. Context

In its current and common usage, context comes from linguistics. It means the parts which are related to a particular text, passage, or speech and that are connected to its meaning. For example, the statement "cheap dates are great" has very different meaning if preceded by a conversation about social engagements or by a discussion on the characteristics and benefits of the *Phoenix dactylifera* fruits. One could say that the meaning of the statement changes with the context. But, if we take semantics to be valued with respect to truth values of propositions, one could also say that the truth values changed from context to context: in a given context the statement "cheap dates are great" may be considered by someone as true, whereas in the other context it may be considered by the same person as false.

So, how can we define context based on this idea? This definition would be particularly difficult if we considered that sometimes it is not clear how to assign truth values, particularly as part of speech. For instance, what is the truth value of the statement "Carlos is a good person"? Even if we ignore the ethical assumptions of establishing the truth of such statement, the very definition of "good person" is context dependent, and in a conversation, either with oneself or with someone else, sometimes we are reluctant to provide definite truth values to such propositions. In another paper [13], we argued that a possible general way to deal with context comes from probability theory. In this section, we will present the main idea of representing contexts with probabilities using a theory known as Contextuality by Default (CbD) [21, 22].[b]

Let us start with the basic ideas. First, we need a probability space, $(\Omega, \mathcal{F}, p)$, where Ω is a sample space, $\mathcal{F}$ is a σ-algebra over Ω, and p a function $p : \mathcal{F} \to [0, 1]$ satisfying the following properties [32]:

K1. $p(\Omega) = 1$;
K2. $p(A \cup B) = p(A) + p(B)$ for $A, B \in \mathcal{F}$ and $A \cap B = \emptyset$.

The probability space $(\Omega, \mathcal{F}, p)$ is just a way to formalize the familiar concept of probability, where the sample space Ω is simply the set of possible outcomes. For example, if we throw a die, the set of possible outcomes is $\Omega = \{1, 2, \ldots, 5, 6\}$, and if we throw two dice, the set is $\Omega = \{(1, 1), (1, 2), \ldots, (5, 6), (6, 6)\}$, where each ordered pair represents the outcomes of the first and second dice, respectively. The algebra $\mathcal{F}$ is a way to talk about more than one outcome: in the case of a die, we can talk about single outcomes, such as $\{1\}$, or we

[b]Here we use a simplified version of CbD, which captures some of its main ideas, but is not as precise and general as the formulation presented in the references.

can talk about combinations of outcomes, such as "$\{1\}$ or $\{6\}$" or "not $\{1\}$", and the combination of all the logical propositions we can make over the outcomes of the die would result in an algebra. Finally, p is simply the probability we assign for each element of $\mathcal{F}$. In the case of the die, we have $p(\{1\}) = \cdots = p(\{6\}) = 1/6$, if the die is unbiased. We can immediately compute from these, using K2, that the probability of getting an even number as outcome, $p(\{2, 4, 6\})$, equals $1/2$.

Now that if we have a probability space, we can define a very important tool in using probabilities to describe quantitatively the outcomes of experiments or the truth values of propositions: random variables. Intuitively, random variables are mathematical objects, functions to be precise, that have the same outcomes as an experiment or measurement we wish to model. For example, if we were to model the outcomes of throwing two dice and then summing their values, we could use a random variable that can have as possible outcomes the numbers $1, 2, 3, \ldots, 12$. The random variable corresponding to this example would be a function $\mathbf{R} : \Omega \to \{1, 2, 3, \ldots, 12\}$, where $\Omega = \{(1, 1), (1, 2), \ldots, (5, 6), (6, 6)\}$ (all possible outcomes of throwing two dice), defined by $\mathbf{R}((i, j)) = i + j$. Since the probabilities of each element of Ω is given by p, one could use the properties of probabilities to compute the distribution of $\mathbf{R}$, i.e., the probabilities of each outcomes. As importantly, given a data set for an experiment, we can always ask the reverse question: what are the random variables and their corresponding probability space that exhibit the same statistical properties as the data set.

Random variables are important because they provide a consistent way of talking about experimental outcomes, truth values, beliefs, or even about semantics.[c] What we mean is that, by assuming an algebra of events, random variables require that the underlying atomic outcomes (i.e., elements of Ω) provide a consistent description of the processes behind the observed events. To illustrate this, let us look at

[c]See [11, 13] for a more detailed discussion of random variables and contextuality as related to those areas.

an example. Imagine we have an experiment where we measure two properties of a system. To keep it simple, let us assume that those properties are binary, i.e., either the system has it or not. Examples of binary properties would be the tossing of a coin (either it is heads or it is not), the passing of an exam (either the grade was sufficient or not), or the success of an offensive play in football (either a goal was scored or not). Let us call those two properties A and B. Since they are binary, we can represent the outcomes of those properties by any two numbers, usually $\{0, 1\}$ or $\{-1, 1\}$, where we can assign, if we wish, 0 or -1 to false and 1 to true. The data from observing A and B would give us their individual statistical characteristics, e.g., their expectations, which tells us how often A or B are true, and their joint expectations (if ever measured together), which tells us how A and B are correlated. With such information, we could create a probability space with a sample space $\Omega = \{\omega_{00}, \omega_{01}, \omega_{10}, \omega_{11}\}$ and the following random variables:

$$\mathbf{A}(\omega_{00}) = \mathbf{A}(\omega_{01}) = \mathbf{B}(\omega_{00}) = \mathbf{B}(\omega_{10}) = 0,$$
$$\mathbf{A}(\omega_{10}) = \mathbf{A}(\omega_{11}) = \mathbf{B}(\omega_{01}) = \mathbf{B}(\omega_{11}) = 1.$$

Our notation here should be straightforward: ω_{00} corresponds to A and B being false, ω_{10} to A true and B false, and so on. If we chose a probability p in a $(\Omega, \mathcal{F}, p)$ that reproduces all the observed expectations and correlations, we created a model of the experimental outcomes in terms of random variables. This model would allow us to infer all possible logical questions about elements of Ω, such as what is the probability of ω_{00}, or "not ω_{00}", or "ω_{10} and ω_{01}", etc. In other words, all possible logical consequences of the possible values of A and B would be accounted in the statistical model.

Now that we have necessary tools to talk about empirical observations, let us go back to contextuality, first with a simple example. Imagine we have instead of two, three binary properties, namely X, Y, and Z. Let us model them with ±1-valued random variables, instead of 0 or 1, for symmetry. After several observations, we conclude the following about the statistical properties of the random variables:

$$E(\mathbf{X}) = E(\mathbf{Y}) = E(\mathbf{Z}) = 0 \tag{1}$$

and

$$E(\mathbf{XY}) = E(\mathbf{XZ}) = E(\mathbf{YZ}) = -1. \tag{2}$$

We do not have the expectation of the product $\mathbf{XYZ}$ because we never observe the three variables simultaneously, but only either by themselves or in pairs.

The first set of expectations (1) tells us that each random variable seems to behave randomly, with each outcome $+1$ and -1 appearing in our data table with equal probability for each. The second set of expectations (2), involving the second moment, tells us that three random variables are not at all independent. In fact, since the expectation of their products is -1, it follows that if one of the variables is -1, the other must be $+1$. In other words, if we know $\mathbf{X}$, we also know $\mathbf{Z}$ or $\mathbf{Y}$. However, there is an issue with the above expectations: they lead to a contradiction. To see this, assume that $\mathbf{X} = -1$, which from the first expectation implies that $\mathbf{Y} = \mathbf{Z} = 1$, which contradicts the observation that $\mathbf{YZ} = -1$. A similar contradiction is reached if we assume that $\mathbf{X} = 1$.

The reader may at this point argue that expectations (1) and (2) are an impossibility: one should never have an experiment that leads to contradictions. In a certain sense they would be right, as the contradictions do not come from the expectations, but from the conclusions we are deriving from our model of the expectations. Notice that, as mentioned above, the experimental conditions are such that we never observe all three properties simultaneously, but only two of them. For example, in one observation, properties may be $\mathbf{X} = 1$ and $\mathbf{Y} = -1$, but this tells us nothing about the *unobserved* value of $\mathbf{Z}$. Assuming that the $\mathbf{Z}$ would be the same in this condition as in the other condition leads to the contradiction. In other words, the assumption is that $\mathbf{Z}$ in the context of $\mathbf{X}$ is the same as in the context of $\mathbf{Y}$. This is what we mean by contextuality.

To emphasize this point, consider that, since the binary properties X, Y, and Z, are only observed in pairs, each pair provides a context, as follows:

Context 1: X and Y;
Context 2: X and Z;
Context 3: Y and Z.

Since we know those variables are contextual, we model them using random variables that are indexed by their context (see [22] for a more formal procedure). Thus, for Context 1, we have $\mathbf{X}_1$ and $\mathbf{Y}_1$, for Context 2, $\mathbf{X}_2$ and $\mathbf{Z}_2$, and for Context 3, $\mathbf{Y}_3$ and $\mathbf{Z}_3$. It should now be clear that we have no contradiction, as the $\mathbf{X}_1\mathbf{Y}_1 = -1$ tells us nothing about the values of $\mathbf{X}_2$ or $\mathbf{Y}_3$. Contextuality is, in this point of view, manifest by the fact that if we assign for the properties in different contexts the same random variable — i.e., if we impose $\mathbf{X}_1 = \mathbf{X}_2$, $\mathbf{Y}_1 = \mathbf{Y}_3$, and $\mathbf{Z}_2 = \mathbf{Z}_3$ — we will reach a contradiction.

In the above example, we worked with perfect correlations. But we do not need to have perfect correlations to reach contradictions. For example, consider the assumption that the properties, X, Y, and Z, are *not* contextual. What this means is that, if we were to observe two of those variables, we would be able to assign also a value to the third one in a way that is consistent with the observations. In fact, if we construct a data table with all the possible values, we get something like Table 1. A quick glance at this table shows us that the sum $XY + XZ + YZ$ is never less than -2. However, in

Table 1. All possible values of properties X, Y, and Z and their products.

X	Y	Z	XY	XZ	YZ
−1	−1	−1	1	1	1
−1	−1	1	1	−1	−1
−1	1	−1	−1	1	−1
−1	1	1	−1	−1	1
1	−1	−1	−1	−1	1
1	−1	1	−1	1	−1
1	1	−1	1	−1	−1
1	1	1	1	1	1

the contextual example shown above, the sum of the joint pairwise expectations $E(\mathbf{XY})+E(\mathbf{XZ})+E(\mathbf{YZ})$ equals -3. This observation is at the core of the Suppes–Zanotti inequalities [44]

$$\begin{aligned}-2 &\leq E(\mathbf{XY}) + E(\mathbf{XZ}) + E(\mathbf{YZ}) \\ &\leq 1 + 2\text{Min}\{E(\mathbf{X}), E(\mathbf{Y}), E(\mathbf{Z})\},\end{aligned} \tag{3}$$

which are necessary and sufficient conditions for the non-contextuality of $\mathbf{X}$, $\mathbf{Y}$, and $\mathbf{Z}$ when we assume them to be ± 1-value random variables with zero expectation.[d]

A useful way to think about the non-contextual bounds of the random variables $\mathbf{X}$, $\mathbf{Y}$, and $\mathbf{Z}$ is the following. Imagine that we set, for simplicity, the following constraints: $E(\mathbf{X}) = E(\mathbf{Y}) = E(\mathbf{Z}) = 0$, $E(\mathbf{XY}) = E(\mathbf{XZ}) = E(\mathbf{YZ}) = \epsilon$, and $E(\mathbf{XYZ}) = \beta$. Then, it is straightforward to compute that a joint probability exists if and only if ϵ and β are within the bounds of a polytope on the plane (ϵ, β) with the following vertices: $(1, 0)$, $(0, -1)$, $(0, 1)$, and $(-1/3, 0)$. Any correlations outside of this polytope will not allow for a proper joint probability distribution.

It is reasonable at this point to ask whether there are some concrete examples of contextuality. As we mentioned, one could argue that language is contextual [13]. The type of contextuality defined above, in fact, also exists in physics [1, 30, 39] and in psychology [8]. Here, we want to focus on examples from physics, which is what we will do in Section 3.

3. Quanta and Context

The world of microscopic phenomena seems to be contextual. But what do physicists mean when they say this? In this section, we will investigate the idea of context in physics, and its relationship to the intuitive idea of context expressed in Section 2. Let us start with

---

[d]If the expectations are not zero, then the right-hand side of (3) needs to be modified.

the famous Kochen–Specker theorem. To understand its origin, let us examine how properties are described in quantum theory.

Quantum theory, in its most basic interpretation, is about how to compute outcomes of experiments. This computation is done in the following way. First, we start with the fact that experiments, or, more precisely, measurements done within an experiment, have numerical outcomes (as the length of a rod, or the temperature of an object). Those possible outcomes are represented in quantum theory by real numbers associated to vectors: the numbers are the value of the outcome itself, whereas the vector describes a system with exactly this outcome. We should not think of those vectors as existing in the real three-dimensional space we live in, but in an abstract representational space of physical systems; sometimes this space can be quite large, even infinite dimensional. The vector space in quantum theory is a Hilbert space, and it has additional properties, such as completeness and metric, but these technical details are not relevant for our purposes.

The simplest types of properties, either in quantum physics or outside of it, are binary properties. Those are properties that are either true or false, such as, informally, whether "this brand of ice cream is tasty", "today is cold", or "my car's battery is dead". A physical property can be "the mass of a body is 1.2 ± 0.1 kg", and if the body's actual mass lies within the given range, this property is true, otherwise it is false.

Binary properties in quantum theory are represented by projection operators who project any vector in the quantum abstract space to its associated vector. If, for a unit vector $\mathbf{w}$ and a projector $\hat{P}$, $\hat{P}\mathbf{w} = \mathbf{w}$, then $\mathbf{w}$ is an eigenvector of $\hat{P}$ with eigenvalue 1, and we can say that $\mathbf{w}$ has the property P associated to $\hat{P}$. For the same projector $\hat{P}$, any unit vector $\mathbf{v}$ on the orthogonal space spanned by $\mathbf{w}$ will have the property $\hat{P}\mathbf{v} = 0$, which means that it is an eigenvector of $\hat{P}$ with eigenvalue 0, and we can say that $\mathbf{v}$ does not have the property P. More interestingly, any linear combination $a\mathbf{w} + b\mathbf{v}$, where a and b are complex numbers such that $\sqrt{|a|^2 + |b|^2} = 1$, is not an eigenvector of $\hat{P}$, and we cannot tell whether the system has property P or not if we do not measure it. However, once we measure the system for property P, we will either get that P is true or false, and if true the state of the system will "collapse" to $\mathbf{w}$, and if false to $\mathbf{v}$.

For a more concrete example, imagine a four-dimensional vector space $\mathbb{C}^4$. The set of vectors $\mathbf{v}_i$, $i = 1, 2, 3, 4$, is a possible orthonormal basis for this space. Given a vector $\mathbf{v}_i$, we can construct an operator $\hat{P}_i$, such that for any other vector $\mathbf{w}$, $\hat{P}_i\mathbf{w} = (\mathbf{w} \cdot \mathbf{v}_i)\mathbf{v}_i$, where "$\mathbf{w} \cdot \mathbf{v}_i$" represents the inner product of $\mathbf{w}$ and $\mathbf{v}_i$. Another way to write $\hat{P}_i$ is by using the dual of $\mathbf{v}_i$, which leads to $\hat{P}_i = \mathbf{v}_i\underline{v_i}$, where $\underline{v_i}$ denotes the dual of $\mathbf{v}_i$. In the quantum formalism, each $\mathbf{v}_i$ corresponds to a property: if the physical system we are describing is in a state represented by one of such vectors, then it has the property associated to it, but not to the others. Again, we could also have a system represented by any vector that is a normalized linear combination of the base vectors, i.e., $\sum_i c_i\mathbf{v}_i$. Such vectors do not have any of the associated properties to $\mathbf{v}_i$'s unless we actually measure them, causing them to collapse into one of the basis vectors.

This brings us to an important point. Take the vector $\sum_i c_i\mathbf{v}_i$. There is a projector associated to this vector, and therefore a property as well. And since the space is four dimensional, there are three other properties that are compatible with this vector, such that if the system has its property, it will for sure not have the others. This new projectors correspond to new axis for the splitting of the three-dimensional subspace, creating new properties. This idea is behind the proof of the Kochen–Specker theorem [31], which we will present here in a form shown by Cabello *et al.* [7].

Let us again consider a four-dimensional vector space. Following [7] notation, let us consider the projectors $\hat{P}_i$ as indexed with the subscript i corresponding to the four components of the vector in a given canonical basis. For example, $\hat{P}_{0,0,0,1}$ corresponds to the projector operator associated to the vector $(0, 0, 0, 1)$. Consider now the following set of equations:

$$\hat{P}_{0,0,0,1} + \hat{P}_{0,0,1,0} + \hat{P}_{1,1,0,0} + \hat{P}_{1,-1,0,0} = 1, \quad (4)$$

$$\hat{P}_{0,0,0,1} + \hat{P}_{0,1,0,0} + \hat{P}_{1,0,1,0} + \hat{P}_{1,0,-1,0} = 1, \quad (5)$$

$$\hat{P}_{1,-1,1,-1} + \hat{P}_{1,-1,-1,1} + \hat{P}_{1,1,0,0} + \hat{P}_{0,0,1,1} = 1, \quad (6)$$

$$\hat{P}_{1,-1,1,-1} + \hat{P}_{1,1,1,1} + \hat{P}_{1,0,-1,0} + \hat{P}_{0,1,0,-1} = 1, \quad (7)$$

$$\hat{P}_{0,0,1,0} + \hat{P}_{0,1,0,0} + \hat{P}_{1,0,0,1} + \hat{P}_{1,0,0,-1} = 1, \quad (8)$$

$$\hat{P}_{1,-1,-1,1} + \hat{P}_{1,1,1,1} + \hat{P}_{1,0,0,-1} + \hat{P}_{0,1,-1,0} = 1, \quad (9)$$

$$\hat{P}_{1,1,-1,1} + \hat{P}_{1,1,1,-1} + \hat{P}_{1,-1,0,0} + \hat{P}_{0,0,1,1} = 1, \quad (10)$$

$$\hat{P}_{1,1,-1,1} + \hat{P}_{-1,1,1,1} + \hat{P}_{1,0,1,0} + \hat{P}_{0,1,0,-1} = 1, \quad (11)$$

$$\hat{P}_{1,1,1,-1} + \hat{P}_{-1,1,1,1} + \hat{P}_{1,0,0,1} + \hat{P}_{0,1,-1,0} = 1. \quad (12)$$

Because all four projectors in each Eqs. (4)–(12) are associated to orthogonal vectors, it follows that the sum of the four projectors is the identity operator. In other words, projecting a vector into four orthogonal components in a four-dimensional space, gives us the component in each of those directions, and their sum gives us back the vector we started with. This has a natural interpretation: since each projector corresponds to a possible property of the system, only one of those properties can be true at a time for each line.

If we now assume that quantum properties are not contextual for a quantum system, e.g., the property associated to $\hat{P}_{0,0,0,1}$ is true both in (4) and (5), this leads to a mathematical inconsistency. To see this, simply notice that every $\hat{P}_i$ in (4)–(12) appears twice. Since a property can be either 0 or 1, if follows that if we sum all the lines on the left-hand side of (4)–(12) we get an even number. However, if we add the right-hand side, we get the number 9, which is odd, a clear contradiction. As discussed in Section 2, this contradiction occurs because we assume the properties are the same in a different context. If we did not assume this, then we would not reach a contradiction. So, quantum observables are contextual.

Let us end this section with a few remarks. First, notice that the contextuality of quantum systems is a consequence of the algebra of observables, defined as projectors in a vector space. This contextuality is state independent, in the sense that it does not matter how you describe your physical system, it will be contextual. However, there are sets of observables that exhibit contextuality only for certain physical systems prepared in special ways [3, 4]. This is particularly interesting because, for some especially prepared systems, it is possible to have "contextuality at a distance", in the sense that the value of a property P may depend not only on the local choices of experimental apparatuses, but also on the far away choices of measurement made by other experimenters within a time-like interval. This is why quantum mechanics is though of as a non-local theory.

Our second point is that this type of contextuality seems not only to be unique to quantum systems, but also to bring non-classically reproducible features that may be responsible for the effectiveness of quantum computers [29,45]. For example, it is possible to create a set of inequalities, similar to (3), that shows certain quantum systems as contextual. For those very same systems, it is possible to simulate the quantum correlations that lead to a violation of non-contextuality conditions, even in a non-local way, with classical fields [43]. However, as [34] points out, such violations are not the same as quantum, and they do not imply contextuality, as the quantum systems are made of particles, and not of continuum fields.

Finally, we would like to emphasize that, even though quantum contextuality seems to be special, there exists contextuality outside of quantum physics. As we remarked in Section 2, contextuality is part of linguistics and pragmatics, psychology [8], and it also seems to play an important role in explorations of consciousness [18, 35].

In Section 4, we will investigate the concept of information, and then examine it as it relates to contextuality. We will see that there are some problems associated to contextuality and information that need to be addressed, and we propose a possible route to approach this problem.

4. Measuring Information

How does one define and measure information? There are many different ways to measure information, but here we focus on the mathematical theory developed by Shannon [41]. Shannon started with the idea that the amount of information a signal carries is related to how surprising that signal is. The more surprising a signal, the more information it carries, whereas the more expected and less surprising a signal is, the less information it carries. For example, the statement "classes will be held as scheduled tomorrow" is less informative than "classes are canceled for tomorrow", as the latter is the unusual event, whereas the former is so common as to not require, from most students' points of view (and for most instructors), constant reminder and reassurance. Formally, Shannon's definition of information relies on probability theory, considered in some interpretations as a measure of our rational expectations [25].

In his famous paper [41], Shannon used probabilities to define the average amount of information. Let $\mathbf{X}$ be a source, whose output can be any of the events in a set of N possibilities $\{x_1, x_2, \ldots, x_N\}$, and let $p(x_i)$ be the probability that x_i is observed (i.e., the less probable x_i is, the more surprising it should be to see it). Shannon defined the amount of information associated to an observation of x_i as $\log_2 p(x_i)$, and consequently the average amount of information of the source $\mathbf{X}$ would be

$$H(\mathbf{X}) = -\sum_{i=1}^{N} p(x_i) \log_2 p(x_i). \tag{13}$$

$H(\mathbf{X})$ is known as Shannon's *entropy* of source $\mathbf{X}$.

The use of $\log_2$ in (13) is related to the choice of using as the unit of information a "yes–no" statement that can be coded by a single binary bit, 0 or 1. To understand the intuition behind it, imagine the case where $N = 2$. In this case, we have $\mathbf{X}$ as either x_1 and x_2, and all we needed is a bit to code which was observed. Now, in the case where $p(x_1) = 0$ and $p(x_2) = 1$ (or, similarly, $p(x_1) = 1$ and $p(x_2) = 0$), there are no surprises: we always get x_2 (or x_1).

Since $\log_2 1 = 0$, it follows that in this case the source's information is zero. The more interesting case is when we have no idea which one, x_1 or x_2, will occur, and the maximum information we can gain is when both are equiprobable. For the equiprobable case, $p(x_1) = p(x_2) = 1/2$, and $\log_2 1/2 = -1$, which implies that $H(\mathbf{X}) = 1$. We can do the same reasoning for $N = 4$, $N = 8$, $N = 16$, etc., and we will find that when all outcomes are equally probable the entropy is 2, 3, and 4, respectively, which is exactly the number of bits that are required to code each one of the outcomes. In other words, by using the information as $\log_2$ the entropy in (13) gives us the average information in units of bits.

Shannon's information is extremely useful in the cases mentioned above, as well as for engineering applications. For example, one of Shannon's main result was a theorem showing that, for a source $\mathbf{X}$ with entropy $H(\mathbf{X})$, there exists an optimal coding for each outcome of the source that allows a representation of signals from $\mathbf{X}$ with $H(\mathbf{X})$ bits or more. This is a non-trivial result, and has profound consequences in signal transmission, coding, and signal compression.

Despite its success, Shannon's information cannot be used indiscriminately, without some changes, for all types of collections of sources. To see this, let us consider the case of two sources that are perfectly correlated, e.g., a source $\mathbf{S}_1$ whose outcomes are isomorphic to the tossing of a coin, and another source $\mathbf{S}_2$ whose outcomes are exactly the same as $\mathbf{S}_1$, i.e., $\mathbf{S}_1 = \mathbf{S}_2$. It is clear that if we look at $\mathbf{S}_1$, it will seem that its entropy is 1, and the same for $\mathbf{S}_2$, and we can naively conclude that their collective information is greater than 1. However, because they are correlated, this would be a wrong conclusion: combined their information content is 1. So, for a collection of sources the amount of information they generate does not depend on their individual entropy, but on a different measure based on their joint probability. We shall examine this in more detail later, but we emphasize that there are collections of contextual sources that do not have joint probabilities. This is the case, for instance, of some entangled quantum sources [5, 23].

So, let us consider now quantum systems. A property O of a system S is represented, in quantum theory, by a Hermitian operator $\hat{O}$ on a Hilbert space $\mathcal{H}$, which are generalizations of the projection operators discussed in Section 3. For this reason, Hermitian operators are called *observables.* The state of the system itself is represented by a vector in $\mathcal{H}$ or by a density operator, defined as an observable that is positive semidefinite and has trace one. States that can be represented by vectors are called pure states, whereas states that must be represented by a density operator (thus having no equivalent in terms of a vector representation) are known as mixed states. For example, for the normalized vector $\mathbf{w} \in \mathcal{H}$ representing a pure state, the density operator associated to it would be $\hat{\rho}_w = \mathbf{w}\underline{\omega}$, where $\underline{\omega}$ is the dual to $\mathbf{w}$. It is easy to prove that a density operator $\hat{\rho}$ is in a pure state if and only if it is idempotent, i.e., $\hat{\rho}\hat{\rho} = \hat{\rho}$, with the implication following from $\hat{\rho}_w$.

While all states represented by a vector in $\mathcal{H}$ are pure states, the density operator allows for other states that are not pure. Consider two linearly independent and normalized vectors, $\mathbf{w}_1$ and $\mathbf{w}_2$. The density operator defined by $\hat{\rho}_M = c_1\mathbf{w}_1\underline{\omega}_1 + c_2\mathbf{w}_2\underline{\omega}_2$ is not idempotent, but it is positive semidefinite with trace one when we choose $c_1 + c_2 = 1$. As such, $\hat{\rho}_M$ has an important physical interpretation: it is how we represent the state of a physical system when we do not know whether it is in the pure state $\mathbf{w}_1$ or $\mathbf{w}_2$, with c_i being the probability of the system to be in state $\mathbf{w}_i$. In other words, the density operator representation allows us to include an uncertainty about the state of the system we are describing, and systems that are not in a pure state (i.e., no uncertainty about the state) are said to have *mixed states.*

Thus, the density operator formalism provides a more general framework to describe quantum systems than the vector formalism: it allows the computation of all quantities available in the vector formalism, plus the same quantities where the state is not known, in a way consistent with classical probability theory. As an example, imagine a three-dimensional Hilbert space where a possible basis for it is $\mathbf{e}_i$, $i = 1, 2, 3$. In the vector formalism, the expectation

$\langle \hat{P}_i \rangle = |\hat{P}_i \mathbf{w}|^2$ of the observable $\hat{P}_i$, where $\hat{P}_i = \mathbf{e}_i \underline{\epsilon}_i$ is a projector, and where $\underline{\epsilon}_i$ is the dual of $\mathbf{e}_i$, gives us the probability of observing a system in the pure state $\mathbf{w}$ as having the property P_i. This same expectation can be computed by using the expression $\langle \hat{P}_i \rangle = \mathrm{Tr}(\hat{\rho}_w \hat{P}_i)$. For the case where we have $\hat{\rho}_M$, as defined above, the linearity of the trace gives us that $\langle \hat{P}_i \rangle = \mathrm{Tr}(\hat{\rho}_M \hat{P}_i) = c_1 \mathrm{Tr}(\hat{\rho}_1 \hat{P}_i) + c_2 \mathrm{Tr}(\hat{\rho}_2 \hat{P}_i)$, or, equivalently, $\langle \hat{P}_i \rangle = c_1 \langle \hat{P}_i \rangle_1 + c_2 \langle \hat{P}_i \rangle_2$. We can interpret this last expression as saying that the expectation of P_i is the weighed expectation for each of the two states composing the mixture. Similarly, for any observable $\hat{O}$, it follows, by the linearity of the trace, that $\langle \hat{O} \rangle = \mathrm{Tr}(\hat{\rho}\hat{O})$.

Let us now go back to the idea of measuring information for quantum systems. The quantum system itself can be thought as being a source of information, equivalent to a source $\mathbf{X}$ in classical communications theory. For a state $\hat{\rho}$, its von Neumann entropy [46] is defined as

$$S = -\mathrm{Tr}(\hat{\rho} \log \hat{\rho}). \tag{14}$$

To see the relationship between (14) and (13), let us examine the case where we have three binary properties, represented by a three-dimensional Hilbert space with an orthonormal basis $\mathbf{e}_i$, $i = 1, 2, 3$. The observable $\hat{P}_i = \mathbf{e}_i \underline{\epsilon}_i$ has binary outcomes 0 or 1. If the state of the system is given by $\hat{\rho} = \sum_i c_i \mathbf{w}_i \underline{\omega}_i$, it follows that $S = -\sum c_i \log c_i$, which is exactly the same as the Shannon entropy formula if we remember that c_i is a proper probability for having the system with state $\mathbf{w}_i$.

Despite their similarities, there are important differences between quantum and classical entropies. For instance, Shannon's entropy was recovered for orthogonal projection measurements, but if $\hat{\rho}$ was a proper mixture of non-orthogonal projectors, this conclusion would not follow. In this chapter, we will not explore in detail these differences, and the interested reader is referred to [40]. However, we will point out that it is possible to prove an equivalent to Shannon's coding theorem for von Neumann's entropy [40], showing that the similarities between those two measures of information are not coincidental. In fact, in [28] the authors argued that for a general orthomodular lattice, which includes more restrictive algebras such as a Boolean one, the natural measure of informational content is von Neumann's entropy, with Shannon's entropy emerging as the measure for classical-like situations.

In this section, we presented two ways of measuring information: Shannon and von Neumann entropies. We also discussed the similarities between then, and how we can think of Shannon as a particular case of von Neumann in special cases. In Section 5, we will explore the concept of context-dependent properties, and how they create a problem for descriptions using classical probability theory. We will use the example in Section 5 to argue about issues on the definition of information for contextual systems.

5. Information and Context

Let us examine the difficulty of defining information for contextual random variables by looking at some examples. Consider two random variables, $\mathbf{X}$ and $\mathbf{Y}$, valued as either $+1$ or -1, and both having equal probability of yielding as outcomes either $+1$ or -1. In other words, each of them looks random. If we were to use Shannon's entropy [41] to compute the amount of information H for each one of them separately, we would find that

$$\begin{aligned} H(\mathbf{X}) &= -p_x \log p_x - p_{\overline{x}} \log p_{\overline{x}} \\ &= -\frac{1}{2}\log\frac{1}{2} - \frac{1}{2}\log\frac{1}{2} = 1, \end{aligned} \tag{15}$$

and similarly, $H(\mathbf{Y}) = 1$. In (15) we are using the simplifying notation that $p_x = p(\mathbf{X} = 1)$, $p_{\overline{x}} = p(\mathbf{X} = -1)$, where the line over x means a negative outcome and no line means a positive outcome. The values of H in (15) are telling us that each source has one bit of information, and we might want to conclude that both $\mathbf{X}$ and $\mathbf{Y}$ would have, combined, two bits of information.

However, the conclusion that $\mathbf{X}$ and $\mathbf{Y}$ carries two bits of information may be incorrect. This is because examining each source separately could be misleading. If we have two sources, $\mathbf{X}$ and $\mathbf{Y}$, and they are independent, then the information carried by both sources is simply the sum of each of the sources. But if they are not independent, then their combined information is smaller. For example, if $\mathbf{X}$ and $\mathbf{Y}$ were perfectly correlated, i.e., $E(\mathbf{XY}) = 1$, then knowing $\mathbf{X}$ implies knowing $\mathbf{Y}$ and vice versa, and the information contained in the pair would not be two bits, but instead one bit, because of the redundancy. The same conclusion would be arrived if $\mathbf{X}$ and $\mathbf{Y}$ were perfectly anti-correlated, i.e., $E(\mathbf{XY}) = -1$.

But how can we compute the amount of information from two sources that are neither independent nor perfectly correlated? To do so, we can consider $\mathbf{X}$ and $\mathbf{Y}$ as a pair, and use their joint probabilities. In this case, instead of using as outcomes -1 and 1 for each variable separately, we can use all possible outcomes for the pairs, namely xy, $x\overline{y}$, $\overline{x}y$, and $\overline{xy}$ (following the notation introduced above). Assuming that $E(\mathbf{XY}) = \alpha$, it follows that

$$p_{xy} + p_{x\overline{y}} + p_{\overline{x}y} + p_{\overline{xy}} = 1, \tag{16}$$

$$p_{xy} + p_{x\overline{y}} - p_{\overline{x}y} - p_{\overline{xy}} = 0, \tag{17}$$

$$p_{xy} - p_{x\overline{y}} + p_{\overline{x}y} - p_{\overline{xy}} = 0 \tag{18}$$

and

$$p_{xy} - p_{x\overline{y}} - p_{\overline{x}y} + p_{\overline{xy}} = \alpha. \tag{19}$$

Equation (16) is simply the statement that the probabilities of all possible events needs to sum to one (Axiom K1 in Section 2). Equations (17) and (18) are a consequence of the zero expectations for $\mathbf{X}$ and $\mathbf{Y}$, since each outcomes $+1$ and -1 are equiprobable, whereas (19) is a consequence of $E(\mathbf{XY}) = \alpha$. The solution to (16)–(19) is

$$p_{xy} = p_{\overline{xy}} = \frac{1}{4}(1 + \alpha), \tag{20}$$

$$p_{\overline{x}y} = p_{\overline{x}y} = \frac{1}{4}(1 - \alpha), \tag{21}$$

and is non-negative for all values of $-1 \leq \alpha \leq 1$. We can use (20) and (21) to compute the Shannon entropy of those two variables,

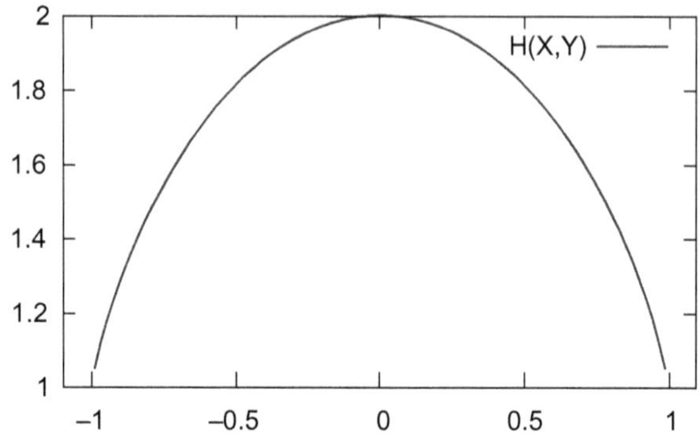

Fig. 1. Joint entropy of $\mathbf{X}$ and $\mathbf{Y}$ as a function of the correlation α.

resulting in

$$H(\mathbf{X},\mathbf{Y}) = -\frac{1}{4}(1+\alpha)\log\left(\frac{1}{4}(1+\alpha)\right) - \frac{1}{4}(1-\alpha)\log\left(\frac{1}{4}(1-\alpha)\right).$$

The behavior of $H(\mathbf{X},\mathbf{Y})$ is shown in Fig. 1. From Fig. 1, we can see that if the correlation α is zero, we get maximum information, as expected, whereas the minimum information is when $\mathbf{X}$ and $\mathbf{Y}$ are perfectly correlated ($\alpha = 1$) or anti-correlated ($\alpha = -1$).

Let us now examine the case where we have three ± 1-valued random variables $\mathbf{X}$, $\mathbf{Y}$, and $\mathbf{Z}$ with zero expectation and with correlations $E(\mathbf{XY}) = E(\mathbf{XZ}) = E(\mathbf{YZ}) = \epsilon$. The corresponding equations set by the expectations are

$$p_{xyz} + p_{xy\overline{z}} + p_{x\overline{y}z} + p_{\overline{x}yz} + p_{\overline{xy}z} + p_{\overline{x}y\overline{z}} + p_{x\overline{yz}} + p_{\overline{xyz}} = 1, \quad (22)$$

$$p_{xyz} + p_{xy\overline{z}} + p_{x\overline{y}z} - p_{\overline{x}yz} - p_{\overline{xy}z} - p_{\overline{x}y\overline{z}} + p_{x\overline{yz}} - p_{\overline{xyz}} = 0, \quad (23)$$

$$p_{xyz} + p_{xy\overline{z}} - p_{x\overline{y}z} + p_{\overline{x}yz} - p_{\overline{xy}z} + p_{\overline{x}y\overline{z}} - p_{x\overline{yz}} - p_{\overline{xyz}} = 0, \quad (24)$$

$$p_{xyz} - p_{xy\overline{z}} + p_{x\overline{y}z} + p_{\overline{x}yz} + p_{\overline{xy}z} - p_{\overline{x}y\overline{z}} - p_{x\overline{yz}} - p_{\overline{xyz}} = 0, \quad (25)$$

$$p_{xyz} + p_{xy\overline{z}} - p_{x\overline{y}z} - p_{\overline{x}yz} + p_{\overline{xy}z} - p_{\overline{x}y\overline{z}} - p_{x\overline{yz}} + p_{\overline{xyz}} = \epsilon, \quad (26)$$

$$p_{xyz} - p_{xy\overline{z}} + p_{x\overline{y}z} - p_{\overline{x}yz} - p_{\overline{xy}z} + p_{\overline{x}y\overline{z}} - p_{x\overline{yz}} + p_{\overline{xyz}} = \epsilon, \quad (27)$$

$$p_{xyz} - p_{xy\overline{z}} - p_{x\overline{y}z} + p_{\overline{x}yz} - p_{\overline{xy}z} - p_{\overline{x}y\overline{z}} + p_{x\overline{yz}} + p_{\overline{xyz}} = \epsilon. \quad (28)$$

We now have seven equations set by the correlation and eight variables, which makes this system of equations underdetermined. An additional equation can be set, without loss of generality, by fixing the (sometimes unobservable) triple moment $E(\mathbf{XYZ}) = \beta$, $-1 \leq \beta \leq 1$, which leads to the extra equation

$$p_{xyz} - p_{xy\overline{z}} - p_{x\overline{y}z} - p_{\overline{x}yz} + p_{\overline{x}\overline{y}z} + p_{\overline{x}y\overline{z}} + p_{x\overline{y}\overline{z}} - p_{\overline{x}\overline{y}\overline{z}} = \epsilon. \tag{29}$$

The solution to Eqs. (22)–(29) is

$$p_{xyz} = \frac{1}{8}(1 + \beta + 3\epsilon),\ p_{xy\overline{z}} = p_{x\overline{y}z} = p_{\overline{x}yz} = \frac{1}{8}(1 - \beta - 3\epsilon), \tag{30}$$

$$p_{\overline{x}\overline{y}z} = p_{\overline{x}y\overline{z}} = p_{x\overline{y}\overline{z}} = \frac{1}{8}(1 + \beta - 3\epsilon),\ p_{\overline{x}\overline{y}\overline{z}} = \frac{1}{8}(1 - \beta + 3\epsilon), \tag{31}$$

and we can see that some of the p's may be negative for certain values of ϵ and β. The non-negative solutions correspond to non-contextual cases, whereas negative solutions are contextual.

Before we compute the entropy based on the joint probability for $\mathbf{X}$, $\mathbf{Y}$, and $\mathbf{Z}$, it is useful to consider some special cases. First, let us examine the perfectly correlated system. In this case, $\epsilon = 1$, and knowing one of the random variables, say $\mathbf{X}$, is sufficient to completely determine all other variables. Therefore, we should expect the entropy for this system to be one bit. The other interesting case is when $\epsilon = 0$, and $\mathbf{X}$, $\mathbf{Y}$, and $\mathbf{Z}$ are uncorrelated. At first, it may seem that, because the three variables are uncorrelated, the total entropy for $\mathbf{X}$, $\mathbf{Y}$, and $\mathbf{Z}$ should be three. However, the entropy also depends on the triple moment. For example, if $E(\mathbf{XYZ}) = \pm 1$, even though the variables are pairwise uncorrelated, there is a global correlation between them that reduces the amount of information from three to two bits, since knowing $\mathbf{X}$ and $\mathbf{Y}$ gives us $\mathbf{Z}$. But if, in addition to the pairwise correlations, we have $E(\mathbf{XYZ}) = 0$, then we need three bits.[e] The third interesting example is when $\epsilon = -1/3$, the lower bound of (3). For this case, from (30) and (31) it follows

[e]We point out that a system $\mathbf{X}$, $\mathbf{Y}$, and $\mathbf{Z}$ of random variables where the pairwise expectations are zero and the triple moment is not can be physically realized in quantum mechanics [15].

that non-negative solutions exist only if $\beta = 0$, i.e., there is no global correlation, and the triple moment does not give us any extra information. Given that the correlations allow us to explain at most 1/3 of the other variables, it is reasonable to assume that the information should be more than two bits, but not as high as 3, when all variables are uncorrelated.

A quick examination of Shannon's entropy applied to the probabilities in (30) and (31) shows that our conclusions for $\epsilon = 1$, $\epsilon = 0$ and $\beta = 0$, and $\epsilon = -1/3$ are corroborated, as the entropy for $\epsilon = 1$ is 1, for $\epsilon = \beta = 0$ it hits a maximum at three bits, and for $\epsilon = -1/3$ is approximately 2.6 bits. A plot of the entropy H as a function of ϵ (for the maximal entropy case when $\beta = 0$) is shown in Fig. 2. Figure 2 shows the informational content of $\mathbf{X}$, $\mathbf{Y}$, and $\mathbf{Z}$ only for the interval $-1/3 \leq \epsilon \leq 1$, as outside of this, interval Eqs. (22)–(29) have negative solutions, and Shannon's entropy is not defined. But just because we cannot apply Shannon's entropy, it does not mean that $\mathbf{X}$, $\mathbf{Y}$, and $\mathbf{Z}$ have no informational content outside of the

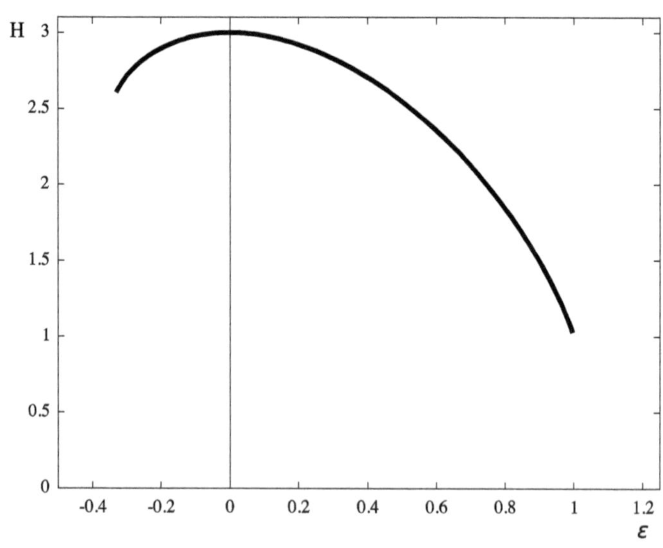

Fig. 2. Entropy $H(\mathbf{X}, \mathbf{Y}, \mathbf{Z})$ for $E(\mathbf{XY}) = E(\mathbf{XZ}) = E(\mathbf{YZ}) = \epsilon$ and $E(\mathbf{XYZ}) = 0$ as a function of ϵ. The maximum of three bits occurs when $\epsilon = 0$.

probability polytope. It just means that Shannon's entropy, as currently defined, cannot measure it.

How can we approach this problem? First, let us examine the case when $\epsilon = -1$. A casual analysis could conclude, as in the perfectly correlated example, that the amount of information should be just one bit: if we know $\mathbf{X}$ we know $\mathbf{Y}$ and $\mathbf{Z}$. However, as we saw before $\mathbf{X}$ in the context of $\mathbf{Y}$ cannot be the same as in the context of $\mathbf{Z}$. So, because of this contextuality, the information content should be at least two bits, perhaps more.

To investigate this in more detail, let us imagine that, instead of $\mathbf{X}$, $\mathbf{Y}$, and $\mathbf{Z}$, due to the contextuality we have the following random variables: $\mathbf{X_Y}$, $\mathbf{X_Z}$, $\mathbf{Y_X}$, $\mathbf{Y_Z}$, $\mathbf{Z_X}$, and $\mathbf{Z_Y}$. The observed expectations are $E(\mathbf{X_Y}) = E(\mathbf{X_Z}) = E(\mathbf{Y_X}) = E(\mathbf{Y_Z}) = E(\mathbf{Z_X}) = E(\mathbf{Z_Y}) = 0$ and $E(\mathbf{X_Y Y_X}) = E(\mathbf{X_Z Z_X}) = E(\mathbf{Y_Z Z_Y}) = -1$. We can see that the index for each variable indicates their context. For example, $\mathbf{X_Y}$ is the variable $\mathbf{X}$ in the context of being observed with $\mathbf{Y}$. So, we moved from an initial system with three properties to a new system with six. Furthermore, because we have six variables, we would need $2^6 = 64$ equations to uniquely determine the joint probability distribution for those variables. But such equations are unavailable, as we only observe the marginals given by the individual and pairwise expectations.

Because of the correlations, we can reduce the $\mathbf{X_Y}$, $\mathbf{X_Z}$, $\mathbf{Y_X}$, $\mathbf{Y_Z}$, $\mathbf{Z_X}$, and $\mathbf{Z_Y}$ system of random variables from six to three. This suggests that the informational content would be three bits. However, as in the case of three variables, higher moments are not defined. If we impose different values of higher moments, ones that are not consistent with the observed marginals, we could imagine an expansion to the necessary number of random variables, thus increasing the amount of information of the system even beyond three bits.

In other words, if we were to draw from the case of two variables, it is not unreasonable to expect that the maximum value for the entropy would be achieved when the higher moments were less correlated. Thus, highly pairwise correlated sources that are contextual may have more information than non-contextual correlated ones. In fact, a highly correlated contextual set of sources can have more information than the same number of sources when completely uncorrelated.

5.1. *Information and negative probabilities*

We now end this discussion with a possible description of contextual information in terms of non-standard probabilities. For the above three-variable example, we may always construct a joint pseudo-probability distribution that is consistent with the observed expectations. This can be done either with upper and lower probabilities (as in [16,27,42]) or with negative probabilities (see [10,11,14]). Here, we will use negative probabilities, as they provide a more convenient computational tool than upper probabilities, in particular for quantum physics (see [36, 37]).

"Negative probabilities" were introduced by Dirac in his famous Bakerian Lectures [19], in the context of quantum field theory. Though the name *negative probabilities* may be misleading, as only some probabilities of atomic events may be negative, but no observable probabilities may be negative, this terminology is widely used in physics.

Let $(\Omega, \mathcal{F}, p^*)$ be an extended probability space, defined as satisfying axioms K1 and K2 (see Section (2)) but with $p^* : \mathcal{F} \rightarrow \mathbb{R}$. We call p^* negative probabilities, as it satisfies the same axioms as Kolmogorov probabilities, except that the non-negativity requirement is relaxed, which means that probabilities measures of the atomic elements can take negative values. A more formal definition of negative probabilities, one that takes into account only the observable marginal probabilities, can be found in [11], and the interested reader is referred to it.

As above mentioned, the term, negative probabilities, may be misleading, for the following reasons. First, as mentioned, negative probabilities are not necessarily negative. It is usually the case that, for all atomic elements in Ω, some probabilities are non-negative and others are negative. A quick examination of (30)–(31) shows this to be the case for $\epsilon = -1$. Second, it is never possible to observe a negative probability. Though for $\omega_i \in \Omega$ it is possible to have $p^*(\omega_i) < 0$, the observable elements of $\mathcal{F}$, corresponding to the experimental random variables, never have negative probabilities. In fact, the Boolean subalgebras of observables are always compatible with a proper probability measure. So, even though the term used in physics is *negative probabilities*, it may be more appropriate to call them *signed probabilities*, as those mathematical objects are known in measure theory [26]. Here, we will follow the physics tradition and call p^* negative probabilities.

Let us use negative probabilities to define an extension of Shannon's entropy for contextual systems. We start with the extended entropy S_{NP} for p^* as given by the following expression[f]:

$$S_{\mathrm{NP}} = -\sum_{\omega_i \in \Omega} |p^*(\omega_i)| \log_2 |p^*(\omega_i)|. \tag{32}$$

The idea of using the absolute value of the negative probabilities comes from using the expression $A = \sum_{\omega_i \in \Omega} |p^*(\omega_i)|$ as a measure of how contextual a probabilistic system is [10]. If $A = 1$, this means that there is no contextuality, since $A = \sum_{\omega_i \in \Omega} p^*(\omega_i)$, which implies that p^*'s are non-negative. However, if $A > 1$, then the system is contextual, and the further A departs from one, the stronger the contextual relations between the variables. So, it is not unreasonable to use $|p^*(\omega_i)|$ as a measure of how "surprising" some outcome is, given its close connections to proper probabilities.

From Eq. (32) we can see that S_{NP} is exactly the same as Shannon's entropy for cases when p^* is non-negative. In other words, when a proper joint probability distribution exists, $S_{\mathrm{NP}} = H$. It should also be clear that, whenever we have contextuality, the quantity $\sum |p^*(\omega_i)|$ is greater than one, which means that the entropy may increase, as expected from our examples.

In Fig. 3 we show a graph of three ± 1-valued random variables $\mathbf{X}$, $\mathbf{Y}$, and $\mathbf{Z}$ with zero expectation, correlations $E(\mathbf{XY}) = E(\mathbf{XZ}) =$

[f]The idea of using something similar to (32) was first suggested informally to the author by Patrick Suppes on the context of quantum information, but he did not pursue this idea.

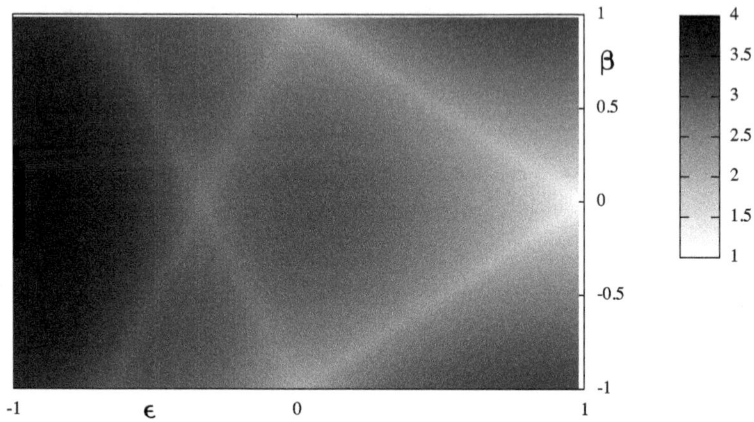

Fig. 3. Surface plot of S_{NP} as a function of ϵ (horizontal axis) and β (vertical axis). Lighter regions correspond to less entropy, whereas darker regions to more entropy.

$E(\mathbf{YZ}) = \epsilon$, and triple moment $E(\mathbf{XYZ}) = \beta$. There are several characteristics of this graph that are interesting. First, we can clearly see the non-contextual polytope as the quadrilateral defined by the vertices $(-\frac{1}{3}, 0)$, $(0, 1)$, $(0, -1)$, $(1, 0)$, as the boundary has lower entropy, as expected. We can also see from the figure that the minimum value for the entropy is 1, which occurs when $\epsilon = 1$ and $\beta = 0$, as discussed above. Still within the classical bounds, the region shown has maximum entropy when $\epsilon = \beta = 0$, consistent with our analysis above. More interestingly, as can be seen from Fig. 4, the maximum entropy is reached when $\epsilon = -1$ and $\beta = 0$, as was predicted by our random variable argument. So, it seems that the entropy with negative probabilities, as defined by (32), may provide a consistent measure of informational content for contextual sources.

6. Final Remarks

In this chapter, we discussed issues related to measuring the informational content of contextual sources. This is an important subject, as it has been argued, for example, that contextuality is a key resource for quantum computation [29, 45]. But, more interestingly,

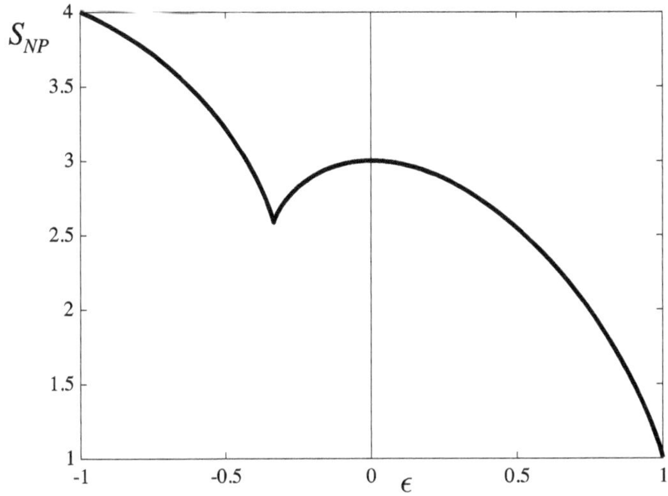

Fig. 4. Cross-section of S_{NP} as a function of ϵ and for $\beta = 0$. We see a local maxima at $\epsilon = 0$ and a global maxima at $\epsilon = -1$, where $S_{\mathrm{NP}} = 4$.

contextuality is a key concept in quantum physics, and perhaps what makes quantum mechanics different from classical mechanics [6].

In our discussion, we provided a simple example of a contextual set of sources, and we showed that this set of sources may provide more information when it is contextual then when it is not. For instance, when two of the three sources are perfectly anti-correlated, this correlation reduces their informational content to at most two bits, if they are non-contextual. However, if they are contextual, the informational content of the three variables may be as high as three bits. More interestingly, the maximum informational content that three non-contextual sources can have is three bits, whereas, as we argued above, contextual sources can have upwards of four bits. In other words, contextuality leads to more information. This result should not be surprising. If we think about the meaning of contextuality, it is exactly that a source depends on the context; what we thought were the same sources in two different contexts (or more) were actually two different (and perhaps uncorrelated) sources.

We also showed that we can provide a measure of informational content by extending Shannon's entropy to negative probabilities.

This new entropy reduces to the classical Shannon entropy for systems with a proper joint probability distribution, but can be used for regions outside of the classical probability polytope. Our analysis showed that the entropy S_{NP} based on negative probabilities give results that are consistent with a simplified contextuality-by-default description of the sources.

We believe our result unveils interesting applications of negative probabilities to the study of contextual information, not only of quantum information. Furthermore, it opens up several questions that are worth exploring. For example, if S_{NP} is a good representation of the informational content of a system, then it should be possible to prove a Shannon coding-type theorem for negative probabilities. Also, it would be interesting to explore the relationship between S_{NP} and von Neumann's entropy, which we know is related to Shannon's, but can also be applied to certain contextual systems. Finally, as we mentioned, von Neumann's entropy cannot be used in situations such as our three random-variable example, as in a Hilbert space description, three pairwise commuting observables have a common base, and therefore a joint probability distribution. Therefore, it is an open question whether S_{NP} can be used for applications outside of physics, such as in psychology, linguistics analysis, or consciousness studies. For example, in consciousness studies, a popular approach, known as Integrated Information Theory [38], faces issues with contextuality [12, 35].

We emphasize that negative probabilities do not provide the most general treatment for measuring information of contextual sources. For example, negative probabilities require that random variables in different contexts are consistently connected (what is called in physics "no-signaling") [36]. A more general approach should probably come from an expanded set of contextual random variables, as it is proposed in Contextuality by Default (CbD). CbD can deal with random variables whose mean values vary from context to context (i.e., they are inconsistently connected) [33]. However, because of the extra variables, it is not clear how to deal with information for contextual variables within this approach, and probably a strategy using CbD will need to impose symmetries or utilize minimization principles. Accordingly, negative probabilities reduce significantly the number of required equations for describing contextual sources, as compared to CbD, thus imposing lots of additional symmetries that may not be satisfied by certain systems. The connections between information with negative probabilities and information using CbD, as well as the limitations of negative probabilities, are topics that need to be explored further.

Acknowledgments

This chapter benefited from conversations with Carlos Montemayor, Gary Oas, Pawel Kurzynski, Federico Holik, and Gregg Jaeger, to whom I am indebted. I would also like to thank the editors of this book, Francisco Doria and Shyam Wuppuluri, for their encouragement and support. It is a great pleasure to dedicate this paper to Professor Chaitin, and it is an honor to have it appear on his Festschrift.

Bibliography

[1] A. Aspect, P. Grangier and G. Roger. Experimental tests of realistic local theories via Bell's theorem. *Physical Review Letters*, 47(7):460–463, 1981.
[2] R. Balian. *Du microscopique au macroscopique: cours de physique statistique de l'Ecole Polytechnique*. Ecole Polytech. Palaiseau, Palaiseau, 1982.
[3] J. S. Bell. On the Einstein–Podolsky–Rosen paradox. *Physics*, 1(3): 195–200, 1964.
[4] J. S. Bell. On the problem of hidden variables in quantum mechanics. *Rev. Mod. Phys.*, 38(3):447–452, 1966.
[5] S. L. Braunstein and C. M. Caves. Information-theoretic bell inequalities. *Physical Review Letters*, 61(6):662–665, 1988.
[6] A. Cabello. Quantum physics: Correlations without parts. *Nature*, 474(7352):456–458, 2011.
[7] A. Cabello, J. M. Estebaranz and G. C. Alcaine. Bell–Kochen–Specker theorem: A proof with 18 vectors. *Physics Letters A*, 212(4):183–187, 1996, arXiv:quant-ph/9706009.
[8] V. H. Cervantes and E. N. Dzhafarov. Snow queen is evil and beautiful: Experimental evidence for probabilistic contextuality in human choices. *Decision*, 5(3):193–204, 2018.

[9] G. Chiribella, G. Mauro D'Ariano and P. Perinotti. Informational derivation of quantum theory. *Physical Review A*, 84(1):012311, 2011.

[10] J. Acacio de Barros, E. N. Dzhafarov, J. V. Kujala and G. Oas. Measuring observable quantum contextuality. In H. Atmanspacher, T. Filk and E. Pothos (eds.), *Quantum Interaction*, Lecture Notes in Computer Science, Vol. 9535, pp. 36–47. Springer International Publishing, 2015.

[11] J. Acacio de Barros, J. V. Kujala and G. Oas. Negative probabilities and contextuality. *Journal of Mathematical Psychology*, 74:34–45, 2016.

[12] J. Acacio de Barros, C. Montemayor and L. P. G. De Assis. Contextualit in the integrated information theory. In *Quantum Interaction*, Lecture Notes in Computer Science, pp. 57–70. Springer, Cham, 2017.

[13] J. Acacio de Barros, C. Montemayor, L. P. G. De Assis, P. Skokowski and J. Perry. On a mathematical representation of linguistic contextuality. *Submitted* for publication.

[14] J. Acacio de Barros, G. Oas and P. Suppes. Negative probabilities and counterfactual reasoning on the double-slit experiment. In J.-Y. Beziau, D. Krause and J. B. Arenhart (eds.), *Conceptual Clarification: Tributes to Patrick Suppes* (1992–2014). College Publications, London, 2015.

[15] J. Acacio de Barros and P. Suppes. Strict holism in a quantum superposition of macroscopic states. Arxiv preprint quant-ph/0003046; to appear in the South American Journal of Logic, 2000.

[16] J. Acacio de Barros and P. Suppes. Probabilistic inequalities and upper probabilities in quantum mechanical entanglement. *Manuscrito*, 33(1): 55–71, 2010.

[17] J. Acacio de Barros and P. Suppes. Quantum mechanics, interference, and the brain. *Journal of Mathematical Psychology*, 53(5):306–313, 2009.

[18] J. Acacio de Barros, F. Holik and D. Krause. Contextuality and indistinguishability. *Entropy*, 19(9):435, 2017.

[19] P. A. M. Dirac. Bakerian lecture. The physical interpretation of quantum mechanics. *Proceedings of the Royal Society of London B*, A180:1–40, 1942.

[20] F. Dretske. *Knowledge and the Flow of Information*. MIT Press, Cambridge, MA, 1981.

[21] E. N. Dzhafarov and J. V. Kujala. Context-content systems of random variables: The contextuality-by-default theory. arXiv:1511.03516 [quant-ph], 2015.

[22] E. N. Dzhafarov and J. V. Kujala. Contextuality-by-default 2.0: Systems with binary random variables. In J. Acacio de Barros, Bob Coecke and E. Pothos (eds.), *10th International Conference on Quantum Interaction, QI 2016*, Lecture Notes in Computer Science, Vol. 10106, Springer International Publishing, 2017, arXiv: 1604.04799.

[23] A. Fine. Hidden variables, joint probability, and the bell inequalities. *Physical Review Letters*, 48(5):291–295, 1982.

[24] C. A. Fuchs and R. Schack. Quantum-Bayesian coherence. *Reviews of Modern Physics*, 85(4):1693–1715, 2013.

[25] M. C. Galavotti. *Philosophical Introduction to Probability*, CSLI Lecture Notes, Vol. 167. CSLI Publications, Stanford, CA, 2005.

[26] P. R. Halmos. *Measure Theory.* Springer-Verlag, New York, 1974.
[27] S. Hartmann and P. Suppes. Entanglement, upper probabilities and decoherence in quantum mechanics. In M. Suárez, M. Dorato and M. Rédei, editors, *EPSA Philosophical Issues in the Sciences*, pp. 93–103. Springer, The Netherlands, 2010.
[28] F. Holik, A. Plastino and M. Sáenz. Natural information measures in Cox's approach for contextual probabilistic theories. *Quantum Information and Computation*, 16(1–2):0115–0133, 2016.
[29] M. Howard, J. Wallman, V. Veitch and J. Emerson. Contextuality supplies the 'magic' for quantum computation. *Nature*, 510(7505):351–355, 2014.
[30] G. Kirchmair, F. Zahringer, R. Gerritsma, M. Kleinmann, O. Guhne, A. Cabello, R. Blatt and C. F. Roos. State-independent experimental test of quantum contextuality. *Nature*, 460(7254):494–497, 2009.
[31] S. Kochen and E. P. Specker. The problem of hidden variables in quantum mechanics. *Journal of Mathematics and Mechanics*, 17:59–87, 1967.
[32] A. N. Kolmogorov. *Foundations of the Theory of Probability.* 2nd edition. Chelsea Publishing Co., Oxford, 1956.
[33] J. V. Kujala, E. N. Dzhafarov and J.-A. Larsson. Necessary and sufficient conditions for an extended noncontextuality in a broad class of quantum mechanical systems. *Physical Review Letters*, 115(15):150401, 2015.
[34] M. Markiewicz, D. Kaszlikowski, P. Kurzynski and A. Wojcik. From contextuality of a single photon to realism of an electromagnetic wave. *npj Quantum Information*, 5(1):5, 2019.
[35] C. Montemayor, J. A. de Barros and L. P. G. De Assis. Implementation, formalization, and representation: Challenges for integrated information theory. *Journal of Consciousness Studies*, 26(1):107–132, 2019.
[36] G. Oas, J. Acacio de Barros and C. Carvalhaes. Exploring non-signalling polytopes with negative probability. *Physica Scripta*, T163:014034, 2014.
[37] Gary Oas and J. Acacio de Barros. A survey of physical principles attempting to define quantum mechanics. In E. Dzhafarov, R. Zhang and S. M. Jordan (eds.), *Contextuality From Quantum Physics to Psychology.* World Scientific, 2015.
[38] M. Oizumi, L. Albantakis and G. Tononi. From the phenomenology to the mechanisms of consciousness: Integrated information theory 3.0. *PLoS Comput Biol*, 10(5):e1003588, 2014.
[39] M. H. M. Passos, W. F. Balthazar, J. Acacio de Barros, C. E. R. Souza, A. Z. Khoury and J. A. O. Huguenin. Classical analog of quantum contextuality in spin-orbit laser modes. *Physical Review A*, 98(6):062116, 2018.
[40] B. Schumacher. Quantum coding. *Physical Review A*, 51(4):2738–2747, 1995.
[41] C. E. Shannon. A mathematical theory of communication. *Bell System Technical Journal*, 27(3):379–423, 1948.
[42] P. Suppes and M. Zanotti. Existence of hidden variables having only upper probabilities. *Foundations of Physics*, 21(12):1479–1499, 1991.

[43] P. Suppes, J. Acacio de Barros and A. Sant'Anna. Violation of Bell's inequalities with a local theory of photons. *Foundations of Physics Letters*, 9(6):551–560, 1996.

[44] P. Suppes and M. Zanotti. When are probabilistic explanations possible? *Synthese*, 48(2):191–199, 1981.

[45] V. Veitch, C. Ferrie, D. Gross and J. Emerson. Negative quasi-probability as a resource for quantum computation. *New Journal of Physics*, 14(11):113011, 2012.

[46] J. von Neumann. *Mathematical Foundations of Quantum Mechanics.* Princeton University Press, Princeton, NJ, Translated by Robert T. Beyer from the 1932 German Edition, 1983.

[47] J. A. Wheeler. Information, physics, quantum: The search for links. In W. H Zurek (ed.), *Complexity, Entropy, and the Physics of Information*, Vol. 8. CRC Press, Boca Raton, FL, 1990.

Chapter 13

A New Bifurcation in the Universe*

A. E. S. Hartmann[†,‡] and M. Novello[†,§]

The combined system of general relativity with a non-minimally coupled electromagnetic field presents a bifurcation in a cosmic framework. In order to simplify our exposition in this chapter, we limit the proof to the case of an universe driven by a cosmological constant. In the same vein, we show the existence of states such that the resulting combined energy (the sum of the minimally and the non-minimally coupled energy momentum tensor of the electromagnetic field) vanishes in a sort of violation of the action-reaction principle.

1. Introduction

In the early 1984, Novello and Rodrigues [1] analyzed the presence of a bifurcation in the universe driven by a viscous fluid. In this kind of classical description, the isotropic pressure of the stress–energy tensor is written as a polynomial of the expansion factor θ,

$$\widetilde{p} = p + \alpha\theta + \beta\theta^2. \tag{1}$$

As they showed, dissipative processes can provoke the appearance of bifurcation in autonomous nonlinear differential equations, as is the case of the equations of general relativity in a spatially isotropic

*We dedicate this chapter to Greg Chaitin.

[†]Centro de Estudos Avançados de Cosmologia/CBPF
Rua Dr. Xavier Sigaud 150, Urca 22290-180 Rio de Janeiro, Brazil.
[‡]aehartmann91@gmail.com.
[§]mnovello42@gmail.com.

and homogeneous universe. This chapter constitutes the return to the open questions of this kind of indeterminacy in cosmological solutions in a very distinct context.

In the standard cosmological description of the electromagnetic fluid as the main source of primordial gravitational field, a sort of average procedure must be made once the geometry is identified to a spatially homogeneous and isotropic metric. This led to describe the electromagnetic effect as nothing but a cosmical fluid with density of energy $\rho = 1/2(E^2 + B^2)$ and pressure $p = 1/3\rho$ given by the traceless property.

The exam of the global properties of the space–time suggests the question: What is the effect on this description if the coupling of the electromagnetic field to gravity is non-minimal and contains functions of the curvature? We shall see that with the same sort of average procedure as in the standard minimal coupling, there is a sea of bifurcations when general relativity is combined with non-minimal interaction of gravity with linear electromagnetic field.

2. Duality Rotation in Minimal and Non-minimal Couplings

Consider the Einstein and Maxwell theories described by the minimally coupled Lagrangian

$$L = \frac{1}{2k}R - \frac{c^2}{4}F, \tag{2}$$

in which $F := F_{\mu\nu}F^{\mu\nu} = 2(B^2 - E^2)$. The field equations for the metric are given by (see the Appendix I for definitions and conventions)

$$R_{\mu\nu} - \frac{1}{2}Rg_{\mu\nu} = -\kappa T^{\gamma}_{\mu\nu}, \tag{3}$$

where the curvature tensor is defined as

$$v_{\alpha;\mu;\nu} - v_{\alpha;\nu;\mu} = R^{\alpha}_{\beta\mu\nu}v^{\beta} \tag{4}$$

and the Maxwell energy-momentum tensor is

$$T^{\gamma}_{\mu\nu} = \Phi_{\mu\nu} + \frac{1}{4} F g_{\mu\nu}, \tag{5}$$

with $\Phi_{\mu\nu} := F_{\mu\alpha} F^{\alpha}{}_{\nu}$.

It is well known that the Maxwell equations are invariant under the global duality rotation $\mathscr{R}(\alpha)$ of the electromagnetic field, stated by the relations

$$\begin{aligned} F_{\mu\nu} &\overset{\mathscr{R}(\alpha)}{\longmapsto} F'_{\mu\nu} = \cos\alpha F_{\mu\nu} + \sin\alpha \overset{*}{F}_{\mu\nu}, \\ \overset{*}{F}_{\mu\nu} &\overset{\mathscr{R}(\alpha)}{\longmapsto} \overset{*}{F}{}'_{\mu\nu} = -\sin\alpha F_{\mu\nu} + \cos\alpha \overset{*}{F}_{\mu\nu}, \end{aligned} \tag{6}$$

with

$$\overset{*}{F}{}^{\mu\nu} := \frac{1}{2} \eta^{\mu\nu\kappa\lambda} F_{\kappa\lambda}. \tag{7}$$

Moreover, the Lagrangian (2) is invariant in first order on the dual angle,

$$\begin{aligned} F'_{\mu\nu} &\approx F_{\mu\nu} + \alpha \overset{*}{F}_{\mu\nu}, \\ \overset{*}{F}{}'_{\mu\nu} &\approx -\alpha F_{\mu\nu} + \overset{*}{F}_{\mu\nu}, \end{aligned} \tag{8}$$

from which it follows, up to a total divergence,

$$F' \approx F. \tag{9}$$

Furthermore, there is no effect on the dynamics of the metric once the Maxwell's energy–momentum tensor $T^{\gamma}_{\mu\nu}$ is invariant under the global map $\mathscr{R}(\alpha)$,

$$T^{\gamma'}_{\mu\nu} = T^{\gamma}_{\mu\nu}. \tag{10}$$

The proof of this is immediate and one should use the algebraic properties of the electromagnetic tensor shown in the Appendix A.

Now comes the question: Is it possible to generalize the interaction between these two fields for a non-minimal coupling in which the curvature tensor appears in the interacting term with the electromagnetic field in such a way that the theory remains invariant under $\mathscr{R}(\alpha)$? The answer is yes and the corresponding interacting Lagrangian is provided by [2, 3]

$$L_{\text{int}} = \zeta C_{\mu\nu} \Phi^{\mu\nu}, \tag{11}$$

where ζ is the non-minimal coupling constant with dimensionality of $length^2$, and

$$C_{\mu\nu} := R_{\mu\nu} - \frac{1}{4} R g_{\mu\nu}. \tag{12}$$

Note that this extra term L_{int} is invariant even in the case the rotating angle depends on space–time. In other words, gravity is responsible for recovering the global invariance of electromagnetic field and introducing a new symmetry without counterpart in the Maxwell's theory. The Lagrangian of interaction (11) can be written in the equivalent form

$$C_{\mu\nu} \Phi^{\mu\nu} = R_{\mu\nu} T^{\mu\nu}_{\gamma}. \tag{13}$$

One can go a step further and instead of using the contracted Riemann tensor $R_{\mu\nu}$ one can consider the conformal Weyl tensor $W_{\alpha\beta\mu\nu}$ to describe non-minimal coupling. In this case, there is a generalized dual rotation that acts both on the Weyl and Faraday tensors that leaves the interaction of the combined electromagnetic and gravitational fields invariant. It is shown in the appendix that such duality invariance occurs if the rotating angle for the spin-2 field, represented by the Weyl conformal tensor, is precisely twice the spin-1 rotating angle, that acts on the electromagnetic field. In this chapter, we limit our analysis to the above L_{int}.

Traditionally, the discussion on the duality rotation of the electromagnetic field concerns to the exam of compatibility of magnetic monopoles with the quantum theory, a question introduced by the seminal paper of Dirac in 1931 [4]. Despite this, the present work follows another possible path.

3. The Field Equations

Let us start by choosing the action principle as given by

$$\delta \int \sqrt{-g} \left(\frac{1}{2k} R - \frac{c^2}{4} F + \frac{\zeta}{2} C_{\mu\nu} \Phi^{\mu\nu} + L_m \right) = 0, \tag{14}$$

where L_m is the rest of matter contribution. The equation of the metric is given by

$$R_{\mu\nu} - \frac{1}{2} R g_{\mu\nu} = -T^{\gamma}_{\mu\nu} - \zeta Z_{\mu\nu} - T^{m}_{\mu\nu}, \tag{15}$$

where

$$\begin{aligned} Z_{\mu\nu} = &-\frac{1}{2} g_{\mu\nu} \left(R_{\rho\sigma} \Phi^{\rho\sigma} + \frac{1}{4} R F \right) + R^{\varepsilon}{}_{(\mu} \Phi_{\nu)\varepsilon} \\ &+ R^{\rho\sigma} F_{\rho\mu} F_{\nu\sigma} - \frac{1}{2} R \Phi_{\mu\nu} + \frac{1}{4} F R_{\mu\nu} \\ &+ \frac{1}{2} \Phi^{\varepsilon}{}_{(\mu;\nu);\varepsilon} + \frac{1}{4} F_{;\mu;\nu} - \frac{1}{2} g_{\mu\nu} \left(\Phi^{\rho\sigma}{}_{;\sigma;\rho} + \frac{1}{2} \Box F \right) - \frac{1}{2} \Box \Phi_{\mu\nu}. \end{aligned} \tag{16}$$

The equation for the electromagnetic field is

$$F^{\mu\nu}{}_{;\nu} + \zeta (C^{\mu}{}_{\epsilon} F^{\epsilon\nu} + F^{\mu}{}_{\epsilon} C^{\epsilon\nu})_{;\nu} = 0. \tag{17}$$

4. The Average Procedure

We are interested in analyze the effects of the non-minimal coupling in the standard spatially isotropic and homogeneous metric. To be consistent with the symmetries of this choice of the metric, an averaging procedure must be performed if electromagnetic fields are to be taken as a source for the gravitational field, according to the standard procedure [5]. The definition of the average of a quantity Q at a instant t is given by

$$\langle Q\rangle = \lim_{V\to\infty}\frac{1}{V}\int\sqrt{-g}Qd^3x, \tag{18}$$

with $V = \int\sqrt{-g}d^3x$. As a consequence, the components of the electric E_i and magnetic B_i fields must satisfy the following relations:

$$\langle E_i\rangle = 0, \quad \langle B_i\rangle = 0, \quad \langle E_iB_j\rangle = 0, \tag{19}$$

$$\langle E_iE_j\rangle = -\frac{1}{3}E^2g_{ij}, \tag{20}$$

$$\langle B_iB_j\rangle = -\frac{1}{3}B^2g_{ij}, \tag{21}$$

where E and B depend only on time. Besides we assume that the time derivative operation commutes with the average procedure, that is we set

$$\partial_t\langle E_iE_j\rangle = \langle\partial_t(E_iE_j)\rangle, \tag{22}$$

$$\partial_t\langle B_iB_j\rangle = \langle\partial_t(B_iB_j)\rangle, \tag{23}$$

$$\partial_t\langle E_iB_j\rangle = \langle\partial_t(E_iB_j)\rangle. \tag{24}$$

Using the above average values, it follows that the Maxwell energy–momentum tensor $T^m_{\mu\nu}$ reduces to a perfect fluid configuration with energy density ρ_γ and pressure p_γ given by

$$\langle T^\gamma_{\mu\nu}\rangle = (\rho_\gamma + p_\gamma)v_\mu v_\nu - p_\gamma g_{\mu\nu}, \tag{25}$$

where

$$\rho_\gamma = 3p_\gamma = \frac{1}{2}(E^2 + B^2). \tag{26}$$

Note that

$$\langle\Phi_{\mu\nu}\rangle = \frac{2}{3}(\sigma^2+1)Xv_\mu v_\nu + \frac{1}{3}(\sigma^2-2)Xg_{\mu\nu}, \tag{27}$$

where we have set $E^2 = \sigma^2B^2$. For simplicity, from now on we will write $B^2 = X$.

Let us look for the other part $\langle Z_{\mu\nu}\rangle$.

4.1. $Z_{\mu\nu}$ interpreted in terms of a perfect fluid

In the case of non-minimal coupling with gravity, the extra term for the energy–momentum tensor $Z_{\mu\nu}$ is rather involved once it contains terms depending on the curvature. Let us set for the metric the form

$$ds^2 = dt^2 - a^2(t)(dx^2 + dy^2 + dz^2). \tag{28}$$

In this case

$$R_{00} = \dot{\theta} + \frac{1}{3}\theta^2, \tag{29}$$

$$R_{ij} = \frac{1}{3}(\dot{\theta} + \theta^2)g_{ij}, \tag{30}$$

$$R = 2\dot{\theta} + \frac{4}{3}\theta^2, \tag{31}$$

where we have defined the expansion factor $\theta = 3\dot{a}/a$.

Using these results, one obtains

$$\begin{aligned}\langle Z_{\mu\nu}\rangle = {} & Mg_{\mu\nu} + Nv_\mu v_\nu + \frac{1}{3}(\sigma^2+1)X_{,\epsilon;(\mu}v_{\nu)}v^\epsilon \\ & + \frac{2}{9}(\sigma^2+1)\theta X_{,(\mu}v_{\nu)} + \frac{1}{3}(\sigma^2+1)XR^\alpha_{(\mu}v_{\nu)}v_\alpha \\ & - \frac{1}{6}(\sigma^2+1)X_{,\mu;\nu} + \frac{1}{2}\left(\frac{\sigma^2}{3} - 1\right)XR_{\mu\nu}\end{aligned} \tag{32}$$

with M and N defined by

$$M = -\frac{1}{6}(\sigma^2+1)\ddot{X} - \frac{5}{18}(\sigma^2+1)\theta\dot{X} + \frac{1}{18}(1-5\sigma^2)X\dot{\theta} - \frac{2}{9}\sigma^2\theta^2X \tag{33}$$

and

$$\begin{aligned}N = {} & -\frac{1}{3}(\sigma^2+1)\ddot{X} - \frac{1}{3}(\sigma^2+1)\theta\dot{X} \\ & - \frac{2}{9}(\sigma^2+1)X\dot{\theta} - \frac{2}{9}(\sigma^2+1)X\theta^2.\end{aligned} \tag{34}$$

4.2. *The perfect fluid*

From what we have shown we can analyze the effects of the non-minimal coupling of the electromagnetic field with gravity in a spatially homogeneous and isotropic geometry in terms of the standard equation of general relativity and a mixed fluid as in Eq. (15). In order to compatibilize the theory with a spatially homogeneous and isotropic metric, one must analyze its corresponding heat flux q_μ and the anisotropic pressure $\pi_{\mu\nu}$ which are defined for an arbitrary energy–momentum tensor $T_{\mu\nu}$ as

$$q_\lambda = T_{\alpha\beta} v^\beta h^\alpha{}_\lambda, \tag{35}$$

$$\Pi_{\mu\nu} = T^{\alpha\beta} h_{\alpha\mu} h_{\beta\nu} + p h_{\mu\nu}, \tag{36}$$

where

$$h_{\mu\nu} = g_{\mu\nu} - v_\mu v_\nu \tag{37}$$

and the pressure

$$p = -\frac{1}{3} T_{\mu\nu} h^{\mu\nu}. \tag{38}$$

In the case of tensor $Z_{\mu\nu}$ it is a rather long although direct calculation to show that both q_μ and $\pi_{\mu\nu}$ vanish. This yields the remarkable consequence that we can write the coupled energy–momentum tensor $Z_{\mu\nu}$ in the form of a perfect fluid

$$Z_{\mu\nu} = (\rho_z + p_z) v_\mu v_\nu - p_z g_{\mu\nu},$$

where

$$\rho_z = -\frac{1}{6}(\sigma^2 + 1)\theta \dot{X} + \frac{1}{3}\sigma^2 \dot{\theta} X - \frac{1}{6}(\sigma^2 + 1)\theta^2 X \tag{39}$$

and

$$p_z = \frac{1}{6}(\sigma^2 + 1)\ddot{X} + \frac{1}{3}(\sigma^2 + 1)\theta \dot{X} + \frac{1}{9}(2\sigma^2 + 1)\dot{\theta} X + \frac{1}{6}(\sigma^2 + 1)\theta^2 X. \tag{40}$$

We will now show that this system admits a bifurcation point. Before this, let us make a short presentation of the mathematical scheme.

5. Conditions for the Bifurcation Point

Let p and q be two variables which characterize a physical system — described in the phase plane (p, q) — whose evolution in time t is given by the planar autonomous system of differential equations

$$\begin{aligned} \dot{p} &= f(p, q, \lambda), \\ \dot{q} &= h(p, q, \lambda), \end{aligned} \tag{41}$$

in which f and h are nonlinear functions. The system is called autonomous because the right-hand side of (41) does not contain explicitly the time variable t. We have add to (41) a parameter λ which has a given range $\mathcal{M}$ on the real axis and distinguishes special interactions among parts of the physical system.

The states of equilibrium of the system (41) are given by the points in the phase plane (p, q) for which f and h are simultaneously annihilated. A multiple equilibrium state is called a *bifurcation point* if for a given value of the parameter, say λ_o, the topological behavior of the integral curves changes discontinuously when passes the value λ_o.

This physical situation of instability of the singular point in the phase plane, coupled to the aleatory character of the fluctuations which the system can undergo, extinguishes almost completely the possibility of predictions. In other words, the system arrived at the vicinity of a bifurcation point evolves in a non-deterministic way, which is a situation already implicitly contained in the equations used to describe the system. The proof of this is based on Bendixson's theorem [6] which states that the Poincaré index — which is a kind of measure of the topological properties in the neighborhood of the singular point — of a multiple equilibrium state is given by the relation

$$I_p = \frac{E - H}{2} + 1, \tag{42}$$

in which E and H represent, in the phase plane, the number of elliptical and hyperbolic sectors, respectively.

The sudden modification of the topological properties of the integral curves of the system in the phase plane represents an abrupt change of behavior of the physical system in the vicinity of the unstable point. The crucial consequence of this is the appearance of non-deterministic features in the metrical properties of the universe, even in a classical and non-statistical description.

6. An Example of Cosmological Solution with Bifurcation

Start by assuming that the expansion factor is a constant driven by L_m identified to a cosmological constant and the electromagnetic fluid is just a test field (in Section 7, we will show that there is a state such that the energy of the fluid interacting non-minimally with gravity, the $Z_{\mu\nu}$ contribution is suppressed). We look for the fundamental state of the cosmic fluid, that is we set

$$p + \rho = 0, \tag{43}$$

where

$$\rho = \rho_\gamma + \zeta \rho_z,$$

$$p = p_\gamma + \zeta p_z,$$

which yield the differential equation

$$\zeta \ddot{X} + \zeta \theta_0 \dot{X} + 4X = 0. \tag{44}$$

The analytical solution is immediate and no further investigation should be required from the mathematical viewpoint. However, the richness of the cosmological questions leaves us for searching another property. In this vein we will investigate the global behavior of such dynamics by asking: can the topology of the dynamical system describing such cosmological solution change in time? The most simple and direct qualitative analysis [6, 7] reveals the important properties of the planar autonomous system

$$\dot{X} = Y, \tag{45}$$

$$\dot{Y} = -\frac{4}{\zeta}X - \theta_0 Y. \tag{46}$$

There is a critical point at the origin $(X, Y) = (0, 0)$. The corresponding matrix $\Delta = \partial(\dot{X}, \dot{Y})/\partial(X, Y)$ is given by

$$\Delta = \begin{pmatrix} 0 & 1 \\ -4/\zeta & -\theta_0 \end{pmatrix}. \tag{47}$$

The determinant is positive when $\zeta > 0$. In this case, the equilibrium point is a focus, stable for θ_0 positive and unstable for θ_0 negative (as shown by the Fig. 1(a)). For $\zeta < 0$ yields $\det\Delta < 0$ and consequently the equilibrium point is a saddle (see the Fig. 1(b)). If $\det\Delta \neq 0$, the only one singular point is the origin [6]. The case

$$\lim_{\zeta \longrightarrow \infty} \det\Delta = 0$$

implicates entire lines of singular points.

The index of Poincaré of a simple equilibrium state is $+1$ in the case of a node or focus and is -1 in the case of a saddle point. Thus, this system contains potentially a change of topology that is the necessary condition for the existence of a bifurcation point [6].

Let us emphasize the dependence of the bifurcation point on the value of the non-minimal coupling constant. Besides, the property of existence of bifurcation points in the dynamical systems describing cosmological solutions of the action principle (14) is valid for almost all relationship between p and ρ. The exception occurs for the case of a "pure" electromagnetic radiation, that is, for the equation of state $p = 1/3\rho$.

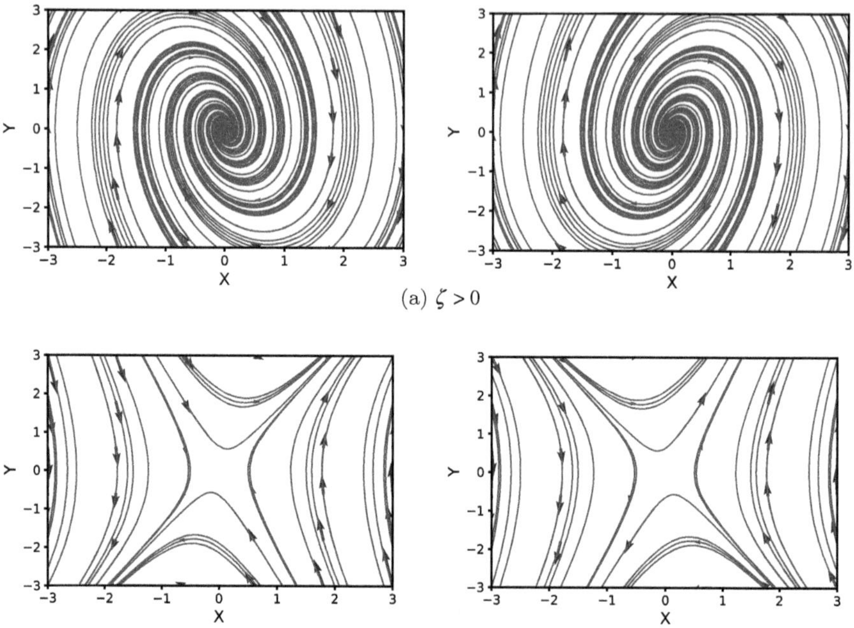

(a) $\zeta > 0$

(b) $\zeta < 0$

Fig. 1. Dynamical system with $\theta_0 > 0$ (to left) and $\theta_0 < 0$ (to right).

7. Energy Suppressed by Curvature or States that Violate the Action-Reaction Principle

A very unexpected result of the property of reduction of the energy distribution of the non-minimal coupling of the electromagnetic field to gravity to a perfect fluid as showed above concerns the possibility of a state such that the combined energy–momentum tensor of the free field plus the interacting term, interpreted as the equations of General Relativity (see Eq. (15)), does not drive the gravitational field. How is this possible? Let us show it. From the average procedure, the coupled system that remains to be solved reduces to the conservation of the energy $\rho_m + \rho_z$ and the evolution equation for the expansion factor. Let us analyze the case in which the gravitational field is due only to the extra matter represented by the term L_m in the action (14) by imposing the two conditions: (i) the sum of the density of energy of the part of the electromagnetic field coupled minimally to gravity ρ_γ with the density of energy of the part of the electromagnetic field coupled non-minimally to gravity ρ_z vanishes; (ii) the sum of the pressure of the part of the electromagnetic field coupled minimally to gravity p_γ with the pressure of the part of the electromagnetic field coupled non minimally to gravity p_z, vanishes, that is,

$$\rho_\gamma + \zeta\rho_z = 0, \tag{48}$$

$$p_\gamma + \zeta p_z = 0. \tag{49}$$

which led to the equations

$$3X - \zeta(\theta_0\dot{X} + \theta_0^2 X) = 0, \tag{50}$$

$$X + \zeta(\ddot{X} + 2\theta_0\dot{X} + \theta_0^2 X) = 0. \tag{51}$$

Solving this system yields an explicit example of a state of the electromagnetic field that is acted by gravity but that does not modify the gravitational field. We thus obtain

$$B^2(t) = \text{constant}\exp\left(-\frac{4}{3}\theta_0 t\right). \tag{52}$$

8. Concluding Remarks

The combination of fields interacting with gravity in both, minimal and non-minimal way allows the possibility to the existence of states such that in a curved geometry the corresponding energy can be annihilated. We have presented here a specific example for the electromagnetic field. However, this property is not restrict to this case. It can appears also in other contexts, for other types of matter source. A scalar field coupled minimally and non-minimally to gravity also exhibit states that violate the action–reaction principle. We will come back to these cases elsewhere [8]. Furthermore, the presence of bifurcation points in the energy tensor under inspection is an example of physical situations where local causality could destroy the global causality of space–time. Maxwell's electromagnetic field, considered the prototype of classical field theory and therefore the most prominent test-field of "deterministic" physical systems, could be responsible, when interacting directly with gravity, for generate non-deterministic cosmological situations even in the representation of perfect fluid. Therefore, the electromagnetic field, in the realm of General Relativity, may imply at least one information about the topology of the dynamical system describing the universe, namely, that it may be not unique.

Appendix A. Mathematical Compendium

Tensor $T_{\mu\nu}$ is obtained from Lagrangian L_m by

$$T_{\mu\nu} = \frac{2}{\sqrt{-g}} \frac{\delta\sqrt{-g}L}{\delta g^{\mu\nu}}. \tag{A.1}$$

This tensor admits a representation under the form

$$T_{\mu\nu} = \rho v_\mu v_\nu - p h_{\mu\nu} + q_{(\mu} v_{\nu)} + \pi_{\mu\nu}, \tag{A.2}$$

where the 10 independent components are the energy density ρ, the pressure p, the heat flux q^α and the anisotropic pressure $\pi_{\alpha\beta}$ given in the frame of a time-like normalized vector

$$v^\mu v^\nu g_{\mu\nu} = 1 \tag{A.3}$$

by

$$\rho = T^{\alpha\beta} v_\alpha v_\beta, \tag{A.4}$$

$$p = -\frac{1}{3} h_{\alpha\beta} T^{\alpha\beta}, \tag{A.5}$$

$$q^\alpha = h^{\alpha\beta} v^\gamma T_{\beta\gamma}, \tag{A.6}$$

$$\pi^{\alpha\beta} = h^{\alpha\mu} h^{\beta\nu} T_{\mu\nu} + p h^{\alpha\beta}. \tag{A.7}$$

The tensors q_μ and $\pi_{\mu\nu}$ satisfy the constraints

$$q_\mu v^\mu = 0, \tag{A.8}$$

$$\pi_{\mu\nu} v^\mu = 0, \tag{A.9}$$

$$\pi_{\mu\nu} g^{\mu\nu} = 0, \tag{A.10}$$

$$\pi_{\mu\nu} = \pi_{\nu\mu}. \tag{A.11}$$

A.1. *Decomposition of any anti-symmetric tensor like Faraday tensor*

Consider an observer endowed with normalized 4-velocity v^μ. This decomposes $F_{\mu\nu}$ into electric and magnetic parts under the form:

$$F_{\mu\nu} = -v_\mu E_\nu + v_\nu E_\mu + \eta_{\mu\nu\rho\sigma} v^\rho B^\sigma, \tag{A.12}$$

where vectors electric (E_μ) and magnetic (B_μ) are given by

$$E_\mu = F_{\mu\alpha} v^\alpha, \tag{A.13}$$

$$B_\mu = F^*_{\mu\alpha} v^\alpha = \frac{1}{2}\eta_{\mu\alpha\rho\sigma} F^{\rho\sigma} v^\alpha \tag{A.14}$$

with

$$\eta^{\alpha\beta\mu\nu} = -\frac{1}{\sqrt{-g}}\varepsilon^{\alpha\beta\mu\nu}. \tag{A.15}$$

It then follows that these vectors are orthogonal to v^μ

$$E_\mu v^\mu = B_\mu v^\mu = 0. \tag{A.16}$$

The six degrees of freedom of $F_{\mu\nu}$ become represented by $(3+3)$ quantities associated to vectors E_μ and B_μ. We can construct two invariants

$$F := F_{\mu\nu} F^{\mu\nu}, \tag{A.17}$$

$$G := F^*_{\mu\nu} F^{\mu\nu}. \tag{A.18}$$

The algebraic identities are satisfied:

$${}^*F^{\mu\alpha*}F_{\alpha\nu} - F^{\mu\alpha} F_{\alpha\nu} = \frac{1}{2} F \delta^{\mu}_{\nu}, \tag{A.19}$$

$$\overset{*}{F}{}^{\mu\alpha} F_{\alpha\nu} = -\frac{1}{4} G \delta^{\mu}_{\nu}. \tag{A.20}$$

Appendix B. The Weyl Tensor

Consider the general dual rotation $\mathscr{R}(\alpha)$ of the electromagnetic field (Section 2) and the similar one $\mathscr{R}(\theta)$ for the Weyl conformal tensor [9]

$$\begin{aligned} \widetilde{W}_{\alpha\beta\mu\nu} &= \cos\theta W_{\alpha\beta\mu\nu} + \sin\theta W^*_{\alpha\beta\mu\nu}, \\ \widetilde{W}^*_{\alpha\beta\mu\nu} &= -\sin\theta W_{\alpha\beta\mu\nu} + \cos\theta W^*_{\alpha\beta\mu\nu}. \end{aligned} \tag{B.1}$$

A direct calculation shows that the quantity

$$L = W_{\alpha\beta\gamma\delta} F^{\alpha\beta} F^{\gamma\delta} \tag{B.2}$$

is invariant under the coupled dual rotation if the rotating angles α and θ satisfy the condition

$$\theta + 2\alpha = 0. \tag{B.3}$$

Acknowledgments

We would like to acknowledge the financial support from Brazilian agencies Finep, Capes, Faperj and CNPq.

Bibliography

[1] M. Novello and L. M. C. S. Rodrigues, Bifurcation in the early cosmos, *Lett. Nuovo Cimento.* **40**(10) (1984), 317, and references therein.

[2] M. Novello, The program of the eternal universe. In: *Proceedings of the Eighth Marcel Grossmann Meeting on General Relativity*. Cingapura: World Scientific, 1997, pp. 1383–1402.
[3] M. Novello and S. E. P. Bergliaffa, Bouncing cosmologies, *Phys. Rept.* **463** (2008), 4.
[4] P. A. M. Dirac, Quantised singularities in the electromagnetic field. *Proc. R. Soc. Lond. A* **133** (1931), pp. 60–72.
[5] R. Tolman, *Relativity, Thermodynamics, and Cosmology* (Clarendon Press, 1934).
[6] A. A. Andronov, E. A. Leontovich, I. I. Gordon and A. G. Maier, *Qualitative Theory of Second-Order Dynamic Systems* (John Wiley & Sons, 1973).
[7] M. Novello and R. A. Araujo, Qualitative analysis of homogeneous universes, *Phys. Rev. D* **22** (1980), 260.
[8] M. Novello and A. E. S. Hartmann, The cosmological constant and a scalar field coupled non minimally to gravity, to be published by *Int. J. Mod. Phys. A*.
[9] R. Debever, La super-énergie en relativité générale, *Bull. Soc. Math. Belg.* **10** (1958), 112–147.

Chapter 14

From the Origins of Life to the Nature of Intelligence

Sara Imari Walker[*,†] **and Paul C. W. Davies**[*,‡]

A remarkable feature of life on Earth is the apparent trend of increasing complexity of the biosphere over the >3.5 billion year history of life on this planet. A striking manifestation of this trend in the last few hundred million years is the growth of intelligence. However, neither the rise of intelligence nor the trend of increasing complexity currently has a satisfactory general explanation in terms of known principles. It is also far from obvious that this trend is truly "open-ended" or if there is a complexity cap that will one day be reached. Indeed, the very concept of *"increasing complexity"* is hard to define and can be challenged, especially as we only have a single sample to go on. In the absence of evidence for any other planet hosting intelligent life, it is unclear at present whether intelligent life like us (or for that matter life at all) is common or rare in the universe. The best path forward, in lieu of discovering alien life, is to better understand the evolutionary mechanisms of known life on Earth. Perhaps the greatest outstanding problem is the transition from non-living to living matter, without which biological evolution could not happen at all. For the past century, attempts to study the origin of life have focused on organic chemistry and the identification of complexifying reaction pathways in plausible prebiotic settings. Inspired by the profound insights of Gregory Chaitin, we here outline an emerging approach, which transcends the chemical substrate of life by attempting to define the non-life to life transition in informational terms. One advantage of this substrate-independent "software" approach is that the mathematical framework applies equally well to life

[*]Beyond Center for Fundamental Concepts in Science, Arizona State University, Tempe, AZ, USA.
[†]sara.i.walker@asu.edu.
[‡]Paul.Davies@asu.edu.

and to intelligence. If there does exist a general directionality principle in the evolution of complexity and intelligence it is most likely to emerge from universal "information laws" that would apply to life here on Earth but also potentially to any other life and intelligence that might exist elsewhere.

1. Are Life and Intelligence Common or Rare?

Gregory Chaitin is one of the most profound thinkers of our age on the nature of the physical world in general, and of complexity in particular. His insights into the conceptual foundations of mathematics and its place in our description of nature have been a source of inspiration to two generations of scientists, these authors included. And no subject is more challenging than life itself. Life and intelligence are the exemplars of emergent complexity, yet nobody knows what life is or how it started. Drawing on Chaitin's ground-breaking ideas, we offer here an attempt to reconceptualize the origin and nature of life.

It is currently fashionable for scientists to declare that the transition from non-life to life cannot be exceedingly improbable because it happened so quickly on Earth [1]. (The oldest traces of life go back at least 3.45 billion years.) If correct, it would follow that life should form readily on the billions of Earth-like planets hypothesized to exist in the Milky Way galaxy alone, implying that the universe is teeming with it. Although the rapid appearance of life on Earth is consistent with an easy transition from non-life to life, it is also equally consistent with an exceedingly hard transition, as pointed out by the cosmologist Brandon Carter over three decades ago [2] (and more recently also by Speigel and Turner [3]). Carter's reasoning, curiously, couples the evolution of intelligence with the origin of life itself.

In outline, Carter's argument is as follows. Any given Earth-like planet will have a finite "habitability window", τ_H, defining the timescale over which life might emerge and eventually evolve intelligence. On Earth, this timescale is $\tau_H \sim 4.5$ billion years — from roughly 3.8 billion years ago when the primordial bombardment abated, to about 800 million years hence, when the Sun will be so hot that the planet will be an uninhabitable furnace. Suppose, reasoned Carter, that life's origin is so improbable that the expectation time for it to occur, τ_L, is many orders of magnitude longer than this habitability window, such that $\tau_L \gg \tau_H$. And further suppose that, in addition to the (improbable) transition from non-life to life, several other very hard (rare) steps are needed before intelligence is attained (for example, eukaryogenesis, sex, multicellularity, evolution of a central nervous system). If in all there are n hard steps, each of which has an expectation time much longer than the habitability window, and each of which is necessary before the next step may be taken, then a simple statistical argument leads to a relationship between n and the duration of the window. Specifically, the window can be divided into $n + 1$ roughly equal intervals (roughly, because this is a statistical argument), with the final interval being the duration between now (the epoch of intelligent life on Earth) and the fiery Armageddon 800 million years hence (predicted based on the finite lifetime of our Sun). Given that the intervals are roughly equal, and given the 800 million year figure, Carter is able to conclude that there should be about 5 or 6 extremely improbable steps involved in attaining intelligent life on Earth [4]. Importantly, the first step is bracketed by a similar interval of 800 million years. That is, if the emergence of life was an exceedingly improbable process (but of course one that had to happen for humans to be here to ponder it) then probability theory predicts it should have happened fairly rapidly — within 800 million years, which is indeed the case. Another way of expressing it is that, unless life had got going quickly, we would not be here to discuss it three-and-a-half billion years later.

A curious corollary of Carter's argument is that the duration between the improbable steps is roughly the same irrespective of just how improbable they are, so long as they are all so improbable that the expectation time for them to occur is far longer than the habitability window. Let us consider a concrete but grossly simplified example. Suppose the expectation time for life to form on a planet like Earth is a billion trillion years, such that $\tau_L \sim 10^{21}$. What this means is that, in an imaginary world in which conditions on Earth remain the same for a billion trillion years, life would very likely arise there at some stage in that (long) window. Alternatively, it means that among a random sample of one trillion Earth-like planets, life would likely arise on one of them every billion years or so. Now, suppose that the expectation time for intelligence to evolve once life is established, τ_I, is much less — say, a mere million trillion years, such that $\tau_I \sim 10^{18}$. That is, the probability for intelligence to evolve, while extremely small, is nevertheless a thousand times greater than the probability for life to emerge from non-life $\frac{\tau_I}{\tau_L} \sim 10^3$. Then you might imagine that the first step — the emergence of life — would on average take a thousand times longer than the evolution of intelligence, since the latter is a thousand times less improbable. But not so: both these steps are predicted by Carter's argument to have taken about the same time to happen, i.e., roughly 800 million years.

The reason that this line of reasoning seems so counterintuitive is because we are used to thinking about probabilities of future events given present uncertainties. But in the case of intelligent life on Earth, events are *post-selected.* That is, scientists attempt to use the fact of the existence of an intelligent species on Earth now (humans) to retrodict about past probabilities. And our intuition collapses. It is entirely possible that our present existence entailed extremely improbable past steps, which nevertheless occurred, against huge odds, and without which we would not be here to reflect on those antecedent events necessary for our existence.

So to summarize Carter's powerful line of reasoning, the rapid appearance of life on Earth is equally consistent with two scenarios. One is that the critical steps leading up to intelligence are all fairly likely, so that if Earth is typical there should be very many other planets with intelligent life (as many optimists involved in SETI — the Search for Extraterrestrial Intelligence — suppose). The other is that there are several critical steps, each of which is so unlikely that the expectation time for them to occur individually is very much longer than the age of the Earth; in which case life will be exceedingly rare in the universe (the SETI pessimist's scenario). Given a sample of only one, we cannot discriminate between them (Carter and McCrea in fact argued that the latter is more probable, based on the odd coincidence in the timescale for intelligence to emerge and the expected lifetime of our Sun, see [2] for discussion).

2. Reasoning One's Way Out of Anthropic Reasoning

The forgoing reasoning joins a set of anthropic arguments often discussed in cosmology that form the basis of the so-called anthropic principle. Expressed in its most general form, it asserts that the observed laws and conditions of the universe must be consistent with life, otherwise we could not be here to observe it. The anthropic principle is saved from being a mere tautology if it is the case that our cosmological vantage point may not be representative of the totality. In one popular scenario, what we have hitherto referred to as "the universe" may in fact be only an infinitesimal fragment of a vast assemblage of universes collectively termed the multiverse. The other universes may have different laws and conditions, most of which would not be consistent with the emergence of biology. It is then no surprise that we are witness to a number of curiously "bio-friendly" features and felicitous coincidences in physics and cosmology: they are simply a selection effect of our existence. Such anthropic arguments have been used to explain (some say explain away) some of the most perplexing questions of our existence, such as why the values of the constants of nature seem surprisingly well suited for life — indicative of a problematic degree of fine-tuning (dubbed the "fine-tuning problem"). While the logic is sound enough, the anthropic principle has been criticized as undermining the search for other explanations of our observations. The origin of life provides one such example. Astrobiologists attempt to determine the steps leading from non-life to life in terms of frequent physical and chemical changes, without appeal to bizarre flukes. But an anthropic argument could proceed by reasoning that it is only necessary for life to have arisen once, by an incredibly rare sequence of random physical and chemical fluctuations, for us to be here to discuss it. While the statement is correct, to settle for this explanation strikes at the very basis of the scientific method.

The multiverse theory has led to a great deal of heated debate in the physics and cosmology communities. Inventing an infinity of universes just to explain the one we observe may seem the antithesis of Occam's Razor. But some multiverse models appeal to a universe-generating mechanism, such as eternal inflation, in which an endless supply of universes arises naturally, on the assumption that general relativity and quantum mechanics apply at the level of the multiverse. Proponents argue that it is simpler to posit a vast selection of universes following naturally from a unified mechanism than to declare that there is a single universe with a large number of ad hoc properties. It seems to us that this is an area in which Chaitin's ideas would find ready application, by permitting (in principle!) the evaluation of the algorithmic information content that attaches to rival cosmological models.

Anthropic arguments may provide an explanation that convinces some at least for why we observe a universe with the life-encouraging features it possesses. There is however at least one feature of the universe as we observe it that cannot be adequately dismissed with these kinds of anthropic selection arguments. The anthropic principle may explain why we observe (some of) the things that we do, but it cannot explain why we also understand them. That is, human beings are more than mere observers; we are also comprehenders (also referred to as "universal explainers", see [5]). There appears to be no necessary reason why the universe should be structured such as to be comprehensible to a subsystem of itself, and no necessary reason that we humans, being a subsystem, should be able to make sense of the whole. Appeals to evolutionary fitness are clearly inadequate; knowledge of quantum physics, for example, is not necessary to our survival (although perhaps in several hundred years it will prove to be, but that does not negate the argument as it stands now). Certainly humans could exist without utilizing our intelligence to comprehend the world in a deep mathematical way, or for that matter making any sense of it at all (we could, for example, instead be zombies in a philosophical sense). While it is true that we would not be having an intellectual debate on the issue if we were not comprehenders, there is nothing about our existence *per se* that depends on our ability to make mathematical sense of the world in the same way that our existence as carbon-based organisms depends rather sensitively on the position of the carbon nuclear resonance.

Note that comprehension-ability is not at all the same as intelligence. Evolution will select for intelligence because of its survival value, and did indeed do so in the case of Homo Sapiens tens of thousands of years before the scientific era. The fitness advantages of human intelligence (e.g., tool-making, social complexity, advanced planning) account for how our species came to be successful long before we began to understand the deep mathematical principles on which the universe runs. While it is true that the surge in population in the last 300 years owes something to the scientific method, that is far too short a time for biological selection to operate. The ability of humans to comprehend the internal workings of atoms or the nature of black holes or Gödel's incompleteness theorem is something that must have been latent in the human brain/mind millennia prior to its successful application.

Usually, the distinction between "observer" and "comprehender" is glossed over. However, a full account of the world must take the difference into account. For example, a perfectly good scientific account exists for how Earth and other planetary bodies and natural satellites were formed via accretion from the solar nebula, some 4.5 billion years ago. We do not have a similar explanatory framework that encompasses the "anti-accretion" happening in the Earth system today,[a] where artificial satellites are being launched into orbit at an increasing pace. The most important difference is that "anti-accretion" requires comprehenders — specifically, the existence of physical systems with knowledge of Newton's laws. To state this distinction more explicitly, the problem here does not arise in our descriptions of the orbital mechanics of the International Space Station (ISS) versus that of the Moon. Both would be much the same,

[a]The term "anti–accretion" was coined by David Grinspoon.

with minor variation (e.g., could be explained using Newtonian gravity, or more precisely general relativity). If instead, however, you were to ask how each was caused to enter its orbit, in the case of the Moon you could provide a plausible account based on the initial conditions of the Solar System and the laws of physics. In the case of the ISS, that account would necessarily also include physical systems with *knowledge* of the laws of gravitation [6].

A second example concerns the element plutonium. It is made during supernova explosions and would have been present in the solar system's proto-planetary nebula, but due to its short half-life naturally-formed plutonium has long since disappeared from our planet. Yet there is plutonium on Earth, made in nuclear reactors. Even if all other traces of human civilization were wiped out, an alien visitor finding this element would correctly deduce that our planet must once have had a technological civilization with an understanding of the principles of nuclear physics. Without the existence of such comprehenders, the presence of plutonium on Earth would be an utterly inexplicable anomaly.

Separating the existence of anthropic reasoners from anthropic arguments may seem an unnecessary or even trivial step. However, the fact that intelligent physical systems (with humans as a pertinent example) can comprehend reality is perhaps the only feature that cannot be readily explained away with anthropic arguments — as stated, nothing about our existence is contingent on this fact (see [6] for discussion). While we may be a long way from formulating a plausible explanation for why the universe permits comprehenders, we can at least aim for a more modest goal. A necessary, though far from sufficient, condition for understanding the world is the ability to not only observe it, but to process the information thereby acquired and act upon it — that is, for information to include causal efficacy. The study of information, causation and agency is a burgeoning field of research, and may provide key insights into the nature of human intelligence and comprehension. While existing theory may serve to explain some basic facts about simple biological behavior in straightforward mechanistic terms, it seems very likely that a full account of human-level intelligence and comprehension will require a radical departure from the existing conceptual framework for physical theory. In the sections that follow, we outline some possible ideas for advancing in this manner inspired by research into the transition from non-living to living matter.

3. Breaking the Rules: Laws ↔ States

In the foregoing discussion, "knowledge" need not be anything nearly as sophisticated as the laws of gravitation, but can include simple rules or algorithms encoded in physical systems. An example is bacterial chemotaxis, in which an organism "programming" permits it to follow a chemical gradient to a food source. Even the seemingly simple narrative of the bacterium is not readily accommodated in our current formulation of physics because it necessitates an informational narrative that appears to assign a fundamental role to information. To see why, we can consider a straightforward example of what it means for a physical system to carry information. The minimal requirement is that a system must have at least two possible states, e.g., to say something carries 1 bit of information means that it could have been in one of two possible states. In physics, we are accustomed to thinking of our most fundamental descriptions of reality in terms of initial conditions and deterministic, fixed laws of motion (the "prevailing conception" as discussed by Deutsch [7]).[b] But, fixed deterministic laws only allow *one* outcome at any given time t for a given initial state at $t = 0$, and therefore do not allow the possibility for more than one possibility to be physically realized. That is, they do not allow for the physical existence of *counterfactuals* — roughly described as situations that do not in fact happen, but could. The known laws of physics permit an infinite number of

[b]Note this is even true of quantum mechanics where "indeterminism" is only introduced in collapse versions of quantum theory, where it is also necessary to invoke some sort of agency as a "measurer" (perhaps suggestive that departures from initial conditions and fixed laws are somehow connected to the physics of observers).

unrealized universes that are consistent with the laws but are not in fact instantiated. (This statement remains true even in most multiverse theories, which do not exhaust all possible universes, but merely all those within a restricted class. In fact, multiverse theories could be cast as an attempt to make counterfactuals physical by assuming they exist in other universes.) Starting from a given initial condition, the future state of the universe as described by fixed dynamical laws is set for all time to realize one, and only one, among the set of all possible trajectories through state space. Thus, while information is useful to externally describe physical systems (as an observer), it is difficult to conceptualize how it might be considered to be *intrinsic* to their operation (as a comprehender) in our current formulations of physics.

We should contrast the situation in physics — where only one history is realizable, posing a challenge for a fundamental status for information — with what we know of biology, which appears to be incredibly path-dependent and riddled with informational descriptions. A good example is biological evolution. Starting from the same initial state, biological systems trace out an enormous array of alternative trajectories through the process of evolution, acquiring genomic "information" about their environment. Thus, a common statement in evolutionary biology such as "we share a common ancestor with X", where X is a chimpanzee, fungus, or Archaeon, is intended to convey how far in the past we once shared a common initial state (to rough approximation, a genome) with species X. Each separate species evolved its unique features due to the peculiar and specific selection pressures it has seen through its evolutionary history; both due to the environment and competition with other organisms. Thus, in biology we are well acquainted with the fact that the current state is a function of (evolutionary) history. It is difficult, if not impossible, to write an equation of motion for such historical processes. The challenge is the "state-dependent" or "self-referential" nature of this kind of dynamical system: the manner in which biological systems evolve through time is a function of their state, such that *the dynamical rules themselves change as a function of the information encoded in the state* [8–10]. Since at least the time of Newton, physics has been founded on assuming that a fundamental asymmetry exists separating states and, in particular initial conditions, from the laws that operate on them. Thus, on the one hand, physical states are generally time dependent and contingent, whereas the laws of physics are regarded as timeless, immutable and universal. So while the world evolves, the laws themselves are assumed to remain fixed, even in conditions as extreme as total gravitational collapse, making it difficult to reconcile what we observe in biology with how we think about physics.

Very little attention has been given to the theoretical development of state-dependent dynamics as a novel way to generate complexity from simplicity, partly because of its unfamiliar nature and because of mathematical difficulties, but also because it is often not regarded as a fundamental property in the same sense as fixed, Newtonian-like laws are. Although the immutability of the laws of physics is occasionally challenged, it remains the default assumption among the vast majority of theoretical physicists. Nonetheless, there exist explicit examples in dynamical systems that are beginning to be considered. One such example of self-referential dynamics is provided by Kaneko and colleagues in their theory of "functional dynamics", which considers discrete maps of self-referential functions [11]. Here, we consider a different model for studying self-referential dynamics, by considering state-dependent cellular automata (CA). CA are widely utilized for studying how complex patterns can emerge from simple, local rules [12]. Some of the simplest CA are one-dimensional linear arrays of cells, each of which can be either empty ("0") or filled ("1"), and whose update depends only on nearest-neighbor interaction [13]. Such CA are called "Elementary Cellular Automata", or ECA for short. Since there are eight possible nearest-neighbor interactions, each of which may update to one of two possible states, there are in total $2^8 = 256$ unique ECA rules. An example of an ECA rule table for Rule 30 and a trajectory generated from applying Rule 30 to the initial state with one filled (black) cell are shown in Fig. 1.

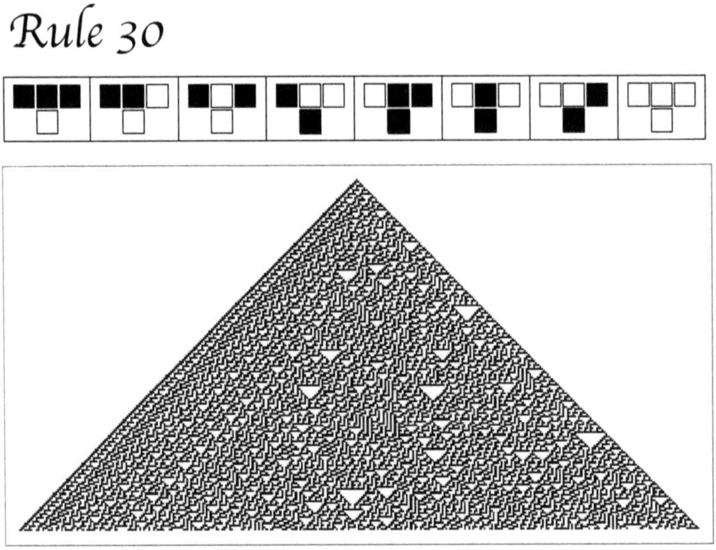

Fig. 1. Elementary Cellular Automaton (ECA) Rule 30. Rule table (top) and execution of Rule 30 starting from an initial state with one "1" cell (bottom).

Like many other classes of dynamical system, CA are typically implemented with fixed, immutable "laws " — that is, in traditional modeling of ECA the rule is fixed for all time. As such, even ECA evolving in time according "complex" rules (such as Wolfram Class IV rules, regarded as those CA that can generate the most complexity [13]) are limited in the patterns that can be generated. To demonstrate an explicit example of this limitation, we recently introduced a new variant of cellular automata with state-dependent update rules [14]. The simulation experiments consider a CA "organism" coupled to an "environment", where the organisms utilize a state-dependent update rule, rather than evolving according to a fixed rule. The phenotype and genotype can be roughly equated to the trajectories of states and of rules, respectively, in the execution of an organism. At time step t, the expressed phenotype and expressed genotype therefore correspond to the state and rule of the organism CA at time t, respectively. Expressible genes are permitted to be any of the 256 possible ECA rules. Each of these rules can be represented as an 8-bit binary string, enumerated by converting the decimal rule number binary, e.g., the binary representation for rule 30 is $G_{30} = 00011110$ and can be read directly from the rule table shown in Fig. 1. The rule table itself is here regarded as the genotype of the CA. At each time step t, "mutations" are calculated by comparing the frequency of all eight triplets states 000, 100, 010, 001, 110, 101, 011, 111, which appear in the rule table, in the organism state (phenotype) and in the environment. A mutation then occurs in a genotype $G_i(t)$ at time step t if the frequency measured in the organism state meets or exceeds the frequency of that same triplet pattern in the environment; if this condition is met the bit in the rule table corresponding to that triplet is flipped in the gene at $G_j(t+1)$. For example, an organism CA using rule 30 at time step t, where only the first triplet 000 is more frequent in the organism state than in the environment state, would mutate to rule 158 with $G_{158} = 10011110$. In this way, the time evolution of an organism is history-dependent, depending on both the organism's current state and rule, and the state of the environment at each time step. Due to the coupling to the environment, organisms are open systems. By contrast to the organism CA, the environment is a one-dimensional nearest-neighbor ECA, which evolves in time according to one of the 256 possible fixed local rules.

In order to demonstrate why state-dependent laws, as implemented in this model, are a potentially more powerful approach to studying complexity than fixed dynamical laws, a criterion for comparison is necessary. For this, we return to the opening question of this essay — namely, what principles might explain the apparent trend of increasing complexity in our biosphere? That is, how might we explain open-ended complexity? A criterion for identifying open-ended evolution is necessary, which has been a notoriously difficult problem in its own right [15]. In Adams *et al.* [14], we provide a formal definition.

Definition 1. A system $o \in O$ interacting with an environment $e \in E$ according to a function f exhibits open-ended evolution if there exists an n such that $f^n(o) = \{o_1, o_2, o_3, \ldots, o_n\}$ is non-repeating for $n > N$, where N is the Poincaré recurrence time for an isolated system $o \in O$ and $\{o_i\}$ is the set of all possible states in O or the set of all possible rules in O.

That is, an organism exhibits open-ended evolution (OEE) *if and only if* its pattern of states (phenotype) *or* its pattern or rules (genotype) is non-repeating with the Poincaré recurrence time for an isolated organism of the same width, w.

An explicit example of two open-ended organism CA of width $w_O = 5$ coupled to environments of width $w_E = 5$, are shown in Fig. 2. These represent the two longest organism-state (phenotype) trajectories discovered over an experiment exploring 104,858 different initial states (consisting of tuples of organism initial state and rule and environment state and rule pairs) — roughly 1% of all possible $w = 5$ organisms (considering only the 88 non-symmetric ECA rules [13]). Both are non-repeating in their phenotype until $t_{\text{recur}} = 237$ time steps, or 7.4 times longer than the Poincaré time, $t_{\text{P}} = 2^5$, of an equivalent isolated system. These two examples are also open-ended in their pattern of rules visited, with non-repeating patterns for $t = 69$ time steps each (2.15 times t_{P}).

More insights can be gained by studying the statistics of recursion times for state-dependent CA as compared to those of ECA according to local, fixed rules (isolated organisms), as shown in Fig. 3, for CA of width $w = 5$. Also shown for comparison are ECA evolved according to random mutation of the rule table, where bits in the rule table are flipped at a fixed probability of 1/2. By definition a closed ECA evolving according to a fixed, local rule cannot satisfy the above definition of open-endedness, as these systems will necessarily repeat within their Poincaré recurrence time (as will any closed, deterministic system). The frequency distribution of recurrence times for all ECA of $w = 5$ evolved according to a fixed rule, with periodic boundary conditions are shown in gray circles in Fig. 3, where all initial states are included for all 88 non-symmetric rules (2816 total experiments). In the figure, it is clearly evident that the vast majority

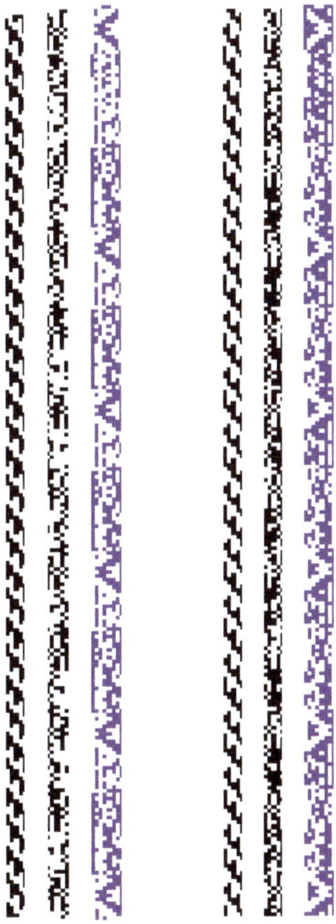

Fig. 2. Example of the two longest state-dependent organism trajectories (phenotypes) for width $w_O = 5$ CA, coupled to environments of width $w_E = 5$. The first panel in each is the environment CA (fixed rule), with the organism phenotype (states) trajectory shown in the middle and the organism genotype (rule) trajectory shown on the left (purple). Both do not repeat for their first 237 time steps, ~ 7.4 times the Poincaré recurrence time of an isolated system (only the early time evolution is shown due to space limitations). *Right*: Initial organism rule and environment rule are 24 and 88, respectively. *Left*: Initial organism rule and environment rule are 231 and 229, respectively.

of ECA recur long before the complexity cap at their Poincaré time (marked by a black vertical line in the figure). These results should be contrasted with that of the state-dependent organisms. The frequency distribution of recurrence times for $w_O = 5$ state-dependent CA, coupled to environments of width $w_E = 5$ and $w_E = 10$ are

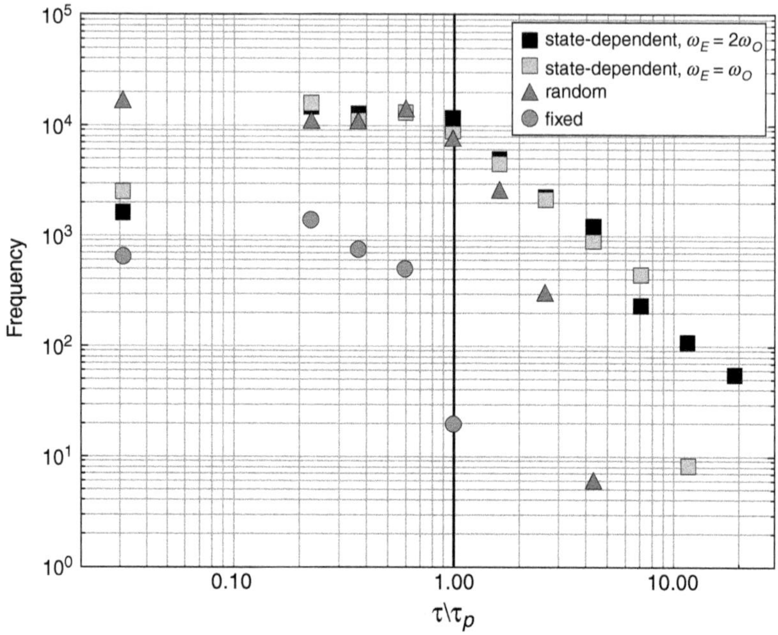

Fig. 3. Frequency distributions of recurrence times for width $w = 5$ CA evolved according to a fixed ECA rule (gray circles) and deterministic but state-dependent rules (squares). The black vertical lines indicate the Poincaré recurrence time for an isolated system of $w = 5$. Recurrence times for state-dependent CA of width $w_O = 5$, coupled to environments of $w_E = 5$ and $w_E = 10$, are shown by the gray and black squares, respectively. Also shown is the distribution of timescales to reach the final attractor for CA evolved with a randomly mutating ECA rule, with mutation rate 0.5, shown by the gray triangles.

shown by the gray and black squares in Fig. 3, respectively (both distributions include 104,858 individual experiments). Here, it is evident that there are many cases that do satisfy the above definition of open-endedness, lying to the right of the black vertical line in the figure.

It is perhaps not too surprising that the state-dependent organisms outperform the isolated, fixed rule CA given that they are open to an environment. More interesting is to compare the case of the (deterministic) state-dependent organisms to CA evolved according to the random mutation of their rule table. For these systems, the final attractor state is a random flipping between the all-0 state and the all-1 state, since every rule either maps 00000 → 00000 or 00000 → 11111 (plus their compliments under the inversion 0 ↔ 1). Thus, the frequency distribution plotted is the time until this final attractor state is achieved, rather than the recurrence time (the latter of which could only be approximate for any given simulation since the random evolution leads the system to never precisely recur). The distribution is illustrated by the gray triangles in Fig. 3, for similar statistics as generated for the state-dependent cases (roughly 104,000 cases). Comparison with the the two state-dependent cases demonstrates that, for comparable statistics between the two systems, the deterministic state-dependent systems outperform the random systems in their ability to generate open-endedness, with more numerous open-ended cases exhibiting longer recurrence times.

There are two further points that should be made with regard to the comparison between the deterministic, state-dependent systems and the random system. The first is that the state-dependent systems provide a mechanism for sustained generation of complexity that is lacking from the randomly evolved CA. That is, for the randomly evolved systems, the final attractor state will always be a random oscillation between the all-0 state and the all-1 state (effectively a random coin-toss), since no new information can enter the system. Whereas for the state-dependent systems, their attractors are in general complex as measured by their compressibility and Lyapunov exponent (see [14]) and therefore the long-time evolution of these systems will still exhibit complex patterns. Additionally, the state-dependent systems exhibit a mechanism for *increasing complexity* by potentially coupling to ever larger environments — in Fig. 3, it is clear that the cases with $w_E = 10$ environments generate more open-ended cases than the $w_E = 5$ environments, with the tail of the distribution achieving longer recurrence times for larger environments. Thus, organisms that evolve to encode more "information" about their environment in their rule table will in this simple system evolve greater complexity and longer recurrence times.

4. Defining Life (and Intelligence?) as Informational and Causal Structure

If state-dependent dynamics are indeed a hallmark feature of life and intelligence, then biological systems should exhibit distinctive informational properties in order to mediate the kind of non-trivial trajectories through state-space illustrated in Section 3. A necessary step to demonstrate this kind of "causal efficacy" for information, as it could operate in biology, is to quantitatively demonstrate that living systems satisfy two criteria [16]:

(1) Biological systems must be demonstrated to somehow be unique in their informational architecture, as compared to other classes of physical systems.
(2) Information must be shown necessary to the execution of biological function — that is to say, information must be shown to matter to matter, such that the dynamics can truly be said to be "state-dependent".

The first criterion could of course be used as a definition of life, that is, a system may be defined as living if its informational architecture exhibits a specific structure. At this time, our understanding of the patterns of information flow and storage in living systems is still rudimentary, so we do not here venture to hypothesize a particular architectural structure. Rather we report on a comparative study we conducted of the informational architecture of biological networks contrasted with the results for random networks to address the above criteria [16, 17]. For illustrative purposes, we focus herein on the Boolean model for the network regulating the cell cycle of the fission yeast *Schizosaccharomyces pombe*, with the goal to identifying any informational properties that might distinguish the biological network from random, by the above two criteria. A motivation for focusing on regulation of cellular division is that it is central to cellular function. The cell-cycle network articulates the cellular division process to create two daughter cells from a single parent. The cell passes through a number of phases, $G1$–S–$G2$–M, which collectively constitute the cell cycle. During the $G1$ stage, the cell grows, and if conditions are favorable, then it can commit to division. During the S stage, DNA is replicated. In the $G2$ stage, there is a "gap" between DNA replication (in the S phase) and mitosis (in the M phase) where the cell continues to grow. In the M stage, the cell undergoes mitosis, and two daughter cells are produced. After the M stage, the daughter cells enter $G1$ again, thereby completing the cycle.

The Boolean network model for the regulation of the cell-cycle process in the fission yeast *S. pombe* is shown in Fig. 4, as first worked out in [18]. Each node corresponds to one gene among the small subset of key regulatory proteins involved in regulating cell-cycle function. The state of node i, $S_i(t) \in \{0, 1\}$, indicates whether the given gene is present (1) or not (0). Biochemical causal interactions between proteins are denoted by edges in the network. Time advances through the cycle via synchronous updating, with the states of all

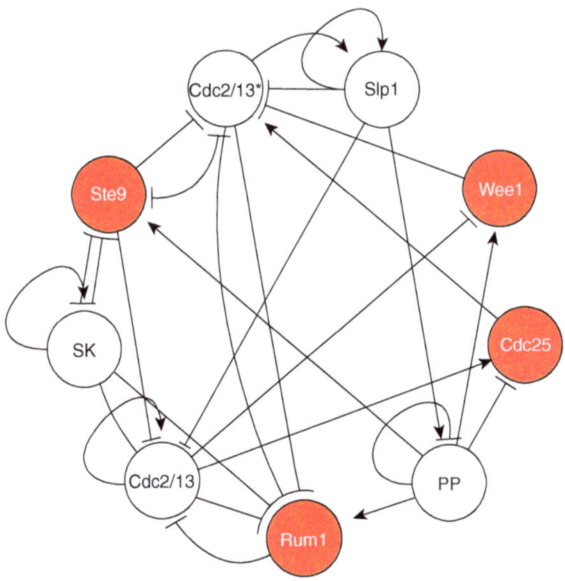

Fig. 4. The Boolean network model governing cell-cycle regulation in the fission yeast *Schizosaccharomyces pombe*. Nodes represent genes and edges their causal interactions. Figure adapted from [16].

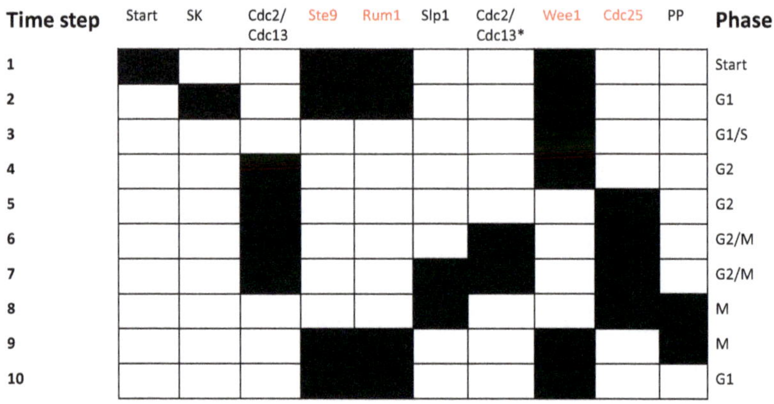

Fig. 5. Pattern of gene-expression states corresponding to phases of the cell cycle for the Boolean network model of gene regulation in fission yeast. Nodes are listed across the top, time runs from top to bottom, and the corresponding cellular phases are labeled on the right. Filled squares indicate that a gene is expressed.

nodes updated at discrete time steps according to the following rule:

$$S_i(t+1) = \begin{cases} 1, & \sum_j a_{ij} S_j(t) > \theta_i, \\ 0, & \sum_j a_{ij} S_j(t) < \theta_i, \\ S_i(t), & \sum_j a_{ij} S_j(t) = \theta_i, \end{cases} \tag{1}$$

where a_{ij} is the edge weight between node i and j ($a_{ij} = -1$ for inhibition links and $a_{ij} = 1$ for activation links), and θ_i is the threshold for node i. The threshold for all nodes is 0, with the exception of Cdc2/13 and Cdc2/13*, which have thresholds of -0.5 and 0.5, respectively. Iterating these rules for the network in Fig. 4 generates a cellular automaton with pattern states matching the expression levels of genes through the $G1$, S, $G2$ and M phases of cellular division in fission yeast, as shown in Fig. 5.

To quantify characteristics of the informational architecture of the fission yeast cell-cycle network and contrast to that of random networks, a relevant, quantifiable concept of "information" is necessary. In Kim *et al.* [16], we focused on the concept of "information flows" as highlighted in a seminal paper by Nobel laureate Paul

Nurse as a concept of information relevant to uncovering new biological insights [19]. In Kim *et al.*, this was interpreted as "information processing" following Lizier *et al.* [21] and measured by Schreiber's transfer entropy (TE) [20]. In brief, transfer entropy (TE) from a *source* node Y to a *destination* node X is defined as the reduction in uncertainty due to knowledge of state of Y about the future state of X, above the reduction in uncertainty provided by knowledge about the past states of X. TE from Y to X can be written as the mutual information between y_n and x_{n+1} conditioned on k previous states of X:

$$T_{Y\to X}(k) = \sum_{(x_n^{(k)}, x_{n+1}, y_n)\in\mathcal{A}_0} p(x_n^{(k)}, x_{n+1}, y_n) \log_2 \frac{p(x_{n+1}|x_n^{(k)}, y_n)}{p(x_{n+1}|x_n^{(k)})}, \tag{2}$$

where $\mathcal{A}_0$ indicates the set of all possible patterns of sets of states $(x_n^{(k)}, x_{n+1}, y_n)$, $x_n^{(k)}$ denotes $(x_n, \ldots, x_{n-k+1})$ the vector of k previous states of destination X at time step $n+1$, and y_n and x_{n+1} represent the state of Y at the current time step and the state of X at the next time step, respectively. Directionality, important for quantifying "flows" of information, arises due to the asymmetry between time steps for the state of the source and the destination in Eq. (2).

Information transferred between pairs of nodes can be calculated through the execution of the function of the cell-cycle network (defined in the Boolean network model as the dynamics of the fission yeast cell-cycle network). The results can then be contrasted with the same analysis performed on ensembles of random networks. In Kim *et al.*, information processing in the fission yeast cell-cycle network was compared to that of networks randomly sampled from two different classes: Erdös — Rényi (ER) networks and scale-free (SF) networks (see [16] for discussion). The ER random networks are defined such that they share no structural bias with the cell-cycle networks other than the size of the networks. The SF random network ensembles, by contrast, maintain the same number of activation and inhibition links for each individual node as the cell-cycle networks. The SF networks therefore share important topological properties with the biological networks, including degree distribution (the rank ordering of the number of edges per node), often cited as a distinguishing feature of many real-world networks including biological networks [22].

Comparing analysis of the fission yeast cell-cycle network to large ensembles of random networks reveals that the biological network exhibits distinctive patterns in the distribution of information transferred between nodes. This was reported in [16, 17] (the paper [16] also considered the network for budding yeast cell-cycle regulation). The results of comparing information processing in the fission yeast cell-cycle network to that of random network ensembles reveals statistically significant differences in the distribution of information processing that distinguish the biological network from random, as reproduced in the top panel of Fig. 6. The scaling distributions reveal a nonlinear relationship between the information transferred between pairs of nodes (y-axis) and their relative rank (x-axis), for each of the network classes studied — biological, SF and ER. The scaling relation is most striking for the biological networks (red), which are significant outliers with respect to either of the ER (green) or SF (blue) ensembles. Figure 6 shows that ER networks have much less information transfer on average than SF networks or the biological network. This result suggests that scale-free structure — as is the case for the SF and biological networks, but not the ER networks — plays as significant role in increasing information transfer within a network. Nonetheless, SF topology alone is not sufficient to explain the pattern observed for the biological network.

The regime where the difference between the biological network and the ensemble of SF networks is most pronounced is highlighted between the red-dashed vertical bars in the top panel of Fig. 6. Detailed inspection of the node pairs in this regime reveals that it is associated with the presence of control kernel nodes. The control kernel consists of a small subset of nodes (highlighted in red in Figs. 4 and 5) that regulate the global function of the network by canalizing all network states toward the primary attractor associated with the biologically functional "resting" state of the cell. Control kernels have been discovered in a number of Boolean models for gene

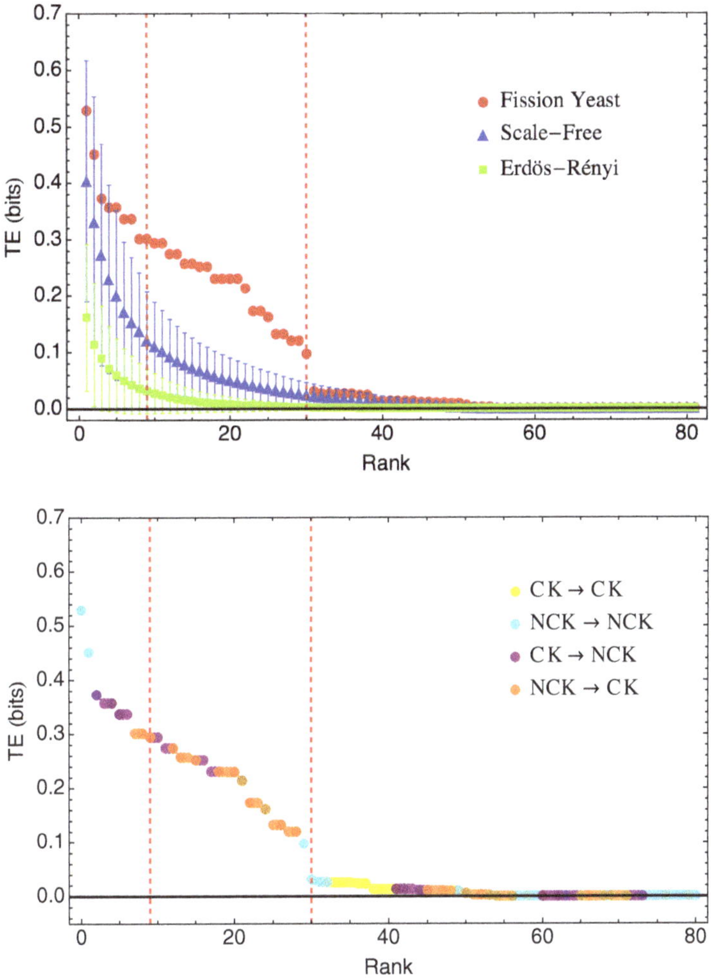

Fig. 6. *Top*: Scaling relation for information transfer among pairs of nodes for the fission yeast cell-cycle (red), and ensemble averages of ER (green) and SF (blue) networks. The history lengths for computing TE was $k = 2$. Averages and standard deviation for each of the random network ensembles were computed for a sample of 1000 networks. *Bottom*: Data shown is the same as the right panel for the fission yeast network, with data points divided into four classes of information transfer: $CK \rightarrow CK$ (yellow), $CK \rightarrow NCK$ (purple), $NCK \rightarrow CK$ (orange) and $NCK \rightarrow NCK$ (aqua). Figure adapted from [16].

regulatory networks [23], suggestive that they may be a generic feature. In the right panel of Fig. 6, control kernel nodes are labeled "CK" and non-control kernel nodes are labeled "NCK": node pairs where information flows to the control kernel are thus labeled NCK $\rightarrow$ CK (orange) and those where information flows out from the control kernel are labeled CK $\rightarrow$ NCK (purple). The figure shows that the biologically distinctive regime of the scaling relation for information transfer in the cell-cycle network (highlighted between the red-dashed vertical lines) is dominated by information transfer both to and from the control kernel.

These features may be common to biological networks of different function, and in particular, the scaling relation in information transfer may be a universal hallmark of biological organization characteristic of life here on Earth and also potentially elsewhere (it has been demonstrated for the Boolean model of at least one other biological network — the cell-cycle regulatory network of the budding yeast *Saccharomyces cerevisiae* [16], with others currently under investigation). The results of Fig. 6 are suggestive of information-based organizational principles that go beyond topological considerations such as scale-free structure, and may be critical to biological function. They thus open a new quantitative framework in support of the view that information is indeed intrinsic to the operation of living systems, by demonstrating how the informational architecture of biologically evolved networks can distinguish them from other complex physical systems that do not exhibit the same informational properties. This suggests that new "informational laws", with biological informational architecture at their core, might be necessary to account for life and perhaps even the emergence of life. It is an open question whether this architecture opens a pathway for realizing state-dependent laws of the variety explored in Section 3.

5. Conclusions

Although the origin of life and the nature of intelligence are two distinct problems, Carter's argument curiously connects the likelihood of intelligence to the likelihood of life itself. This may be a mere artifact of our current vantage point as observers with only one biosphere (our own) to attempt to understand it. While it is difficult to reason one's way out of anthropic reasoning, the existence of comprehenders (which cannot readily be explained by anthropic arguments) suggests that an explanatory framework for life and intelligence is likely to include a concept of information that is causal at its core. In particular, a key facet of comprehenders making a difference to the world with which they interact is that their rules of interaction depend not just on fixed immutable laws, but are also somehow "programmed" into states of the world. We have shown an explicit example of how such state-dependent dynamics can generate novel pathways to complexity, utilizing elementary cellular automata as a toy model system. As far as we are aware, nobody has attempted to characterize state-dependent dynamics and complexity growth in terms of algorithmic information theory. In effect, we are prescribing a generalized type of algorithm (involving coarse-graining) to determine up update algorithm for the system of interest at each time step. Whether Chaitin's ideas may be extended to encompass such a construction is a fascinating open question. It would be a significant advance if state-dependent dynamical pathways to complex states could be systematically classified in terms of some measure of computational complexity.

If it is indeed true that such dynamics play out in real biological systems, then we should expect to observe patterns of information flows in living systems (or intelligent systems more broadly) that are distinct from other complex physical systems. While our current understanding of what these defining patterns might be is still in its infancy, we have shown at least one concrete example where this approach does indeed distinguish biological systems from random. Taken together, the toy models explored here suggest new pathways for understanding state-dependent dynamics and how biological systems might "break the rules" as a result of their distinctive informational architecture. It may be the case that the emergence of these attributes is the very emergence of life itself, and in turn they could provide important clues to the fundamental nature of intelligence.

Bibliography

[1] C. H. Lineweaver and T. M. Davis, Does the rapid appearance of life on Earth suggest that life is common in the universe? *Astrobiology* **2**(3) (2002), 293–304.

[2] B. Carter and W. H. McCrea, The anthropic principle and its implications for biological evolution and discussion, *Philos. Trans. Roy. Soc. A* **310** (1983), 347–636.

[3] D. S. Speigel and E. L. Turner, Bayesian analysis of the astrobiological implications of life's early emergence on Earth, *Proc. Natl. Acad. Sci. USA* **109** (2012), 395–400.

[4] B. Carter, Five-or six-step scenario for evolution? *Int. J. Astrobiol.* **7**(2) (2008), 177–182.

[5] D. Deutsch, *The Beginning of Infinity: Explanations that Transform the World* (Penguin, UK, 2011).

[6] S. I. Walker, The descent of Math, in *Trick of Truth: The Mysterious Connection Between Physics and Mathematics*? eds. A. Aguirre, B. Foster and Z. Merali (Springer, 2016).

[7] D. Deutsch, Constructor theory, *Synthese* **190**(18) (2013), 4331–4359.

[8] D. R. Hofstadter, *Gödel, Escher, Bach: An Eternal Golden Braid* (Basic Books, 1979).

[9] N. Goldenfeld and C. R. Woese, Life is physics: Evolution as a collective phenomenon far from equilibrium, *Ann. Rev. Cond. Matt. Phys.* **2** (2011), 375–399.

[10] S. I. Walker and P. C. W. Davies, The algorithmic origins of life, *J. Roy. Soc. Interface* **10** (2013), http://doi.org//10.1098/risf.2012.0869.

[11] N. Kataoka and K. Kaneko, Functional dynamics. I: Articulation process, *Physica D* **138**(3) (2000), 225–250.

[12] J. P. Crutchfield and M. Mitchell, The evolution of emergent computation, *Proc. Nat. Acad. Sci. USA* **92**(23) (1995), 10742–10746.

[13] S. Wolfram, *A New Kind of Science* (Wolfram Media, Champaign, 2002).

[14] A. Adams, H. Zenil, P. C. W. Davies, and S. I. Walker, Formal definitions of unbounded evolution and innovation reveal universal mechanisms for open-ended evolution in dynamical systems. *Sci. Rep.* **7**(1) (2017), 997.

[15] M. A. Bedau, *et al.*, Open problems in artificial life, *Artificial Life* **6**(4) (2000), 363–376.

[16] H. Kim, P. C. W. Davies, and S. I. Walker, New scaling relation for information transfer in biological networks, *J. R. Soc. Interface* **113**(12) (2015), http://doi.org/10.1098/risf.2015.0944.

[17] S. I. Walker, H. Kim, and P. C. W. Davies, The informational architecture of the cell, *Proc. Roy. Soc. Phil. Trans. A*, in press.

[18] M. I. Davidich and S. Bornholdt, Boolean network model predicts cell cycle sequence of fission yeast, *PLoS ONE* **3** (2008) e1672.
[19] P. Nurse, Life, logic and information, *Nature* **454** (2008), 424–426.
[20] T. Schreiber, Measuring information transfer, *Phys. Rev. Lett.* **85** (2000), 461–464.
[21] J. T. Lizier, M. Prokopenko, and A. Y. Zomaya, A framework for the local information dynamics of distributed computation in complex systems, in *Guided Self-Organization: Inception* (Springer, Berlin, 2014), pp. 115–158.
[22] R. Albert, Scale-free networks in cell biology, *J. Cell Sci.* **118** (2005) 4947–4957.
[23] J. Kim, S.-M. Park, and K.-H. Cho, Discovery of a kernel for controlling biomolecular regulatory networks, *Sci. Rep.* **3** (2013), 2223.

Chapter 15

Meta+phenomenology: Primer Towards a Phenomenology Formally Based on Algorithmic Information Theory and Metabiology

Michael Winter*

1. Introduction

To start, let us assume that our world is made of information whose dynamics are governed by algorithms. This idea, often called digital philosophy or digital physics, is quite established. As Gottfried Leibniz put it over 300 years ago: "Omnibus ex Nihilo ducendis sufficit Unum" (in order to produce everything from nothing, one is sufficient). In the last century, such a world view has been strengthened by the theories of Alan Turing, John von Neumann, Claude Shannon, and somewhat more recently, people such as Gregory Chaitin and Stephen Wolfram.

The implications of digital philosophy invoke vexing ontological and epistemological questions regarding the nature of our universe. If everything is computable, how can we reconcile the seemingly complex phenomena that we observe? While physics has primarily been

*Composer and Independent Researcher; mwinter@unboundedpress.org.

explained via theoretical mathematics, other intellectual domains such as biology and phenomenology often remain unquantifiable and therefore irreconcilable with a digital philosophy.

However, starting from 1931, there was a paradigm shift marked by Gödel's incompleteness theorems and then later by Turing's 1936 publication "On Computable Numbers, with an Application to the Entscheidungsproblem". Gödel was the first to show that there exist truths that cannot be proven true. This was a major blow to traditional mathematics.

Gödel's proof was complicated, algebraically based on Georg Cantor's diagonal argument. Turing simplified the proof of incompleteness through the concept of a universal computer: a theoretical machine that can compute any computable function. It is also more dangerous. Turing's proof showed that uncomputable problems, and thus incompleteness, are ubiquitous.

Finally, in the lineage of incompleteness, comes Chaitin with an even more dangerous form of incompleteness. One showing the existance of maximally complex axioms that are not self-evident. This finding was the final nail in the coffin of the traditional world view of formalized mathematics.

Both Turing and Chaitin would go on to consider mathematical worlds that would model the complex phenomena of biology. Turing's investigations would rely on partial differential equations and not the more universal power of algorithms. This can perhaps be attributed to a few factors. The biological inquiry was rather focused on a particular phenomenon, morphogenesis, and it preceded the discovery of the structure of DNA. Also prior to the work of James Watson and Frances Crick, von Neumann, in a series of lectures published posthumously as the "Theory of Self-Reproducing Automata", outlined a computable system that exhibited the fundamental biological phenomenon of reproduction. In von Neumann's self-reproducing automata, a universal constructor takes as input a set of instructions for reproduction. For this reason, biologist and Nobel Laureate Sydney Brenner would recognize von Neumann's work as fundamental to his own DNA research. In particular, Brenner's understanding of the phylogenic mechanism of DNA. That DNA was essentially a computer program which contained the biological instructions for production but not the means to execute those instructions. This concept is now understood as the differences between genotype and phenotype; the difference between the genetic code and the phenomenological expression of the genes.

Von Neumann not only explored the digital mechanism of self-reproduction, but also how complexity evolves in such a system. To do so, he proposed using Shannon's recently published theories of information as a measure of complexity.

At the time of von Neumann's lectures, Chaitin had yet to formalize Algorithmic Information Theory; a theory of program size complexity formally equivalent to Information Theory. Both Shannon classical Information Theory and Chaitin Algorithmic Information Theory focus on the complexity of objects through compression: the former statistically through measuring entropy and the latter by comparing the size of the smallest (most elegant) program to the size of the object that program outputs.

Algorithmic Information Theory might have been a more useful measure of complexity for von Neumann. This has been made clear in the last 10 years by Metabiology, Chaitin's toy-model of Darwinian evolution. Through an algorithmic information theoretic approach, Chaitin's metabiological model shows how a single software organism (a Turing machine) undergoing random algorithmic mutations will compute larger and larger integers (known as the Busy Beaver function) at evolutionary rates.

One of the hallmarks of Chaitin's work is his ability to understand a problem in the most minimal, essential way. The metabiological model does not contain a population or anything like sexual reproduction; just a single mutating organism. Nonetheless, Chaitin's elegant model has significant philosophical implications beyond the domain of evolutionary biology. For example, Chaitin's wife and collaborator, Virginia, uses Metabiology as an ontology for the human self-image and his student, Felipe Abrahao, has studied the effects of metabiological models extended such that a population of software organisms shares information along edges in networks.

Chaitin himself recognizes that Metabiology goes beyond simply proving conditions in which continuous evolution occurs in toy mathematical worlds. A fundamental tenet of Metabiology is the relation between mathematical and biological creativity in terms of how new information enters the system of the metabiological model. He has also recently started to discuss theories of consciousness with respect to Algorithmic Information Theory drawing from David Chalmer's ideas set forth in "The Conscious Mind: In Search of a Fundamental Theory" in which Chalmers proposes that any physical system able to represent and process n bits of information contains n bits of consciousness. The fact that Chaitin's metabiological model seems to intrinsically incorporate notions of creativity and consciousness suggests a new understanding of experience; a new phenomenology. One based on a digital philosophy.

Historically, phenomenology has aimed to understand, in an objective manner, the subjective response to an experience. "Logical Investigations" by Edmund Husserl is often cited as establishing the field. His arguments are almost completely predicated on, as suggested by the title, logic. Not surprisingly, readings of Leibniz are also integral to his theses. Following Husserl, philosophers such as Martin Heidegger and Maurice Merleau-Ponty continued to deal with overarching concepts of phenomenology: consciousness, perception, intuition, intentionality, and subjectivity. Later, more focused subdomains were created to investigate specific perceptual mechanisms. For example, Gestalt psychology tries to understand the perceptual mechanism of gestalt formation; how several objects may or may not be perceived as one. This was initially studied in the visual scene by Kurt Koffka and then extended to the auditory scene by composer James Tenney and researcher Alfred Bregman.

Even though Husserl's phenomenology maintains a logical basis, it does not completely reconcile a digital philosophy or the epistemological consequences of incompleteness. This is another justification for a new phenomenology based on digital philosophy. One that would further cement the arguments set forth in "Logical Investigations" by incorporating computation, Algorithmic Information Theory, and Metabiology.

On the surface, Chaitin's model may seem void of concepts central to phenomenology: experience, perception, information exchange, subjective response, and by extension art and music. However, the metabiological model can be analyzed to show that the mechanism of the algorithmic mutation of Metabiology — an algorithm that takes as input a software program and outputs another (its mutated self) — can actually serve as a conceptual framework for a mathematical understanding of experience and subjectivity.

In order to lay the foundation for such an extended phenomenology, I will address three fundamental questions:

(1) What is the digital mechanism of the experience?
(2) How can a phenomenology based on digital philosophy account for subjectivity?
(3) What is the ultimate objective of creativity?

As the title suggests, these expository thoughts aim to function as a primer towards a mathematical phenomenology based on digital philosophies and filtered through the lens of a composer seduced into epistemology. A primer towards the more ultimate goal of understanding, through mathematics and with a minimal set of conceptual assumptions, why beauty is the subjective response to an experience and not intrinsic to or characteristic of an object.

2. A Personal Note and Acknowledgments

As I am not a mathematician, I do not feel bound to the rigors of mathematics. I imagine this text will easily reveal that fact to the reader. Yet when one sheds the cloak of specialization all too common in our current world, we venture into territories that bear the fruit of new possibilities. And even if I fail in the process, at least I tried.

No other person has encouraged that mentality in me more than Greg Chaitin. Greg is a true polymath and intellectual wanderer. He deserves the ranks of those great thinkers who exhibit pure, unexplainable creativity (here I am thinking of other great mathematicians such as Leonhard Euler, Srinivasa Ramanujan, and von Neumann). I am grateful for his friendship and inspiration.

Three others that I feel compelled to thank are composers Larry Polansky, Curtis Roads, and James Tenney. Most of all for their friendship and mentorship, but also for the fact that they instilled in me an interest exploring other domains and a phenomenological approach to making music. The title, Meta+phenomenology, with the somewhat strange "+" sign, is a reverent nod to Tenney, who at the young age of 25, wrote a seminal thesis called "Meta+Hodos", which explicates a phenomenological approach to music theory.

Naturally, I saw it fitting to merge these two interests: epistemology and phenomenology. The following ideas are my humble attempts to lay a general framework for phenomenology based on epistemology and digital philosophy.

3. Creativity, Inspiration and a Brief Overview of Metabiology

At the core of this new phenomenology, like Chaitin's metabiological model upon which it is based, is incompleteness. It gives rise to the most fundamental aspect of all the post-Gödelian theories: that there exist problems that require infinite creativity. Chaitin's findings are published in great detail in his paper *Life as evolving software*, but I will give an overview of aspects germane to the following discussion.

Metabiology is a toy mathematical model of Darwinian evolution. The fitness test of the model is an alternate version of Tibor Rado's Busy Beaver Function in which $\mathrm{BB}(n)$ is defined as the computer program of $\leq n$ bits that computes the largest whole number. This is an uncomputable function that, as Chaitin would say, requires an infinite amount of mathematical creativity because you can always define a larger and larger number: adding a number to itself, multiplying a number to itself, exponentiation, hyperexponentiation, etc.

Chaitin shows that approaching BB(n) is also the same as getting a better and better lower bound on the Halting Probability, defined as the probability that a random computer program will halt.

In Metabiology, mathematical creativity equates to biological creativity. Evolutionary progress is measured by the rate in which a single mutating software organism calculates bigger and bigger numbers. There are ostensibly no bodies, no sex, and the environment is a software space. Time is measured by the number of mutations tried up to a given point.

The mutation functions are not point mutations or crossovers as in traditional biology. They are actually algorithms that take as input a software organism and output another organism. Such a mutation can be extremely powerful, allowing violent changes and giving an organism the potential to traverse the entirety of the software space as it mutates. The mutations are chosen randomly with a probability based on something akin to relative complexity in Algorithmic Information Theory, which is the size of the minimal program to compute string b from a given the minimal program to compute a. The only difference is that in Metabiology, the mutation algorithm is given a directly instead of the minimal program for a. This is how Chaitin defines distance in the environment of the software space.

Chaitin's proof of cumulative random evolution is predicated on comparing the rate of evolution in the model to two extremes: systematic search and intelligent design. In the former, mutations are tried in order of some computable function. The equivalent of complete randomness and not cumulative at all. In the latter, an Oracle is used to pick the best possible mutations in perfect order reaching the solution in linear time. Chaitin's toy model sits in between these two. Mutations are chosen at random, but with a probability determined by the altered version of relative complexity described above. His ultimate finding is that this version of random evolution is cumulative, progresses faster than systematic search, and has a lower bound that is slower than intelligent design. In intelligent design, the creativity is given absolutely and entirely by an Oracle. In cumulative random evolution, only a limited use of an Oracle is employed in order to throw away non-halting algorithmic mutations as they run endlessly and never produce a new organism. Oracles are the source of inspiration. They are used to inject novelty and new information into the system, but there is no completely free lunch.

For our purposes, the most important aspect of the metabiological model is that creativity and inspiration are central to the model. This thesis simply extends Chaitin's equation between biological and mathematical creativity to *all* creativity and to *all* experiences.

4. The Digital Experience

An experience is transformative; comprised of an event/situation, a subject/observer, and the effect of the former on the latter. The metabiological mechanism of evolution, an algorithmic mutation that takes as input a computer program (a software organism) and outputs an altered version of the input, encapsulates this paradigm quite elegantly. In a digital phenomenology, the experience can be conceived of almost verbatim: an event algorithm takes as input a software organism and outputs a transformed version of it. At any given time, the software organism is defined as the accumulation of these experiences; the sum of these transformations.

Let us take the general notion of the artistic experience for example. Trivially put, an observer perceives an aesthetic object (or situation) created or framed by another person. This suggests that there is something that Metabiology, which only contains a single mutating organism, cannot account for: a population and the cumulative effect of information exchange between members of the population.

Chaitin addressed this issue to some extent in an earlier work, *Towards a Mathematical Definition of Life*. This article can be read as a sort of precursor for Metabiology and demonstrates that Metabiology is some 40 years in the making. In the paper, Chaitin describes how independent structures (which could be members of a population, aesthetic objects, etc.) within a digital scene can be differentiated. The boundaries are determined using mutual information from Algorithmic Information Theory, which is the difference between the size of the minimal program that computes two strings compared to the sum of the sizes of the individual minimal programs that compute each string separately. Two strings that have a minimal amount of a mutual information are considered separate entities in the scene.

Chaitin seemed to jettison considering multiple structures in the formalization of Metabiology in favor of a simpler, more elegant model with a single organism. However, the mutation algorithm can actually contain a representation of entities within itself. This basically reintroduces sex into the equation. The result of shared information among members of a population. It also accounts for the setting/scene itself including aesthetic objects. Therefore, the analog of the mutation algorithm of Metabiology to the event algorithm of Meta+phenomenology does not necessarily preclude a population or sex and is powerful enough to represent the multiplicities that we are accustomed to in real-world artistic experiences, from artists to aesthetic objects to shared experiences among multiple members of the population (the sex of Meta+phenomenology). As such, there does not necessarily need to be a redefining of Metabiology in order to fit it to Meta+phenomenology. All that is required is an interpretation of the model through the alternative lens of experience as opposed to biology.

5. Towards a Proof of Subjectivity

A proper phenomenology must take into account subjectivity: how entities will react differently to similar experiences both between the entities and over time. Let's start with the first case as it seems simpler/clearer to explain. While Chaitin's model is instantiated with a single software organism, one could track multiple instances of the model starting with the same or different organisms. It is completely possible that the two organisms can go through mutations with a high level of mutual information at the same step in time. This would be the equivalent of a shared experience. The two mutation algorithms need not be the same, only similar. The differences between the organisms themselves and between the mutation functions (no matter how similar they might be) will have varying effects. The metabiological model accounts for subjectivity between organisms in this way. Organisms are transformed to different extents as they pass through similar event/mutation algorithms.

The second case, varying experiences to similar stimuli over time might seem more complex, but is equally straightforward. There is a spectrum of experiences from those that are common and less transformative to those that are rare and more transformative. Further, once a rarer, more formative experience occurs, an experience of a similar nature may have a lesser effect. Simply put, what was once new, often becomes old; becomes normalized and canonized. Say for example, the first time you listen to a piece of music that truly changes you. You are likely to listen to it again and while it still has effect, that effect may wane over time. It seems reasonable to conjecture that something similar occurs in the metabiological model. That mutations similar to a highly transformative mutation that has already occurred will have a lesser effect over time. Also, the software organism itself will have changed between the two mutations (or, in the case Meta+phenomenology, the two experiences).

Novelty is also relative. An experience that is novel to one organism may not be to another who has already experienced a similar event. Leibniz suggests this in the "Monadology" where he posits a world made of monads that have varying scopes of perception. One monad, essentially the God monad, sees everything. So in the metaphenomenological model, monads are in the process of expanding their perceptual views; expanding their horizons.

This sketch proof of subjectivity also suggests a well-defined concept and measure of beauty based on the transformative effect of an event/mutation algorithm. One that reflects the equivalent of an evolutionary leap in the metabiological model. By definition, the effect of a particular event/mutation is largely based on the current state of the organism — the sum of the cumulative experiences up to that point — and likely more so than any entities, such as aesthetic objects, that may be represented within the mutation algorithm.

Another important issue is how status quo biases arise. To address this, we will make another conjecture. Organisms that have a high level of mutual information are more likely to be transformed a similar amount by mutations that have a high level of mutual information. This would show how status quo biases arise among similar members of a population and refute the greatest misunderstanding of some phenomenologies: that an aesthetic object possesses the universal quality of goodness or badness.

6. Conclusion: Transcendence and the Metaphenomenological Objective

Finally, we come to the discussion of whether or not there is a metaphenomenological objective; especially for intentional acts of transformation such as creative/artistic experiences. Is there a reason for doing something that often seems so impractical? What is the point of a novel experience?

In Darwinian evolution, survival itself is the biological imperative reached by overcoming obstacles that test the fitness of an organism. In Metabiology, the organisms are essentially mathematicians/computer scientists that are in an arms race to name larger and larger integers. But let us examine a somewhat hidden meaning of such a quest and then extend that to novel, artistic experiences.

Solving Chaitin's alternative Busy Beaver Function is equivalent to getting better and better approximations of the halting probability. The halting probability, in a way, can be seen as a representation of all truths because knowing n bits of the halting probability means knowing how many programs of size $\leq n$ halt. It is pure, compressed knowledge.

The software organisms of Metabiology, and by extension Meta+phenomenology, are striving to become Oracles, but since that is a never-ending process that requires infinite creativity, it is a never-ending pursuit. A pursuit towards total transcendence. Towards becoming gods.

Dedicated to *Charis von Ristok*, with whom I had the great opportunity to discuss the ideas in this text extensively. She listened to me rant about digital philosophy, constructively challenging me throughout, and always doing so elegantly and gracefully. I now realize more than ever the gravity of a particular type of experience; that of falling in love. In the metaphenomenological sense, the time we spent together as I wrote this was one of the most transformative experiences I have ever had and transcendent on a divine order.

Schloss Solitude
Stuttgart, Germany
November 2018 to January 2019

Addendum: A Few More Thoughts/Treatises Related to Meta+phenomenology

The following treatises are ideas that I considered and/or removed from the original text after deciding the primer on Meta+phenomenology should be as succinct and clear as possible. I am grateful to the editors, Francisco Antonio Doria and Shyam Wuppuluri, for encouraging me to extend the chapter and allowing me to do so in the form of an addendum of treatises. They are intended to be rather episodic reflections on potential implications of Metabiology considered in the process of writing the original text. In particular, I would like to thank Shyam for not only encouraging me to extend the text but also becoming a caring, selfless friend in the process. His kind words and consideration helped fuel the following the ideas.

1. Run-Time Versus Program-Size

As with most of Chaitin's work, his proofs focus strictly on program-size (measured in number of bits) without reconciling run-time (the time it takes a program to halt). In the metabiological model, time is measured by the number of mutations up to a given point in the evolution even though the mutations themselves run for different amounts of computational steps. However, Cristian Calude's findings in *Most programs stop quickly or never halt* suggest that program-size and run-time might be reconcilable. For example, Chaitin's algorithm that approximates Omega, the halting probability, spawns and runs all programs of size less than n bits up until an arbitrary time-limit in order to determine a lower bound on Omega. The longer the time-limit; the better the approximation. Given Calude's finding, a sufficient time-limit related to n should produce a very good lower bound on Omega.

2. Time in Finite Systems

In order to discuss time in finite systems, I will start by focusing on cellular automata. In an automaton on a finite grid (e.g., in the form of a torus), time can be a well-defined constant: the number of computations needed to update all the cells in the system. This is essentially the computational time needed to get from one frame to the next in the moving image that is the universe. Just as with automata, time in the metabiological model can be completely reconciled given a finite system with only halting programs (a version proposed by Chaitin himself). It is a constant equal to the run-time of the program that takes the longest amount of time to halt.

3. Time in Infinite Systems

If there exists an infinite amount of information in the system, it becomes much more difficult to conceptualize time and reconcile program-size with run-time. Assuming that all universal systems are somehow equivalent and that an understanding of time through cellular automata likely has some correlation/equivalence in the metabiological model, I will continue my focus on cellular automata. A cellular automata on an infinite grid can be partitioned into subsets of finite and infinite subgrids (parallel universes). If all the partitions are finite subgrids, time in the system as a whole can be conceptualized as an infinite number of finite parallel universes each of which abides by the well-defined concept of time described in the previous treatise. If any of the partitions are infinite (or there is only one partition, the infinite whole), then there is no well-defined concept of the "current state" of the system as a whole because an infinite grid can only update in infinite time.

4. Classical Versus Quantum Mechanics in A Digital System

Assuming that cells of automata (in both the finite and infinite cases) update serially, there are two topologies of space–time: the Euclidean distance between the cells in the multidimensional grid (the scope of the large scale: a classical topology) and the distance between cells with respect to when they are updated in a one-dimensional order (the scope of the small scale: a quantum topology). Some of the anomalies of quantum mechanics in our observable universe seem interestingly similar to phenomena in automata that result from the difference between these two levels. For example, quantum entanglement could be the particular case where two cells are far apart in the multidimensional grid but are actually neighbors in the one-dimensional update sequence resulting in bits of information that are seemingly correlated over vast distances in space (from the perspective and scope of the large scale). In the case of an automaton on an infinite grid, there would need to be a mechanism that triggers the one-dimensional update sequence at a given interval of time. This would spawn an ever increasing number of parallel computations (yet another notion of parallel universes if the update sequence is non-deterministic or differs upon every instantiation). Whatever the mechanism, by definition, there would be an infinite number of cells whose neighbors only update in infinite-time. It would be impossible to ever see the current states of these cells similar to, say, dark matter.

5. All Universal Systems are Equivalent: Tilings and Aggregation

As suggested in the previous treatises, it seems likely to me that all universal systems are equivalent and can be reduced to one another. Such equivalences are ubiquitous in mathematics (e.g., how an *np*-complete problem is reducible to all others *np*-complete problems). While I have focused primarily on Metabiology and cellular automata, there is another universal system that warrants mention: tilings (specifically Wang-tilings). Through music, I have become fascinated by the history of the domino problem; the decidability problem that asks: given an arbitrary set of square tiles with colored edges, is there an algorithm that will determine if the set tiles the infinite plane such that adjacent squares share the same color on the abutting edges. Hao Wang conjectured that there was such an algorithm, but his student Robert Berger showed that the problem was undecidable by discovering a tile set that tiles the infinite plane strictly aperiodically. Such a construction should not exist if Wang conjectured correctly. Since Berger's proof, several other aperiodic tile sets have been discovered. The most obvious reason why tilings can be useful mathematical objects to explore is that they are toy models of how things "fit together" and aggregate (be it atomic structures, DNA, or higher order molecular constructs). Further, any Turing machine can be converted to a Wang tiling and Chaitin has also shown how Wang tiles can be used to represent the bits of Omega. Of particular interest is the work of Jarkko Kari, who has shown several problems involving tilings as undecidable. This is yet another example/model of incompleteness and infinite creativity in the evolution, aggregation, and construction of (toy, mathematical) objects. In particular Kari has defined a set of "directed Wang tiles" and shown that it is undecidable if an arbitrary set tiles and includes a "snake" (a path) that covers the infinite plane. Kari's paths are always adjacent tiles in the multidimensional tiling. It would be interesting to extend his work such that the snake can be disconnected. Intriguingly, Kari's directed tiles combine a multidimensional scale with a one-dimensional scale in a unified system.

6. Intuition and Inspiration: Further Considerations about Intelligent Design Versus Divine Intervention

The limited use of an Oracle in Metabiology and by extension Meta+phenomenology posits that there is no intelligent designer. However, it does suggest some form of divine intervention which accounts for how new information enters the system. I would suggest that this is the phenomenological equivalent to intuition: the acquisition of new information by an organism in an act of inspiration. A failure to distinguish between a god/intelligent designer and a limited divine intervention is the fundamental flaw of most organized religions wherein beliefs grow out of a need to explain that which is not understood (the seasons, the movement of celestial objects, etc.). But as knowledge is accumulated, the notion of a god gets replaced by an understanding of our world. As concluded in the original text, we are in the process of transcendence and becoming Oracles ourselves. In *Life as evolving software*, Chaitin gives several "open problems". Problems 5 and 6 relate to the rate of evolution: the former asks if there is a way to get a better upper bound on the rate of evolution (e.g., more akin to those that we see in the real-world) while the latter asks if there is a way to get a lower bound that is strictly slower than intelligent design. As Chaitin suggests in Problem 5, the range of potential evolutionary rates of cumulative random evolution should have a mean and variance. Perhaps it is be possible to prove that cumulative random evolution has the potential to result in a system that evolves at a rate equivalent to intelligent design, but that it is *extremely* unlikely. In the case that the mean evolutionary rate in Chaitin's model still reflects an evolutionary rate slower than we see in the real-world, it would be interesting to explore a notion of statistical feedback in the system by further weighting the probability of a mutation algorithm occurring such that the longer it has *not* occurred, the more likely it is *to* occur. This idea is borrowed from an algorithm frequently used by James Tenney in the compositional process to get a more uniform series of selections over a small number of trials (see *A few more Words about James Tenney: dissonant counterpoint and statistical feedback* by Larry Polansky, *et al.* for a detailed analysis of the procedure).

7. Intentionality and Feedback: A Beautiful Soup

By definition, the organisms in Metabiology contain their own history: evolution is cumulative. The current best solution to the Busy

Beaver Function is encoded in the organism itself in order to test the output of the next generation. Therefore, the organisms are always working with the best solution up to that point in the evolutionary process (the current state of the art and technology). This is a fundamental observation that I did not elaborate upon in the initial text. The creative act, in both the real-world and in Metabiology, is the invention. Of equal importance, is that the invention is demonstrated and shared such that the next generation is working with the knowledge accumulated up to that point. The analog in the real-world is as follows. An invention is made by a person, that person puts the invention in the public domain (i.e., publishes a paper, premieres a piece of music, etc.), the observers of those acts then process that information, which sometimes leads to new inventions. This explains an important aspect of phenomenology: intentionality. The act of disseminating knowledge into the public domain. This is elegantly encapsulated in Chaitin's model by the cumulative evolution of knowledge within the evolving organism! It can be seen as a feedback loop of sorts giving rise to more and more knowledge.

Calle Monclova 62
Mexico City, Mexico
April 2019

Chapter 16

Metabiology, Interdisciplinarity and the Human Self-image*

Virginia Maria Fontes Gonçalves Chaitin[†]

1. Introduction to Interdisciplinary Research Methodology

We discuss metabiology at conceptual and methodological levels in light of a permeable pluralistic epistemology (Chaitin, V.M.F.G., 2009). Permeable pluralistic epistemology studies interdisciplinary research and provides the criteria for evaluating whether or not it is fertile to combine different conceptual frameworks, vocabularies, research techniques and methods.[a]

This chapter deals with three such criteria: (a) the very important distinction between locally useful analogies and helpful metaphors in contrast with non-isomorphic migrations of entire conceptual neighborhoods that create a new vocabulary; (b) the ability of this new vocabulary to reformulate existing and propose new research challenges, specific problems and epistemic goals; and (c) the ability of

*Reprinted from *La Nuova Critica* **61–62** (2015) 115–124 with the kind permission of Prof. Arturo Carsetti.
[†]Federal University of Rio de Janeiro (UFRJ), Brazil; virginia.mfgc@gmail.com.
[a]Our epistemological approach was inspired by Paul K. Feyerabend's pluralistic epistemology, which however, is not permeable because it does not emphasize the important role of non-isomorphic concept migration.

specific techniques transported from individual disciplines within this interdisciplinary research setting to be applied to these new problems and goals, and deliver meaningful results for the ongoing effort.

Let us begin by discussing criteria (a) and (b). Conventional analogies and metaphors are widely used in colloquial and scientific communication but tend to have mostly local semantic applicability. We propose that within wide-ranging interdisciplinary studies, local conceptual analogies and metaphors are extended to neighboring concepts and as a result of this process cover significant conceptual neighborhoods in the different disciplines involved in a particular interdisciplinary research setting. These non-isomorphic relationships between conceptual neighborhoods that belong to different disciplines constitute a hybrid semantic territory in which the new interdisciplinary field of research can take shape, together with its initial guiding questions, problems and goals. It is within this amorphous hybrid semantic territory that new concepts may gradually emerge by non-isomorphic concept migration, generating new meanings and novel conceptual frameworks, all of which taken together add up to an entirely new vocabulary, especially tailored for expressing the particular interdisciplinary research context. The significance of this novel conceptual neighborhood is evidenced by its usefulness for formulating meaningful research questions and possible answers as well as new topics for research that could not even be expressed in the conceptual frameworks of previous individual disciplines.

We are now ready to discuss criterion (c). As this expanding non-isomorphic migration process creates a new interdisciplinary vocabulary and a new conceptual framework, it is also being affected by and influencing possible modes of utilization — and maybe even reshaping — of research techniques, methods and epistemic goals coming from the different disciplines. If this process of experimentation with alternate versions delivers meaningful results for the ongoing research, then it is epistemically fertile.

In other words, an epistemically fertile interdisciplinary area of study is one in which the original frameworks, research methods and epistemic goals of individual disciplines are combined and recreated, yielding novel and unexpected prospects for knowledge and understanding. This is where interdisciplinary research really proves its worth.

In Section 3, we shall use this permeable pluralistic epistemology to discuss a crucial moment in the development of metabiology. But first, in Section 2, we discuss metabiology at a conceptual level in the context of Deleuzian post-modern philosophical thought and in light of permeable pluralistic epistemology. As we have seen, permeable pluralistic epistemology addresses interdisciplinary research and provides it specific methodological support by proposing epistemic criteria for analyzing such studies. These criteria involved the presence of non-isomorphic conceptual migrations, transportation of methods and techniques across disciplinary boundaries, and obtaining significant new results, all of which occur in the case of metabiology.

2. Conceptual Overview of Metabiology and Its Possible Impact on the Human Self-image

Now, we shall use permeable pluralistic epistemology to appraise metabiology. Based on this analysis, we shall argue that metabiology is a paradigm-shifting interdisciplinary research area that successfully combines methods, techniques and ideas from the following fields: algorithmic information theory, computability theory, metamathematics and evolutionary biology. We shall argue that the new paradigms proposed by metabiology cover and affect a broad range of academic disciplines and that our interpretation of its conceptual framework includes the reframing of the general philosophical question of what is life and what is it that makes us human.

In order to make our paper reasonably self-contained, we begin by reviewing the interdisciplinary aspects of this new field, then give a brief overview of the mathematics and computer science involved in metabiology, and finally propose a theoretical basis for studying its impact on the contemporary human self-image.

Combining algorithmic information theory, computability theory, metamathematics and evolutionary biology, metabiology starts off with the basic analogy between DNA and programming languages and then proposes considering life to be randomly mutating, evolving software (Chaitin, G., 2012, 2013). In effect, in metabiology, Nature is "programming" its organisms through random mutations subjected to an open-ended selective criterion.

This unfolds into a series of analogies and metaphors which embody and provoke an extensive renovation of the conceptual framework used to perceive evolutionary biological processes, shifting the core conceptions of these processes from metabolism to algorithm, replacing energy with algorithmic information, and proposing creativity measured in bits of algorithmic information rather than adaptation and preserving successful genes as a criterion for evolution.

We emphasize that metabiological evolution is not about adaptation or passing one's genes to the next generation but about coming up with new genes that contain new algorithmic information content.

To operate within this conceptual framework, metabiology replaces continuous pre-Gödelian math (used in population genetics and systems biology) by discrete post-Gödelian math. It is not surprising that the mathematical techniques used to derive results in this field have to be in tune with its conceptual framework. Furthermore, metabiology has an entirely different epistemic goal as compared to other mathematical efforts involving evolutionary biology: it goes from explaining the response of biological systems to selective pressure, to explaining biological creativity. This change in epistemic goal carries with it a change of emphasis from evolutionary simulations to mathematical proofs, giving an entirely different epistemic status to the interplay between mathematics and biology. In fact, the goal of metabiology is to prove mathematically that there are life forms — in this case, algorithmic life forms — that evolve due to natural selection filtering random mutations and to simultaneously explain biological creativity and where new genes come from.

However, explaining the emergence of novelty requires a fundamentally new kind of mathematics capable of dealing with creativity. Metabiology achieves this by adopting a non-reductionist view of mathematics and by taking advantage of the power of post-Gödelian and post-Turing mathematical concepts (Chaitin, G., 1995). Furthermore, since the metabiological evolutionary process applies randomly chosen algorithmic mutations to algorithmic organisms, Turing oracles are required in order to avoid the pitfall of non-halting algorithms, which is an uncomputable step (Turing, 1936, 1939). The oracle solves a technical problem: it prevents metabiological evolution from getting stuck when non-halting algorithms are encountered.

Notwithstanding the importance of this technical role, the oracle also has the more fundamental epistemic role of bringing new information from outside the system that generated the current life form, the current algorithmic organism. The new information coming from the oracle cannot be deduced or calculated from the information already in this system and consists of deciding whether the randomly chosen algorithmic mutation halts, whether the mutated organism halts, and finally whether the mutated organism is fitter than the previous one. The fitness criterion is that an organism is fitter than another if it calculates a bigger number, and jumps in fitness are measured by using the Busy Beaver function. One of the key theorems in metabiology is that, in the long run, the metabiological evolutionary process is accumulating conceptual complexity given by information content, that metabiological creativity is taking place and can actually be measured in bits of algorithmic information. Furthermore, creativity is unending and unbounded.

From this brief description, one can infer that the role of the Turing oracle in metabiology is analogous to the role played by the environment in biology. The following table shows some other analogies between metabiology and the neo-Darwinian new synthesis, which also involve algorithmic information theory, computability theory and metamathematics:

Biology (neo-Darwinian new synthesis)	Metabiology
Natural software: DNA-RNA (base 4: ACGT)	Metabiological software: computer programs (base 2: 0,1)
Genes — complete genome	Subroutines — complete program
Organism results from biological processes involving DNA-RNA	Metabiological organism is the software itself
Metabolism — energy	Algorithm — information
Biological creativity: random mutations that generate new genes — diversity of life forms in the biosphere	Metabiological creativity: random mutations that generate new algorithms — diversity of all possible programs in a software space
Darwinian challenge: competition, survival of the fittest, adaptation to the changing environment, plus passing your genes to the next generation (selfish genes)	Metabiological challenge: solve a mathematical problem that requires a creative, uncomputable step
Empirical theory of evolution	Theorems that prove that metabiological evolution takes place

However, this chart does not tell the whole story; there are additional new meanings that emerge from these analogies within the new metabiological conceptual framework due to the fertility of these conceptual analogies. The fact that this occurs is a distinctive feature of the non-isomorphic processes that results in effective interdisciplinary research. Although metabiology was inspired by Darwinian evolution, in this section, we are mostly interested in the non-Darwinian aspects of metabiology, which move away or escape from the neo-Darwinian conceptual framework in a Deleuzian line of flight (Deleuze and Guattari, 1987) toward a metabiological conception of life as an intrinsically creative process.

We perceive this line of flight in the following terms. The challenge for Darwinian life is competition, survival of the fittest, adaptation to the changing environment, and passing your genes to the next generation (selfish genes). The challenge for metabiological life is to solve a mathematical problem that requires a creative, uncomputable step. Note that there is no competition, no desire for perpetuation. In the metabiological evolutionary process, there is no clear analogy for reproduction; instead, the life form represented by the mutating algorithmic organism is constantly searching for and incorporating new information.

Populated by algorithmic organisms, metabiology seems to be deprived of any corporeality. However, the mathematical proof of the existence of a software life form primarily driven by creativity gives rise to a vigorous metaphoric potential that translates directly to the human being in all his dimensions from our most abstract ideas about ourselves to the most concrete, organic elements of our bodies, because it forces us to reevaluate the neo-Darwinist dogma that we are creatures of egotistical competition for survival.

As the moral philosopher Mary Midgeley (2010) argues, when viewed as mere biological support for neo-Darwinian selfish genes, the human being turns into an impoverished "solitary self." However, based on the new metabiological perspective combining randomness, uncomputability and Turing oracles to generate novelty, we propose providing the metabiological life form with a human body or, conversely, providing a metabiological kind of life to the human body, inaugurating in this manner an unexpected *metabiological body.* This gives rise to a brand-new and more generous conceptual framework for the human being, which now evolves around the idea of a life form motivated by a non-mechanical, lawless, subjective creativity instead of a life form driven by a pre-determined "winner or loser" survival dichotomy.

Based on these considerations, we are currently continuing our work on the conceptual basis for a contemporary reworking of the human self-image by exploring possible analogies between the *metabiological body* and Deleuze and Guattari's *body without organs.*

This is a rather surprising unfolding of metabiology, showing how non-isomorphic conceptual migrations can be powerful in combining different disciplines. Furthermore, metabiology can be seen as a *meta-theory* because in attempting to explain biological creativity, it does not directly address the empirical phenomenon of biological creativity *per se* but does so at a meta level by mathematically expressing principles that engender creativity involving randomness and uncomputability.

3. A Crucial Episode in the Development of Metabiology: The Story of Algorithmic Mutations

We conclude our chapter by giving an example of the usefulness of our interdisciplinary research methodology.

As we have seen, the goal of metabiology is to prove mathematically that there are life forms — in this case, algorithmic life forms — that evolve due to natural selection filtering random mutations and to simultaneously explain biological creativity and where new genes come from. There were two stages in the achievement of this goal.

The first stage consisted of a random walk in software space with localized mutations and with the fitness of the organisms determined by their ability to solve the challenge posed the Busy Beaver Problem, that is to say, to calculate a very large number (Chaitin, G., 2009, 2011). This early version of metabiology was rather awkward because proofs were distressingly complicated, requiring unnatural auxiliary assumptions.

In the second stage in the development of metabiology (Chaitin, G., 2012, 2013), the localized mutations were replaced by global algorithmic mutations. As a result, proofs became simple and straightforward. However, this seemed to further distance metabiology from biology, since current molecular biology deals mostly with localized mutations.

At this point, the above-mentioned criterion (c) for evaluating the fertility of interdisciplinary efforts came into play. Metabiology uses mathematical proofs. In pure mathematics, elegance is an important aesthetic criterion for truth. A mathematician would say: "If it's ugly, it can't be right!"

Consequently, as remarked in (Chaitin, G., 2012), if algorithmic mutations yield a more elegant version of metabiology, and if metabiology is relevant as a meta-theory for biology, then we must ask: Do algorithmic mutations play a role in molecular biology? Can algorithmic mutations shed light on problematic biological phenomena such as missing intermediate forms, the Cambrian explosion or the major transitions in evolution (Maynard Smith and Szathmáry, 1995)? Can viruses give rise to algorithmic mutations?

This insight was later confirmed at a September 2014 meeting at Tegernsee, Baviera, on theoretical biology organized by Eörs Szathmáry and the Parmenides Foundation. At this meeting, algorithmic mutations were not rejected out of hand. Instead, they were considered intriguing and perhaps an interesting topic for further research.

This episode in the development of a new field of theoretical biology, metabiology, shows that a permeable pluralistic interdisciplinary methodology that discusses non-isomorphic concept migrations and proposes criteria for evaluating their fertility may indeed be of value to practicing scientists.

Bibliography

Chaitin, G. J., Randomness in arithmetic and the decline and fall of reductionism in pure mathematics. In: Cornwell, J., *Nature's Imagination.* England: Oxford University Press, 1995, pp. 27–44.

Chaitin, G. J., Evolution of mutating software, *Bulletin of the European Association for Theoretical Computer Science*, vol. 97, pp. 157–164, 2009.

Chaitin, G. J., Metaphysics, metamathematics and metabiology. In: Zenil, H., *Randomness Through Computation.* Singapore: World Scientific, pp. 93–103, 2011.

Chaitin, G. J., *Proving Darwin: Making Biology Mathematical.* New York: Pantheon, 2012.

Chaitin, G. J., Life as evolving software. In: Zenil, H., *A Computable Universe.* Singapore: World Scientific, pp. 277–302, 2013.

Chaitin, V. M. F. G., *Mimetic Conceptual Networks in the History of Ideas: A Proposal for a Pluralistic Epistemology.* Ph.D. Thesis, Federal University of Rio de Janeiro (UFRJ), 2009.

Deleuze, G. and Guattari, F., *A Thousand Plateaus: Capitalism and Schizophrenia.* Minneapolis and London: University of Minnesota Press, 1987.

Maynard Smith, J. and Szathmáry, E., *The Major Transitions in Evolution.* New York: Oxford University Press, 1995.

Midgeley, M., *The Solitary Self.* Durham: Acumen, 2010.

Turing, A., On computable numbers, with an application to the Entscheidungsproblem, *Proceedings of the London Mathematical Society*, (2), vol. 42, pp. 230–265, 1936.

Turing, A., Systems of logic based on ordinals, *Proceedings of the London Mathematical Society*, (2), vol. 45, pp. 161–228, 1939.

Chapter 17

Incompleteness and Complexity in Economic Theory

J. Barkley Rosser, Jr.*

1. Introduction: Chaitin's Way

Kurt Gödel (1931) showed that the formalist program of David Hilbert was doomed by demonstrating that within a consistent formal system that a Peano axiomatized arithmetic can be carried out is incomplete. It will be able to generate statements that can neither be proved or disproved. Likewise, for such a system its consistency cannot be proved within the system itself. Central to the incompleteness theorem was the use of paradox based on self-referencing.[a] "This statement is unprovable." A great irony of the theorem is that Gödel used the system developed by Whitehead and Russell (1910–13) to avoid various logical paradoxes to generate his own paradox, thus hammering home the failure of the formalist program.

This result soon inspired further important developments that would provide the foundation of computer science in the study of

*James Madison University, Harrisonburg, VA, USA; rosserjb@jmu.edu.

[a]For a more extending the theorem to a more generalized view of consistency, see Rosser (1936) in which the key is to show that for any sentence provable in the system a shorter proof of its negation can be found, this known as "Rosser's trick" and also "the Rosser sentence." See Kleene (1967) who provides a Turing machine-based proof of Rosser's version of the incompleteness theorem.

recursive functions (Church, 1936) Especially important would be Alan Turing's conceptualization of a generalized form of computing device that could solve (or not solve) such functions, which would come to bear his name, the Turing machine (Turing, 1936). Indeed, realizing the possibility of not being able to prove a statement arising from a logical system, Turing extended this to the idea of not being able to compute a solution. The self-referencing of the Liar's Paradox ("All Cretans are liars, and I am a Cretan.") sets up the possibility for an infinite do-loop that never ends in a program as one goes back and forth between being a Cretan and not being a Cretan. A program is not computable because it never halts. This would become known as the *halting problem*, and it would be the key to the idea of computational complexity.

While non-computability is clearly the extreme case of computational complexity, the question of defining such complexity and measuring its degree arose out of the work of Turing. Deriving from this case the idea that the length of a program could be the measure of such complexity gained prominence (Kolmogorov, 1965). It was into this discussion and study that Gregory Chaitin (1966) first became involved in these matters and began to influence the nature of the path these ideas followed. Inspired by quantum mechanics and recognizing that one can never reach infinity in a program, Chaitin discovered the number Ω, which is the probability of halting in a program. In the spirit of Gödel and Turing, while this number is definable, it is not itself computable. Study of this number and its implication that randomness is integral to the foundations of mathematics and computer science led to Chaitin's (1987) development of highly influential *algorithmic information theory.*

Chaitin has argued in various places that his randomization of mathematical logic provides a new way of relating underlying axioms to "mathematical facts," which are never fully to be known. "Randomness = incompressibility" (Chaitin, 2002, p. 37). While the traditional notion is that a small set of finite axioms generates mathematical truths, algorithmic information theory implies going the other way to compress mathematical facts into a small set of axioms.

With unproven assertions or programs that run for a very long time, estimating a probability of provability allows one to reach this irreducible relation between the two sides, a relation that at some level turns the mathematical facts into axioms.

With influence pervasive on computer science as well as the foundations of mathematics regarding the nature of undecidability, Chaitin himself would branch out and expand his applications of these ideas to a broader array of disciplines. These would include philosophy and the arts (Chaitin, 2002), generalized complexity (Chaitin, 2008), evolutionary biology (Chaitin, 2012) as well as physics and economics (Chaitin *et al.*, 2012). It can be argued that the strands of thought pulsing through these meditations have laid out a distinctive path of thought, a veritable *Chaitin's Way.* This essay seeks to explore aspects of this multi-disciplinary, or even transdisciplinary, way that manifest themselves more broadly within complexity and economics. These include such matters as incompleteness and general equilibrium theory, how a constructivist mathematics changes economic theory, how this perspective links with complex reflexivity in economics, and how complex emergence arises within economics.

2. Incompleteness and the Limits of Economic Theory

There has long been a pattern of mathematical ideas and issues in economic theory appearing in game theory before they have in general equilibrium theory (GET), with the application of fixed-point theorems by Brouwer, initially done by von Neumann (1928. 1945) for both, followed by the Kakutani version initially by Nash (1951) for game theory to be followed by Arrow and Debreu (1954) for GET. So it has been the case with seriously considering problems arising from self-referencing and the incompleteness theorems.

Thus probably the first recognition of such limits were for game theory and came from Michael Rabin (1957), who found an uncomputability result that there are games in which the player who can always win cannot do so in practice because it is impossible to supply her with effective instructions regarding how to play in order to win. This resembles Chaitin's number Ω that is known to exist but cannot be computed. It can be approached by successive approximation, but not fully reached. It is already the case that even known game theory equilibria often involve mixed strategies that are already probabilistic, with this uncomputability result simply adding another layer of stochasticity on top.

Rabin's result would be extended later to more general cases by Prasad (1991) and Tsuji *et al.* (1998) who showed a more general uncomputability of Nash equilibria. This would be followed by Koppl and Rosser (2002) who showed how self-referencing leads to uncomputability of best response functions in game theory. Their arguments rely on Cantor-style diagonal arguments of the sort used by Turing (1936). Their work drew on Binmore (1987) whose response to this problem was to propose a sophisticated form of Bayesian updating to approach solutions that, ironically, assume greater ignorance by the agents involved. However, this only holds in finite spaces and when the basis is continuous (Diaconis and Freedman, 1986). The paradoxes underlying these non-computabilities involve infinite regresses of agents knowing the other agents' knowing of their own knowing of the other agents' knowing (*ad infinitum*) of each other's expected strategies. A famous example of this is the Holmes–Moriarty problem and has a mixed strategy solution. The problem is how to get to that solution, with it not being possible by trying to use best response functions.

Others began to understand that these problems can appear in a variety of areas of economics.[b] Peter Albin (1982) observed that self-referencing and associated incompleteness with uncomputability arise in many problems where a micro-decision solution depends on a macro-result while at the same time the macro-result depends on the micro-decision solution. It can be thought that this might be

[b]Some more general discussions of limits on economics due to self-referencing include Winrich (1984), Bookstaber (2017), and Landini *et al.* (2019).

solved by a giant handwaving assertion of general equilibrium, but in fact this is more a problem of aggregation, a problem that GET has seriously, as we shall see. A particular example Albin cites is the old problem of aggregation in capital theory that became entangled in the paradoxes of capital theory debates. In any case, the problem as posed by Joan Robinson (1953–54) was that on the one hand in order to aggregate capital one needs to know what the marginal product of capital is so as to aid in determining the discount rate for calculating present values, while at the same time one needs to know what the value of aggregate capital is before one can determine its marginal product. One is potentially stuck in an endless do-loop like Holmes and Moriarty trying to outguess each other as to whether to get off the train from Victoria Station at Canterbury or Dover.

Which brings us to GET itself. Probably the first to raise issues based on self-referencing, incompleteness, and uncomputability emphasizing particularly Turing's work was Alain Lewis (1986, 1992). His work relates to complexity in a curious way in that it reportedly heavily influenced Kenneth Arrow, playing a role in his advocacy of and support for establishing the Santa Fe Institute and getting economists involved with it from the start, which might not have happened otherwise as its main founders and inspiration came from physicists associated with the nearby Los Alamos nuclear physics laboratory. Lewis's work in turn prepared the way for a more general result on uncomputability of Walrasian general equilibria due to Richter and Wong (1999). Their general result also applies to such simplifications of GET as dynamic stochastic general equilibrium (DSGE) models now so widely used in macroeconomic modeling.

There are at least three problems of incompleteness with GET. One not directly tied to Gödel–Turing forms of incompleteness involves aggregation and is due to what is known as the SMD theorem (Sonnenschein, 1972; Mantel, 1974; Debreu, 1974). This theorem states that for a continuous, homogeneous of degree zero function, with Walras' Law holding, there exists an economy with at least as many agents as goods for which the function is an aggregate demand function with prices bounded away from zero. This says effectively that there are no restrictions on aggregate demand functions if no restrictions are based on individual preferences, meaning any aggregate demand function based on a microfoundation is highly arbitrary. Stability and other useful characteristics cannot be guaranteed, with, indeed, the theory as classically conceived incomplete as a way of understanding economic phenomena. To add insult to injury, Kirman and Koch (1986) have shown that the SMD theorem still holds even with identical individual preferences and identical shares of income for agents.

Unsurprisingly this is linked to incompleteness problems more clearly tied to the underlying Gödel theorems, very much in the spirit of Chaitin's algorithmic information theory. This involves the problem of moving from an axiomatic system that allows for proving existence of general equilibrium to establishing uniqueness or stability. Ingrao and Israel (1991) have explained how this problem presented itself to Debreu (1959). Let **FS** be the formal system presented by Debreu (1959) based on axiomatic system **AX**. This system can be arithmetized by a Gödelization procedure on each symbol, statement, and demonstration with natural numbers, thus allowing a direct application of the incompleteness theorem. When it becomes clear that certain outcomes such as stability cannot be established while relying on **AX**, then it can be expanded by adding new axioms, $\mathbf{A}^*$, giving a new axiom system $\mathbf{AX} + \mathbf{A}^*$, which in turn gives a new formal system, **FS***. However, now according to the second part of the Gödel theorem this new system cannot be shown to be consistent from within itself. There is no end to this, and we are left with a probabilistic outcome as argued by Chaitin. Debreu himself was aware to a considerable extent of these problems limiting GET and how hard it was to get uniqueness and stability for his system even when it could prove existence, although his concerns generally focused on local results.

To complete this section, we consider issues more tied directly to computational complexity and unsolved problems whose probability of proof is characterized by a Chaitin Ω that is not computable.

This involves a theorem on efficient markets by Maymin (2011) that involves skirting the boundaries of the still unsolved problem of the relation between P and NP complete problems. While most computer scientists and logicians believe that P does not equal NP, an argument first made by both Gödel and Nash independently and secretly in the 1950s, it remains one of the most important unproven hypotheses in computer science and thus subject to ongoing revision of its Chaitin probability as new efforts proceed.

The Maymin theorem posits that under certain conditions about information, markets are efficient if $P = NP$, which few believe is really the case. In a study that pushes the boundaries of our knowledge of these matters, da Costa and Doria (2016) consider the O'Donnell algorithm in connection with this theorem (O'Donnell, 1979). This algorithm is exponential and thus not P, although it involves an exponential function that is slowly growing and thus "almost P". This holds if $P < NP$ is provable for any theory strictly stronger than Primitive Recursive Arithmetic (PRA) while PRA itself cannot prove it. This is known as the *counterexample function to the* $P = NP$ *problem.* The sorts of economics problems that appear to involve NP solvability problems include such computationally complex ones as the traveling salesman problem. Anyway, da Costa and Doria establish that under these conditions where the O'Donnell algorithm behaves as an "almost P" system, this then implies an outcome of "almost efficient markets". This is a result that walks on the edge of the unknown, if not the unknowable.

3. Incompleteness and Constructivist Economics

Chaitin's irreducible randomness of provability that attempts to generate a balance between axioms and mathematical facts not only has implications for economic theory, but has implications for mathematics itself. Nearly, all current economic theory has been derived from a mathematics that accepts a standard, formalist, or "classic" set of axioms that fit the Hilbert approach and that is indeed used by most mathematicians themselves, this set of axioms having a virtue of providing easy ways of proving various theorems, even if many mathematicians retains doubts regarding the fundamental truth of them. This set of axioms is often labeled as being the Zermelo–Fraenkel-[Axiom of] Choice set, or ZFC. This set has long been viewed as easily consistent with the Peano arithmetic, and many mathematicians do not question this set and view worrying about their truth as an unfortunate distraction from getting at the work of proving theorems and developing new knowledge.

Nevertheless, there has been an ongoing debate within mathematics over this set of axioms that dates back into the 19th century. For a period in the mid-20th century these dissenters were pushed back to far corners, but starting in the 1960s a new group suggested that the alternative approaches should be taken seriously (Kleene and Vesley, 1965; Bishop, 1967). As it is these alternative approaches, which go broadly under the label *constructivist*, emphasize relying on using computation to solve problems rather than questionable formalistic axioms that allow for neat and easy solutions. Thus the increasing power of modern computers, something that Chaitin has celebrated and emphasized, have probably helped encourage the revival of this constructivist approach.

There are varying factions among mathematicians sympathetic to constructivism depending on which and how many of the standard ZFC set of axioms they question, but broadly there are two axioms and a principle that have been questioned by one or another of the constructivists. Probably the first axiom to be questioned starting in the late 19th century was the Axiom of Infinity. The fundamental argument is that infinity simply does not exist in the real world. It is the abstraction beyond abstractions, a complete fantasy. As it is, using sending variables asymptotically to infinity as a method in proving theorems is deeply embedded in large parts of classical mathematics. Of course such methods arguably remain in the world of the finite, and indeed what really upset the constructivist critics was the introduction by Cantor of levels of infinity, the "paradise" that Hilbert praised him for "bringing us [mathematicians] into." Dumping infinity makes proving things more difficult and simply ends substantial discussions in mathematics, but constructivist critics argue that this keeps mathematicians more real and credible, and forces them to lay out computationally how to establish mathematical facts.

The other axiom that has been questioned by many is the Axiom of Choice, which allows the relatively easy ordering of infinite sets. What is involved here may not be as profoundly metamathematically deep as the question of infinity, but in fact getting rid of the Axiom of Choice may be more damaging to classical mathematics and generate even more problems for proving various theorems. It may be less dramatic than the Axiom of Infinity, but the Axiom of Choice is probably a harder workhorse of the classical axiomatic structure, especially important in topology and parts of real analysis, even as its role remains less obvious and visible.[c]

Finally, we come to a deep principle that dates back at least to Aristotle, if not earlier, and which also involves a deep perspective on the nature of truth. This is the *Law of the Excluded Middle.* Statements must be either true or false. It may seem that such approaches as Boolean algebra and fuzzy set theory, which allow for probabilistic statements, may look like they get around this sharp duality of truth versus falsehood. But in fact they do not do so as the very statement that "there is a 10 percent chance of a thunderstorm tomorrow" is itself arguably subject to this same harsh truth versus falsehood duality. Obviously this is a very philosophically profound position, and for classical mathematics it allows for the use of proof by contradiction, which has often been used in many of the most famous and neat theorems of mathematics, ranging from the original proof that the square root of two is not a rational number up to Cantor's proof that there is no highest level of infinity and more.

---

[c]On the one hand, Specker (1953) has shown that the ultimate Axiom of Choice that allows for the well ordering of all infinite sets is false, a sort of ultimate victory of this branch of constructivism. However, non-standard analysis introduced by Abraham Robinson (1966) allows for infinite (and infinitesimal) real numbers, which allows for not needing the Axiom of Choice nearly as much for proving well-known theorems.

The particular branch of constructivist mathematics that emphasizes not accepting the Law of the Excluded Middle is known as *intuitionism*, and may be the most radical of the constructivist schools. The founder and most famous advocate of intuitionism was Luitzen Brouwer (1908), who also proved the first and most famous fixed-point theorem. Ironically, Brouwer used non-intuitionist methods in that proof, with him only much later providing an alternative proof consistent with intuitionism (Brouwer, 1952). Brouwer essentially became the leader of the broader constructivist group in debates with Hilbert and others defending the formalist classical approach in the 1920s and 1930s, when it was widely perceived that the latter managed to get the upper hand (even as Gödel more profoundly undercut all of them in 1931). In any case, intuitionism has seen a revival (Bishop, 1967; Kleene, 1967), and its defenders do so on grounds of ultimate philosophical and ontological principles.

Until quite recently these matters have rarely been even mentioned by economists with the occasional exception (Scarf, 1973). But indeed concern over these matters has become a focus of research among some economists, especially among those concerned with issues of computability of the sort Chaitin has long been concerned with (Velupillai, 2005; Bartholo *et al.*, 2009; Rosser, 2012a). Unsurprisingly much of this has taken a critical turn by pointing out the non-constructivist elements in standard proofs of widely accepted theorems in economic theory. Thus Velupillai (2006, 2008) has pointed out how most proofs of the existence of general equilibrium rely upon fixed-point theorems. However, both the Brouwer and Kakutani versions of this have traditionally relied upon using Sperner's Lemma, which in turn depends on the Bolzano–Weierstrass theorem which in turn was proven using proof by contradiction.

A common problem arising related to these concerns in computational economic models used for policy purposes has involved estimating approximate solutions using only the countably infinite set of rational numbers in programs. That this might lead to problems has been observed by Clower and Hewitt (1978) who posed a peculiar finance function for which rational number solutions are distinctly different from solutions for nearby irrational numbers. Velupillai (2005, p. 186) has noted that for such functions roundoff errors may lead to wildly divergent results that may have real world implications, such as the serious misfiring of a Patriot missile during the First Gulf War in Dhahran, Saudi Arabia that missed its target by 700 m, thereby killing 28 soldiers in "friendly fire".

Some have argued that actual economic variables only take rational values so that therefore there should be no problem regarding this, Clower and Hewitt's peculiar finance function aside. However, it would seem that at least some economic variables may take irrational values, with examples from real estate appearing, such as the diagonal length of a square piece of land. This sort of example may still not be too difficult to deal with ultimately given that polynomial roots are part of the set of algebraic numbers, which are also a countably infinite set like that of the rational numbers. One can still operate with a countably infinite set of numbers if all one adds to the set of algebraic numbers is a small finite set of the transcendental numbers with *pi* and *e* being the obvious candidates.

There have been efforts made to overcome the problems of computing with real numbers (Blum *et al.*, 1998). However, these efforts have generally not been accepted by the advocates of a constructivist approach to building up economic theory. It must be noted that while most of the literature in this area has taken a critical approach, there have been some efforts made attempting to develop just such a mathematical economics based on a fully constructivist approach (Velupillai, 2012).

4. Reflexivity and Incompleteness

Closely related to self-referencing is the idea of *reflexivity*. This is a term with no agreed upon definition, and it has been used in a wide variety of ways (Lynch, 2000). It is derived from the Latin *reflectere*, which is usually translated to mean "bending back," but can refer to "reflex" as in a knee jerking when tapped, not what is meant here, or more generally is linked to "reflection" as in an image being reflected, possibly back and forth many times as in the situation of two mirrors facing each other. This latter is more what the focus is here and more the type that is connected with self-referencing and all that implies. Someone who made that link strongly was Douglas Hofstadter (1979) in his *Gödel, Escher, Bach: An Eternal Golden Braid* as well as even more so in later works (Hofstadter, 2006). For Hofstadter, reflexivity is linked to the foundations of consciousness through what he calls "strange loops" of indirect self-referencing, which he sees certain prints by Maurits C. Escher as highlighting, particularly his "Drawing Hands" and also his "Print Gallery," with many commentators on reflexivity citing "Drawing Hands," which shows two hands drawing each other (Rosser, 2019).[d] Hofstadter argues that the foundation for his theory is the Incompleteness Theorem of Gödel, with its deep self-referencing, along with certain pieces by J.S. Bach, as well as these prints by Escher.

Chaitin himself appears not to have used this term in any of his discussions as near as I can tell. But others have used it in discussing some of his work, with the meaning clearly related to the paradoxes that can arise from self-referencing such as the Cretan Liar's Paradox, an example cited by many who have written about reflexivity. Thus Mlynarski (2018, p. 1) argues "...that a formal system contains a sentence, which introduces reflexivity, does not imply that the same system does not contain a sentence or proof procedure which solves the problem." This paper applies this to diagonal proofs such as that showing that the real numbers are not countably infinite and asserts that this shows a limit to the "Chaitin–Gödel theorem" (Chaitin, 1974). So Chaitin may not have used the term (or at least not very

[d]Examples of reflexivity in art are often thought to involve the *Droste Effect*, in which a work contains an image of itself within itself, clearly a matter of self-referencing. Among the earliest known examples is a painting by Giotto from 1320, *The Stefaneschi Triptych*, in which in the central panel Cardinal Stefaneschi is depicted kneeling before Saint Peter and presenting to him the triptych itself. Needless to say, even if they cease to be depicted after a finite sequence of images, such artworks exhibiting this Droste Effect imply an infinite regress of ever smaller images containing ever smaller images (Rosser, 2019).

much) in connection with these sorts of limits-to-mathematics discussions, but others are doing so in connection with his arguments about these matters.

The term has probably been most widely used, and with the greatest variety of meanings, in sociology (Lynch, 2000)). Its academic usage was initiated by prominent sociologist, Robert K. Merton (1938), who used it to pose the problem of sociologists thinking about how their studies and ruminations fit into the broader social framework, both in how they themselves are influenced by that framework in terms of biases and paradigms, but also in terms of how their studies and how they do their studies might reflect back to influence society as well. Among the sociologists the most radical uses of the concept involved sharp self-criticism wherein one deconstructs the paradigm and influences one is operating in to the point that one can barely do any analysis at all (Woolgar, 1991), with many complaining that this leads to a nihilistic dead end. The earliest usages of the term by economists followed this particular strand of analyzing how particular economists are operating within certain methodological frameworks and how they came to do so from broader societal influences and how their work may then reflect back to influence society, sometimes even through specific policies or even ways of gathering and reporting policy-relevant data (Hands, 2001; Davis and Klaes, 2003).

Merton (1948) would also use the idea to propose the idea of the *self-fulfilling prophecy*, an idea that has been widely applied in economics as with the concept of sunspot equilibria (Azariadis, 1981), with many seeing this as deriving originally from Keynes (1936, Chap. 12) and his analysis of financial market behavior based on the early 20th century British newspaper beauty contests. In those contests newspapers would publish photos of young women and ask readers to rate them on their presumed beauty. The winner of such a contest was not the person who guessed which young woman was objectively the most beautiful, but rather which one received the most votes. This meant that a shrewd player of such a game was really trying to guess the guesses of the other players, with Keynes comparing this to financial markets where the underlying fundamental of an asset is less important for its market value than what investors think it is. This led Keynes even to note that this kind of reasoning can move to higher levels, trying to think what others think others think, and on to still higher levels in a potential infinite regress, a classic infinite reflection in a non-halting program. This beauty contest idea of Keynes has come to be viewed as a centerpiece of his philosophical view, implying ultimately not only reflexivity but complexity as well (Davis, 2016).

Among the first to pick up on Keynes's argument and apply it to self-fulfilling prophecies in financial markets and also bringing in reflexivity as relevant to this was George Soros (1987), who would later also argue that the analysis was part of complexity economics (Soros, 2013). Soros has long argued that thinking about this beauty contest-inspired version of reflexivity has been key to his own decision-making in financial markets. He sees it as explaining boom and bust cycles in markets as in the US housing bubble of the early 2000s, whose decline set off the Great Recession. He first got the term from being a student of Karl Popper's in the 1950s (Popper, 1959), with Popper also an influence on Hayek (1967) in connection with these ideas (Caldwell, 2013). Thus the idea of reflexivity with links to arguments about incompleteness and infinite regresses associated with self-referencing have become highly influential among economists and financiers studying financial market dynamics and other related phenomena.

5. Incompleteness, Complexity, and Emergence

The concept of *emergence* has long been viewed as a central phenomenon in complex systems (Rosser, 2010, 2012b). Chaitin has at times written about it (Chaitin, 2008) usually stressing its role in such matters as creativity and the appearance of unexpected outcomes, with this aspect stressed also by Moore (1990). It is also deeply tied to elements of algorithmic information theory, particularly the matter of levels of computational complexity, with it associated with a transformational function that changes one state to another that is more computationally complex as defined by Chaitin (1987), Kolmogorov (1965) and Lee (2005). Discussing Chaitin's ideas, Requardt (1991, p. 185) argues that emergence involves a "suitable extension of 'algorithmic complexity' to possibly non-recursive... structures in general, [and to] ...possible existence of levels of complexity beyond, e.g., the human brain and the epistemological consequences thereof."

Some argue that the idea of emergence goes all the way back to Aristotle's discussion of situations where "the whole is greater than the sum of its parts," although this has also been identified as being the core concept of *complexity* as opposed to mere "complicatedness" (Israel, 2005). However, many argue that the modern father of the emergence idea was John Stuart Mill (1843) who labeled it as an example of *heteroclinic laws*. For Mill these involved something more along Hegelian lines, a qualitative change arising from some interaction. His original example involved chemistry, in particular how when one combines methane with oxygen one gets out carbon dioxide and water, two qualitatively different things. Certainly emergence involves this, although later many would add in the element of hierarchy, moving from one level to a presumably higher level, somehow defined, as well as a qualitative change.

Following Mill's influence, Lewes (1875) coined the term "emergence" for these heteroclinic laws, and this led to a broad British intellectual movement known as *Emergentism* that emphasized essentially holistic approaches to a variety of subjects, including even theology. Probably the most important strand in this movement involved evolutionary theory, reaching a peak of development and influence in the 1920s with the work of Morgan (1923), whose great concern was explaining how more complex forms of life evolved out of simpler ones, how single-celled organisms developed into multi-celled ones with organs, how eventually consciousness would emerge, and so on. This program fell out of favor in the 1930s as the neo-Darwinian synthesis took hold that combined probability theory with genetics to emphasize a reductionist emphasis on individual genes (Gould, 2002).

However, these concerns would re-emerge later as these problems of qualitative change and leaps in hierarchical levels cried out for answering. Herbert Simon (1962) used such evolutionary emergence as a central example in his development of the idea of *hierarchical complexity* with Kaufmann (1993), Crutchfield (1994), and Bak (1996) tying this directly to computational processes. Boulding (1978) labeled these emergent hierarchical leaps to be forms of *anagenesis*, while Eigen and Schuster (1979) coined the *hypercycle* as the key to the emergence of multi-celled organisms out of single celled ones. Rosser (1992) would identify such transformations as being *hypercyclic morphogenesis.* Rosser *et al.* (1994) identified the moment of such a hierarchical leap as being the *anagenetic moment.*

Chaitin (2012) has also entered these waters with his *Proving Darwin.* Arguably he stays away from talking about emergence or even specifically about the specific nature of hierarchical or qualitative leaps, although they are implicit in his analysis. Rather he draws on the more reductionist neo-Darwinian synthesis approach, however placing it into the context of algorithmic information theory, unsurprisingly. It is a natural application, given the stochastic nature of mutations. He presents his famous number, Ω, as "the organism that evolves" (Chaitin, 2012, p. 65). He poses this as coming from *metabiology*, which emphasizes evolution as a computational process of increasing fitness in response to the random shocks of mutations. He invokes the *Busy Beaver problem* of Rado (1962), which involves naming numbers by first adding them then multiplying them then raising them to powers and to powers of powers, with this process leading to fitness that "increases faster than any computable function" (Chaitin, 2012, p. 42). While again he does not specifically highlight qualitative leaps, they are implicit in the jumps such a Busy Beaver function makes as it goes from one arithmetic operation to another, and he discusses such problems as Darwin's own concern about how "half an eye is useless". Clearly this is all very much up Chaitin's alley.

Let us return then to the idea of levels of hierarchy of computational complexity, with many proposing various systems for this. A common one involves four levels: linear programs, polynomial (P) ones, non-polynomial (NP) ones that still halt (although if one is running one must especially confront Chaitin's Ω), and then the fully uncomputable ones that do not halt, the highest level of computational complexity. This has led many to identify other four level systems of complexity with this one, even if arguably this involves stretching things somewhat. Thus Mirowski (2007) introduced the idea of *markomata*, that markets are evolving forms of algorithms. He sees simple spot markets leading to the emergence of futures markets, which lead to the emergence of options markets, which in turn lead to higher level derivatives markets. He compares this four-level hierarchy to Chomsky's (1959) four-level hierarchy of grammar: from regular language to context-free language to context-sensitive language to recursive enumerable language (that is, a Turing machine). He also sees it as comparable to Wolfram's (1984) four-level hierarchy: from systems converging on a homogeneous state to ones that evolve towards simple separated or periodic structures to ones that evolve towards chaotic dynamical structures to ones that evolve to complex localized structures. Again, in all of these hierarchical systems, moving from one level to another is a matter of complex emergence.

The link of all this to Gödel's Incompleteness theorem goes back to the matter of how one resolves the appearance of incompleteness in a formal system. This involves moving to a higher level system with more axioms, arguably an emergent leap. But as noted above, one then faces the problem of determining whether or not this system is consistent. There is no end to all this.

6. Conclusion: Chaitin's Transdisciplinary Way

It should be quite clear by now that Chaitin's Way is one that follows a transdisciplinary path despite his grounding in mathematics.

Incompleteness implies the halting problem implies uncomputability and a hierarchy of levels of computability. This leads to his random dance between axioms and mathematical facts. But the existence of these levels and possibility of moving up and down them leads to complex emergence, with this phenomenon appearing across many disciplines from physics to chemistry to evolution to economics. This is a path that itself is subject to Chaitin's own limit of its own uncomputable Ω. We do not know how or when it may halt, but the path is clearly one of ever-expanding knowledge and possibilities. May it proceed forward and further.

Bibliography

Albin, Peter S. 1982. "The Metalogic of Economic Predictions, Calculations and Propositions." *Mathematical Social Sciences* 3, 329–358.

Arrow, Kenneth J. and Gérard Debreu. 1954. "Existence of Equilibrium for a Competitive Economy." *Econometrica* 22, 265–290.

Azariadis. Costas. 1981. "Self-Fulfilling Prophecies." *Journal of Economic Theory* 25, 380–396.

Bak, Per. 1996. *How Nature Works: The Science of Self-Organized Criticality.* New York: Columbia University Press.

Bartholo, Robert S., Carlos A.N. Cosenza, Francisco A. Doria, and Carlos T.R. Lessa. 2009. "Can Economic Systems Be Seen as Computing Devices?" *Journal of Economic Behavior and Organization* 70, 72–80.

Binmore, Ken. 1987. "Modeling Rational Players I." *Economics and Philosophy* 3, 9–55.

Bishop, Errett A. 1967. *Foundations of Constructive Analysis.* New York: McGraw-Hill.

Blum, Lenore, Felipe Cucker, Michael Shub, and Steve Smale. 1998. *Complexity and Real Computation.* New York: Springer-Verlag.

Boulding, Kenneth E. 1978. *Ecodynamics: A New Theory of Social Evolution.* Beverly Hills: Sage.

Bookstaber, Richard. 2017. *The End of Theory: Financial Crises, the Failure of Economics, and the Sweep of Human Interaction.* Princeton: Princeton University Press.

Brouwer, Luitzen E.J. 1908. "De Onbetrouwbaarheid der Logische Principes." *Tijdschrif voor wijsbegeerte* 2, 152–158.

Brouwer, Luitzen E.J. 1952. "An Intuitionist Correction of the Fixed-Point Theorem on the Sphere." *Proceedings of the Royal Society London* 213, 1–2.

Caldwell, Bruce. 2013. "George Soros: Hayekian?" *Journal of Economic Methodology* 20, 350–356.

Chaitin, Gregory J. 1966. "On the Length of Programs for Computing Finite Binary Sequences." *Journal of the ACM* 13, 547–569.

Chaitin, Gregory J. 1974. "Information Theoretic Limitations of Formal Systems." *Journal of the ACM* 21, 403–424.

Chaitin, Gregory J. 1987. *Algorithmic Information Theory.* Cambridge, UK: Cambridge University Press.

Chaitin, Gregory J. 2002. *Conversations with a Mathematician: Math, Art, Science and The Limits of Reason.* London: Springer.

Chaitin, Gregory J. 2008. "Irreducible Complexity in Pure Mathematics." In Herbert Hrochavec and Alvin Pichler, eds. *Wittgenstein and Philosophy of Information: Proceedings of the 30th International Ludwig Wittgenstein-Symposium in Kirchberg 2007.* Berlin: de Gruyter, 261–272.

Chaitin, Gregory J. 2012. *Proving Darwin: Making Biology Mathematical.* New York: Pantheon Books.

Chaitin, Gregory J., Newton da Costa, and Francisco A. Doria. 2012. *Gödel's Way: Exploits into an Undecidable World.* Boca Raton: CRC Press.

Chomsky, Noam. 1959. "On Certain Formal Properties of Grammars." *Information and Control* 2, 11–54.

Church, Alonzo. 1936. "A Note on the Entscheidungsproblem." *Journal of Symbolic Logic* 1, 40–41, correction 101–102.

Clower, Robert and Peter W. Howitt. 1978. "The Transactions Theory of the Demand for Money: A Reconsideration." *Journal of Political Economy* 86, 449–465.

Crutchfield, James. 1994. "The Calculi of Emergence: Computation, Dynamics, and Induction." *Physica D* 75, 11–54.

da Costa, Newton C.A. and Francisco A. Doria. 2016. "On the O'Donnell Algorithm for *NP*-Complete Problems." *Review of Behavioral Economics* 3, 221–242.

Davis, John B. 2016. "The Continuing Relevance of Keynes's Philosophical Thinking: Reflexivity, Complexity, and Uncertainty." GREDEG Working Paper 2016-36, University of Nice Sophia Antipolis, http://www.gredeg.cnrs.fr/working-papers/GREDEG-WP-2016-36.pdf.

Davis, John B. and Matthias Klaes. 2003. "Reflexivity: Curse or Cure?" *Journal of Economic Methodology* 10, 329–352.

Debreu, Gérard. 1959. *Theory of Value: An Axiomatic Analysis of Economic Equilibrium.* New York: John Wiley & Sons.

Debreu, Gérard. 1974. "Excess-Demand Functions." *Journal of Mathematical Economics* 1, 15–21.

Diaconis, Persi and D. Freedman. 1986. "On the Consistency of Bayes Estimates." *Annals of Statistics* 14, 1–26.

Eigen, Manfred and Peter Schuster. 1979. *The Hypercycle: A Natural Principle of Self-Organization.* Berlin: Springer-Verlag.

Gödel, Kurt. 1931. "Über Formal Unentscheidbare Satz der Pincipia Mathematica und Verwandte Systeme I." *Monatshefte für Mathematik und Physik* 38, 173–198.

Gould, Stephen Jay. 2002. *The Structure of Evolutionary Theory.* Cambridge, MA: The Belknap Press of Harvard University Press.

Hands, D. Wade. 2001. *Reflection without Rules: Economic Methodology and Contemporary Science Theory.* Cambridge, UK: Cambridge University Press.

Hayek, Friedrich A. 1967. "The Theory of Complex Phenomena." In *Studies in Philosophy, Politics and Economics.* London: Routledge & Keegan Paul, 22–42.

Hofstadter, Douglas R. 1979. *Gödel, Escher, Bach: An Eternal Golden Braid.* New York: Basic Books.

Hofstadter, Douglas R. 2006. *I am a Strange Loop.* New York: Basic Books.

Ingrao, Bruno and Giorgio Israel. 1991. *The Invisible Hand: Economic Equilibrium in the History of Science.* Cambridge, MA: MIT Press.

Israel, Giorgio. 2005. "The Science of Complexity: Epistemological Problems and Perspectives." *Science in Context* 18, 1–31.

Kaufmann, Stuart A. 1993. *The Origins of Order: Self-Organization and Selection in Evolution.* Oxford: Oxford University Press.

Keynes, John Maynard. 1936. *The General Theory of Employment, Interest and Money.* London: Macmillan.

Kirman, Alan P. and Karl J. Koch. 1986. "Market Excess Demand in Exchange Economies with Identical Preferences and Collinear Endowments." *Review of Economic Studies* 53, 457–463.

Kleene, Steven C. 1967. *Mathematical Logic.* New York: John Wiley & Sons.

Kleene, Steven C. and Richard E. Vesley. 1965. *Foundations of Intuitionistic Mathematics.* Amsterdam: North-Holland.

Kolmogorov, Andrei N. 1965. "Combinatorial Foundations of Information Theory and the Calculus of Probabilities." *Russian Mathematical Surveys* 38(4), 29–40.

Koppl, Roger and J. Barkley Rosser, Jr. 2002. "All That I Have to Say Has Already Crossed Your Mind." *Metroeconomica* 53, 339–360.

Landini, Simone, Mauro Gallegati, and J. Barkley Rosser, Jr. 2019. "Consistency and Incompleteness in General Equilibrium Theory." *Journal of Evolutionary Economics*, doi: 10.1007/s00191-018-08580-6.

Lee, Cassey. 2005. "Emergence and Universal Computation." In K. Vela Velupillai, ed. *Computability, Complexity and Constructivity in Economic Analysis.* Oxford: Blackwell, 198–217.

Lewes, George Henry. 1875. *Problems of Life and Mind, First Series: The Foundations of a Creed, Vol. II.* Cambridge, UK: The Riverside Press.

Lewis, Alain A. 1985. "On Effectively Computable Realizations of Choice Functions." *Mathematical Social Sciences* 10, 43–80.

Lewis, Alain A. 1992. "On Turing Degrees of Walrasian Models and a General Impossibility Result in the Theory of Decision-Making." *Mathematical Social Sciences* 24, 143–171.

Lynch, Michael. 2000. "Against Reflexivity as an Academic Virtue and Source of Privileged Knowledge." *Theory,Culture, and Society* 17, 26–54.

Mantel, Rolf R. 1974. "On the Characterization of Aggregate Excess-Demand." *Journal of Economic Theory* 7, 348–353.

Maymin, Philip Z. 2011. "Markets are Efficient if and only if $P = NP$." *Algorithmic Finance* 1, 1–11.
Merton, Robert K. 1938. "Science and the Social Order." *Philosophy of Science* 5, 523–537.
Merton, Robert K. 1948. "The Self-Fulfilling Prophecy." *Antioch Review* 8, 183–210.
Mlynarski, Kajetan. 2018. "Reflexivity and the Diagonal Argument in Proofs of Limitative Theorems." https://arxiv.org/pdf/111.6954.pdf.
Morgan, C. Lloyd. 1923. *Emergent Evolution.* London: Wiliams and Norgate.
Mill, John Stuart. 1843. *A System of Logic: Ratiocinative and Inductive.* London: Longmans Green.
Mirowski, Philip. 2007. "Markets Come to Bits: Evolution, Computation, and Markomata in Economic Science." *Journal of Economic Behavior and Organization* 63, 200–242.
Moore, Christopher. 1990. "Undecidability and Unpredictability in Dynamical Systems." *Physical Review Letters* 64, 2354–2357.
Nash, John. 1951. "Non-Cooperative Games." *Annals of Mathematics* 54, 286–295.
Neumann, John von. 1928. "Zur Theorie der Gesellschaftspiele." *Matematische Annalen* 100, 296–320.
Neumann, John von. 1945. "A Model of General Equilibrium." *Review of Economic Studies* 13, 1–9.
O'Donnell, M. 1979. "A Programming Language Theorem that is Independent of Peano Arithmetic." *Proceedings of the 11th Annual ACM Symposium on the Theory of Computation*, 179–188.
Popper, Karl. 1959. *The Logic of Scientific Discovery.* London: Hutchinson Verlag von Julius Springer.
Prasad, Kislaya. 1991. "Computability and Randomness of Nash Equilibria in Infinite Games." *Journal of Mathematical Economics* 20, 429–442.
Rabin, Michael O. 1957. "Effective Computability of Winning Strategies." In M. Dresher, A.W. Tucker, and P. Wolfe, eds. *Annals of Mathematical Studies, No. 39, Contributions to the Theory of Games, vol. III.* Princeton: Princeton University Press, 147–157.
Rado, Tibor. 1962. "On Non-Computable Functions." *Bell System Technical Journal* 41, 877–884.
Requardt, M. 1991. "Gödel, Turing, Chaitin and the Question of Emergence as a Meta-Principle of Modern Physics: Some Arguments against Reductionism." *World Futures* 32, 185–195.
Richter, M.K. and K.C. Wong. 1999. "Non-Computability of Competitive Equilibrium." *Economic Theory* 14, 1–28.
Robinson, Abraham. 1966. *Non-Standard Analysis.* Amsterdam: North-Holland.
Robinson, Joan. 1953–54. "The Production Function and the Theory of Capital." *Review of Economic Studies* 21, 81–106.
Rosser, J. Barkley [Sr.]. 1936. "Extensions of Some Theorems of Gödel and Church." *Journal of Symbolic Logic* 1, 87–91.

Rosser, J. Barkley, Jr. 1992. "The Dialogue between the Ecologic and Economic Theories of Evolution." *Journal of Economic Behavior and Organization* 17, 195–215.

Rosser, J. Barkley, Jr. 2010. "Constructivist Logic and Emergent Evolution in Economic Complexity." In Stefano Zambelli, ed. *Computability, Constructive and Behavioural Economic Dynamics: Essays in Honour of Kumaraswamy (Vela) Velupillai*. London: Routledge, 184–197.

Rosser, J. Barkley, Jr. 2012a. "On the Foundations of Mathematical Economics." *New Mathematics and Natural Computation* 8, 53–72.

Rosser, J. Barkley, Jr. 2012b. "Emergence and Complexity in Austrian Economics." *Journal of Economic Behavior and Organization* 81, 122–128.

Rosser, J. Barkley, Jr. 2019. "Reflections on Reflexivity and Complexity." James Madison University, http://cob.jmu.edu/rosserjb.

Rosser, J. Barkley, Jr., Carl Folke, and Folke Günther, Heikki Isomäki, Charles Perrings, and Tõnu Puu. 1994. "Discontinuous Change in Multilevel Hierarchical Systems." *Systems Research* 11(3), 77–94.

Scarf, Herbert E. 1973. *The Computation of Economic Equilibria*. New Haven: Yale University Press.

Simon, Herbert A. 1962. "The Architecture of Complexity." *Proceedings of the American Philosophical Society* 106, 467–482.

Sonnenschein, Hugo. 1972. "Market Excess-Demand Functions." *Econometrica* 40, 549–563.

Soros, George. 1987. *The Alchemy of Finance*. Hoboken: Wiley & Sons.

Soros, George. 2013. "Fallibility, Reflexivity, and the Human Uncertainty Principle." *Journal of Economic Methodology* 20, 309–329.

Specker, E.P. 1953. "The Axiom of Choice in Quine's New Foundations for Mathematical Logic." *Proceedings of the National Academy of Sciences of the U.S.A.* 39, 972–975.

Tsuji, Marcelo, Newton C.A. da Costa, and Francisco Doria. 1998. "The Incompleteness Theories of Games." *Journal of Philosophical Logic* 27, 553–568.

Turing, Alan M. 1936. "On Computable Numbers with an Application to the Entscheidungsproblem." *Proceedings of the London Mathematical Society* 2, 230–265.

Velupillai, K. Vela. 2005, ed. *Computability, Complexity and Constructivity in Economic Analysis*. Oxford: Blackwell.

Velupillai, K. Vela. 2006. "Algorithmic Foundations of *Computable* General Equilibrium." *Applied Mathematics and Computation* 179, 360–369.

Vellupillai, K. Vela. 2008. "Uncomputability and Undecidability in Economic Theory." Department of Economics Working Paper 0806, University of Trento, Italy.

Velupillai, K. Vela. 2012. "Taming the Incomputable, Reconstructing the Nonconstructive, and Deciding the Undecidable in Mathematical Economics." *New Mathematics and Natural Computation* 8, 5–51.

Whitehead, Alfred North and Bertrand Russell. 1910-13. *Principia Mathematica, Volumes I-III.* London: Cambridge University Press.

Winrich, J.E. 1984. "Self-Reference and the Incomplete Structure of Neoclassical Economics." *Journal of Economic Issues* 4, 987–1005.

Wolfram, Stephen. 1984. "Universality and Complexity in Cellular Automata." *Physica D* 10, 1–35.

Woolgar, Steve. 1991. "The Turn to Technology in Social Studies of Science." *Science, Technology & Human Values* 16, 20–50.

Chapter 18

The Incompletability of Markets

Sami Al-Suwailem*

If the market is modeled as a formal axiomatic system, then Gödel's incompleteness theorem implies that if the system is consistent, it cannot be formally complete. Formal incompleteness implies the impossibility of complete descriptions of commodities and contracts if such descriptions are to be logically consistent. Market incompleteness explains the fundamental role of free choice and private ownership in market systems. A complete market, technically speaking, cannot be a free market.

1. Introduction

In September 1917, the great mathematician David Hilbert delivered an address to the Swiss Mathematical Society entitled *Axiomatic Thought* [82]. He concluded his talk with a firm belief:

> Anything at all that can be the object of scientific thought becomes dependent on the axiomatic method, and thereby indirectly on mathematics, as soon as it is ripe for the formation of a theory. By pushing ahead to ever-deeper layers of axioms ... we also win ever-deeper insights into the essence of scientific thought itself, and we become more conscious of the unity of our knowledge. In the sign of the axiomatic method, mathematics is summoned to a leading role in science. [79, p. 90].

According to Herman Weyl, Hilbert liked to illustrate the axiomatic method in his lectures through examples taken from various fields, including biology and economics [79, p. 88].

*Islamic Research and Training Institute, Jeddah, Saudi Arabia; sami@isdb.org.

These ideas influenced John von Neumann, who started the axiomatization of various fields of knowledge, including economics. The trend reached its peak during the 1950s, in the hands of Kenneth Arrow and Gerard Debreu, among others, in what is known as the "Formalist Revolution" [15]. Today, the axiomatic approach is an integral part of graduate textbooks on microeconomic theory and general equilibrium models.

Despite the wide use of the axiomatic approach in economic theory, not enough attention is paid to the "price" associated with it. The price was demonstrated by Kurt Gödel in 1930 when he developed his Incompleteness Theorems, which set serious limits to the power of logic and mathematical formalism and brought in deep insights along the way. In his *Microeconomic Foundations* book, Kreps [45, p. xv] asserts that "the foundations of economics are resolutely mathematical." But, if this is the case, then economists should be at least aware of the foundations of mathematics as well, and at the heart of these foundations lie Gödel's theorems. For a variety of reasons, however, the limits imposed by these theorems were ignored or marginalized. They were viewed mostly as intellectual curiosities hovering at the far edge of scientific inquiry (see [44]).

In this chapter, we argue that formal incompleteness implies the impossibility of complete description of commodities and contracts if such descriptions are to be logically consistent. The fundamental role of private ownership and decentralized markets cannot be explained if markets were complete. Gödel's incompleteness theorems show that, in a formal system of the economy, ownership and free choice are of intrinsic value, even in the most idealistic world.

2. Gödel at a Glance

It would be helpful to first present an informal brief of the main ideas behind Gödel's two incompleteness theorems before we discuss the axiomatic approach to choice. More details can be found in [29, 71].

In 1931, Kurt Gödel published his now-infamous paper "On Formally Undecidable Propositions of *Principia Mathematica* and

Related Systems, I." The paper presents, among other things, two fundamental results on formal systems.

2.1. *First incompleteness theorem*

An axiomatic system, S, rich in arithmetic, can be either complete or consistent, but cannot be both.

By "complete," we mean that any well-formed sentence in the system must be either proved or disproved. By "consistent," we mean that the system cannot prove a sentence and its negation. The First Theorem shows that truth extends beyond formal reasoning.

In any axiomatic system like S, there will be many sentences formulated in the system but are not decidable within it. We are able to decide the provability of these sentences only "outside" S, as will be explained shortly.

Is it possible to find out those sentences that are true but cannot be proved within S and simply add them to the axioms of the system? If we do, then new statements will arise. There are infinitely many sentences of this sort and there is no way to systematically identify them in advance. Axiomatic systems can never encapsulate the entirety of true statements. Are these sentences of substantial value? Or are they simply logical puzzles? As we shall argue, these sentences are of critical importance.

2.2. *Second incompleteness theorem*

The most important sentence that cannot be proved within the system is: "System S is consistent." This is the Second Theorem: System S, if consistent, cannot prove its own consistency.

We can show (or assume) that S is consistent only "outside" S. The combination of S and the assertion that S is consistent forms a new system, S'. The consistency of the new system, in turn, cannot be proved or asserted within S'. Every time we try to "complete" the system by adding new axioms, we end up with a new system with yet more true but unprovable sentences. Thus, axiomatic systems are not completable. Mathematical truth is inexhaustible.

2.3. *Gödel and economics*

Gödel was interested in economics since his days in Vienna, according to his pupil, mathematician Hao Wang [78, p. 75]. Dawson [28, p. 232] reports that Gödel also discussed the foundations of economics with Oskar Morgenstern, who was a close friend of Gödel at Princeton.

In response to a proposal to generalize his theorem to human affairs, Gödel commented [78, p. 4]:

> A completely unfree society (i.e., one proceeding in everything by strict rules of "conformity") will, in its behavior, be either inconsistent or incomplete, i.e., unable to solve certain problems, perhaps of vital importance. Both of course, may jeopardize its survival in a difficult situation. A similar remark would also apply to individual human beings.

This shows that, from Gödel's point of view, incompleteness theorems could, in principle, apply to the real world, not only the physical universe but the social sphere as well.

3. Limits to Choice

3.1. *Axioms of choice*

Microeconomic theory defines consumer choice axiomatically, i.e., by stating certain axioms based on which the entire theory of choice can be derived.

The two fundamental axioms of consumer's preference relation $\succsim$ over a consumption set X are as follows:

- *Completeness*: For all $x, y \in X$, either $x \succsim y$ or $y \succsim x$. Thus, for any two bundles of commodities, the rational consumer is able to decide whether he or she prefers one to the other or is indifferent between the two. Put differently, the rational consumer is able to decide *every* choice problem.
- *Transitivity*: For all $x, y, z \in X$, if $x \succsim y$ and $y \succsim z$, then $x \succsim z$. This axiom is basically about consistency. A rational consumer shall not contradict him or herself by being intransitive.

These two axioms, together, imply that the consumer can consistently rank *any* finite number of elements in the consumption set, X, from best to worst, possibly with some ties. See [43, pp. 5–6].

However, Gödel's First Incompleteness Theorem shows that these two properties, completeness and consistency, are impossible to achieve simultaneously in an axiomatic system rich in arithmetic. This obviously includes axiomatic choice theory. If the rational agent is consistent, then he or she cannot *always* decide which is preferred to what. That is, if preferences are consistent, they cannot be complete.

The axiom of completeness was questioned, long ago, by no less than John von Neumann, in *Theory of Games and Economic Behavior* (1953). With Oskar Morgenstern, he writes (p. 19):

> It is conceivable, and may even in a way be more realistic, to allow for cases where the individual is neither able to state which of two alternatives he prefers nor that they are equally desirable.[a]

Moreover (p. 630):

> It is very dubious, whether the idealization of reality which treats [the completeness axiom] as a valid one, is appropriate or even convenient. I.e., one might want to allow, for two utilities u, v, the relationship of *incomparability* ... which means that neither $u = v$ nor $u > v$ nor $u < v$.

This question "amounts to asking whether events, with respect to preference, form a completely or a partially ordered set." (p. 19, footnote 4).

Concerning the appropriate modeling strategy, they remark (p. 29):

> If the general comparability assumption is not made, a mathematical theory ... is still possible. It leads to what may be described as a many-dimensional vector concept of utility. This is a more complicated and less satisfactory set-up, but we do not propose to treat it systematically at this time.[b]

[a] In this case, the authors remark, indifference curves "do not exist" (p. 20).

[b] Both von Neumann and Morgenstern were well aware of the important work of Gödel (see [70]). Hence, one might speculate that questioning the completeness axiom and the

Aumann [6] was even more critical, and constructive, regarding the completeness axiom. He writes (p. 446):

> Of all axioms of utility theory, the completeness axiom is perhaps the most questionable. Like others of the axioms, it is inaccurate as a description of real life; but unlike them, we find it hard to accept even from the normative point of view. Does "rationality" demand that an individual make definite preference comparisons between *all* possible lotteries ...?

Aumann notes the difference between indifference and incomparability, where the former involves a positive decision, while the latter means that no decision is made at all (footnote 6). Aumann's work stimulated an active line of research, starting most notably with Bewley [10–13], followed by Mandler [51], Ok [55], Bade [8], among others (see [32] for a survey).

This rich branch is still absent from economics textbooks. More important, perhaps, is that these works question the completeness axiom from a practical or intuitive perspective. While this is perfectly valid, it would be more appropriate, and more insightful, to consider Gödel's incompleteness theorems as the foundations for incomplete preferences. As we shall see later, this approach is more fruitful and presents a sound framework for rationality in principle.

For now, we might ask: why wouldn't we be able to rank any pair of commodities or bundles of commodities? The rationale for the incompleteness theorems is that, if the system is consistent, then completeness inevitably leads to contradictions and paradoxes. We explore some of these in the following sections.

3.2. *Paradoxes of classification*

Ranking two or more objects requires a system of classification. We cannot compare two objects without first identifying certain properties or attributes based on which we carry out the comparison. But for rich objects like economic goods and services, classification and

---

interest in partial ordering were probably motivated by the incompleteness theorems. However, the difficulty of modeling may have curtailed pursuing this line of research.

Table 1. Paradoxes of classification.

Paradox	Classification	Contradiction
Russell	(1) Sets that are members of themselves (2) Sets that are not	Set of sets of type (2) is/is not a member of itself
Grelling	(1) Auto-logical words* (2) Hetero-logical words**	"Hetero-logical" is/is not hetero-logical
Library	(1) Books that list themselves (2) Books that don't	A book that lists all books of type (2) does/does not list itself
Berry	(1) Numbers definable in less than 60 characters (2) Numbers definable in 60 or more characters	Type (2) numbers are definable in less than 60 characters

*Words that describe themselves, like English and Short.
**Words that do not describe themselves, like German and Long.

comparison of these attributes not only becomes computationally complex, they also tend to produce self-reference and logical paradoxes. These paradoxes, like Russell's paradox, arise mainly in the context of classification; see Table 1.

These paradoxes follow more or less the same pattern[c]: All objects of the concerned universe are classified into two sets, A and B. The function $P(x)$ is supposed to correctly identify members of A and of B:

- $A = \{x \mid P(x) = 1\}$;
- $B = \{x \mid P(x) = 0\}$.

Since A and B are mutually exclusive, no member of either set can belong to the other. However, a problem arises when the property P is true for the set B *as a whole*, so that $P(B) = 1$. This implies that $B \in A$. But this is a contradiction because the two sets are supposed to be mutually exclusive. In this case, the function $P(x)$ cannot exist. That is, there is no general or universal procedure that

[c]For a general formulation of paradoxes, see [16, 60, 61, 81].

can consistently classify all objects among the sets A and B. Hence, the classification cannot be complete.

3.2.1. *Russell's paradox*

For Russell's paradox, the property is self-membership decided by the (hypothetical) function $M(x)$. So, we have the following:

- $A = \{x \mid M(x) = 1\}$, the set of all sets that are members of themselves;
- $B = \{x \mid M(x) = 0\}$, the set of all sets that are not members of themselves.

Since the set B is not supposed to belong to A by construction, then B is not a member of itself. But this means that B must include itself because it is a set that does not belong to itself. This means that $M(B) = 1$. This implies that B belongs to A — a contradiction. It follows that the function $M(x)$ does not exist. Similar reasoning applies to the Grelling and Library paradoxes.

3.2.2. *Berry's paradox*

For Berry's paradox, we define the two sets:

- $A = \{x \mid Y(x) = 1\}$, the set of all numbers definable in less than 60 characters;
- $B = \{x \mid Y(x) = 0\}$, the set of all numbers definable in 60 or more characters.

But the set B itself is defined in less than 60 characters! This means that $Y(B) = 1$. Thus, $B \in A$. But this contradicts the assumption that A and B are mutually exclusive. Suppose instead that we set the cutoff number to be equal to the number of characters defining B, which is 12. Since the definition of A has the same number of characters, we get $A \in B$ — a contradiction again. This shows that the function $Y(x)$ cannot exist.

Gregory Chaitin [21,22] explains how he builds on Berry's paradox to develop the information-theoretic incompleteness theorems. Along the line, he introduces a paradox arising from classifying computer programs into "elegant" and "inelegant" programs. A program is elegant if its output cannot be produced by a shorter program. To apply the above framework, suppose we have the following two sets:

- $A = \{x \mid G(x) = 1\}$, the set of all elegant programs;
- $B = \{x \mid G(x) = 0\}$, the set of all inelegant ones.

Now, $G(x)$ is supposed to be a program that identifies elegant from inelegant programs, and thus each of A and B is a program as well. Are they elegant? Suppose they are. Then A belongs to itself, which causes no issue. But, if B is also elegant, then $G(B) = 1$, and thus $B \in A$ — a contradiction, since the two sets are supposed to be mutually exclusive. Suppose instead that A and B are inelegant. Then $G(A) = 0$, and thus $A \in B$ — a contradiction again. It follows that the program G cannot exist.

3.2.3. *Gödel and Turing*

To see how these paradoxes relate to Gödel's theorem, suppose we want to classify all statements into two groups: true statements and false ones. So, let

- $A = \{x \mid T(x) = 1\}$, the set of all true statements;
- $B = \{x \mid T(x) = 0\}$, the set of all false ones.

The function $T(x)$ can be seen as a formal system that generates or proves all true statements and disproves all false ones. But the statement B must be true, and thus, $T(B) = 1$. This means that $B \in A$ — a contradiction. Therefore, there is no general way to identify true statements from false ones, and so, $T(x)$ cannot exist. Hence, not all true statements are provable, and this is the essence of Gödel's first theorem.

A related result is obtained when we want to classify statements into provable and unprovable ones. So, we have the following:

- $A = \{x \mid V(x) = 1\}$, the set of all provable statements;
- $B = \{x \mid V(x) = 0\}$, the set of all unprovable ones.

But the statement B itself is provable given the function $V(x)$. So, we end up with $B \in A$ — a contradiction. Hence, $V(x)$ cannot exist, and thus, there is no systematic or universal function to identify provable statements from unprovable ones. This is known as the Decision Problem, or *Entscheidungsproblem* in German, proved by Alan Turing (see [9, Chapters 1 and 7]). More on this in Section 6.1.

4. Classifying Commodities

The above paradoxes will arise with almost any scheme of classification of rich objects that aims to be exhaustive. In the following sections, we present some examples of classifying commodities and other economic variables.

4.1. *Substitutability*

Suppose we want to classify commodities or bundles of commodities into two groups or sets based on whether they are unique, i.e., have no substitutes, or not. Let

- $A = \{x \mid Q(x) = 1\}$, the set of all unique commodities, i.e., those with no substitutes;
- $B = \{x \mid Q(x) = 0\}$, the set of all non-unique commodities, i.e., those with substitutes.

The two sets must be mutually exclusive. But, since the set B itself is a bundle of commodities, then this bundle, by construction, is unique because all substitutes are already included in the set, and there is no substitute to the bundle B. Thus, $Q(B) = 1$. It follows that $B \in A$. But this is a contradiction of terms, and thus, the function $Q(x)$ cannot exist.

For the same reason, there will be no systematic way to identify commodities that are commensurable, in terms of other commodities, from incommensurable ones. This indicates that we cannot be sure that all useful objects or goods can be priceable in terms of money or other commodities. More on this is covered in Section 6.2.

Asset specificity is of great importance in economic activities. Without asset specificity, notes Williamson [80, p. 53], there is no reason for firms to exist. The above result shows that asset specificity is undecidable. This contributes to the incompleteness of contracts, as discussed in Section 6.5.

4.2. *Describability*

Commodities can also be classified, in general, into search goods and experience goods [30, 54, Chapter 7]. Search goods are goods that the consumer can evaluate in advance via description or observation. Experience goods, in contrast, are goods that cannot be evaluated without experiencing or consuming them. Examples include restaurants, movies, etc. Almost all new products are experience goods [69, Chapter 1]. Demand for experience goods is especially sensitive to reputation and trust, unlike search goods. So, suppose we have the following:

- $A = \{x \mid D(x) = 1\}$, the set of all search or describable commodities;
- $B = \{x \mid D(x) = 0\}$, the set of all experience, or indescribable, commodities.

But the set B is nothing but a systematic description of experience goods. Thus, $D(B) = 1$. This means that $B \in A$. But this is a contradiction, and so the function $D(x)$ cannot exist. This means it is not possible to provide an accurate description of all types of commodities. Describability is an undecidable property.

4.3. *Trust*

Now, suppose that we want to classify commodities depending on whether their brand or label is trustworthy or not. Let:

- $A = \{x \mid W(x) = 1\}$, the set of all commodities with trustworthy labels;
- $B = \{x \mid W(x) = 0\}$, the set of all commodities with untrustworthy labels.

We obviously must trust the label of the set B itself, i.e., $W(B) = 1$. Thus, $B \in A$. But this is a contradiction, and so, the function $W(x)$ cannot in general exist. This result shows why trust itself cannot become a market commodity. Intuitively, how can we trust someone who claims to "sell" trust? This will obviously devolve into an infinite regress. This shows the contradictory nature of commercializing trust, as Arrow [4] rightly points out.

4.4. *Risk*

Commodities of a given type tend to have variable qualities. These variations might be due to systematic factors, e.g., seasonal or cyclical factors that are generally predictable, or they might be purely random. Suppose we want to identify these two types of variations. So, we have the following:

- $A = \{x \mid R(x) = 0\}$, the set of all systematic variations of a given commodity;
- $B = \{x \mid R(x) = 1\}$, the set of all random variations.

Now, if we are able to define what variations belong to which set, then the set B is identified in a systematic manner. This means that $R(B) = 0$. But this implies that $B \in A$ — a contradiction. Hence, the function $R(x)$ cannot in general exist. This shows that risk, in principle, cannot be systematically quantified, as argued in [3].

Chaitin [23, pp. 123–125] points to the paradoxical nature of randomness: if it can be well defined, it is not random anymore. Byers [20, p. 14] remarks that mathematics is the study of a pattern. But randomness is the absence of pattern, and thus, there can be no general mathematical formula for randomness.

4.5. *Change and motion*

Given that risk is undecidable, what can we say about *change* in general? Change is very important in economic behavior. It includes growth, inflation, depreciation, and many other and many other phenomena. Economists usually formulate "laws of motion" describing the growth of major variables, e.g., population, income, capital, etc. [47, p. 59]. For a complex dynamical system, like the economy, is there a general procedure to identify changing variables from those that stay constant within a certain process? This is an important question because constant properties reflect conserved or invariant characteristics of the system, which preserve its identity throughout the change.

Following the above framework, suppose we have

- $A = \{x \mid \Delta(x) = 0\}$, the set of all constant properties of the system;
- $B = \{x \mid \Delta(x) = 1\}$, the set of all changing ones.

But B itself must be constant because Δ must be so. Thus, $\Delta(B) = 0$. But this implies a contradiction as before. Hence, there is no universal or general function that can identify constant properties from changing ones. Of course, we can know, in principle, which is which by measurement *after* the process is completed. But there is no general method to identify *in advance* the set of changing properties and that of constant ones.

Change, conceptually, has puzzled philosophers since the age of Aristotle [52]. It led to many paradoxes of motion, e.g., Zeno's Arrow paradox. According to Zeno, at each moment in time, a moving object (e.g., an arrow) is supposed to occupy a well-defined space. If this is the case, then, at each moment in time, the arrow is indistinguishable from a motionless one. How could a collection of motionless arrows add up to a moving one? Put differently, if the position of the arrow at each moment in time is well defined, motion will be impossible. This implies that, if motion is possible, then position at any moment of time cannot be well defined. But this tradeoff reminds us of the Uncertainty Principle.

The principle implies, *inter alia*, that the position of a particle (or a wave) cannot be precisely determined, in advance, simultaneously with its momentum. The principle follows from the mathematical formalism of Fourier transformation and is not confined to the quantum world. Economically, position corresponds to level (of capital stock, say), while momentum corresponds to growth. Baaquie [7] applies the uncertainty principle to finance. He argues[d] that, if the price of a stock, say, and its growth rate (capital gain) can be computed simultaneously in advance, this implies an arbitrage opportunity that is not supposed to survive in a competitive market. The established risk–return relationship in economics might be attributed, at least in part, to the uncomputability of change.

Is there a relationship between the uncertainty principle and the incompleteness theorem? Many intellectuals were interested in this question. Gregory Chaitin asked John Wheeler about the relationship between Gödel's theorem and Heisenberg's uncertainty principle. According to Svozil [74, p. 112], Chaitin recalls the following:

> [In 1979] I went up to Wheeler and I asked him, "Prof. Wheeler, do you think there's a connection between Gödel's incompleteness theorem and the Heisenberg uncertainty principle?" ...
>
> Wheeler said, "I went to Gödel, and I asked him, 'Prof. Gödel, what connection do you see between your incompleteness theorem and Heisenberg's uncertainty principle?' ... And Gödel got angry and threw me out of his office!" Wheeler blamed Einstein for this. He said that Einstein had brain-washed Gödel against quantum mechanics and against Heisenberg's uncertainty principle!

Calude and Stay [24] argue that algorithmic randomness is equivalent to a formal uncertainty principle which implies Chaitin's information-theoretic incompleteness. They also show that the formal uncertainty principle applies to all systems governed by the wave equation not just quantum waves. This fact, they argue, supports the conjecture that uncertainty implies algorithmic randomness not only in mathematics but also in physics. There is no reason why the same cannot apply to economics.

[d]Personal communications, 30 June 2014.

4.6. *A generalization*

In general, therefore, there is no systematic procedure to *consistently* classify *all* components of a complex system like the economy. Any proposed procedure of this sort will likely bump into contradictions of the same nature as Russell's paradox. The impossibility of a complete and consistent classification procedure means that the rational consumer will not be able to systematically rank *all* commodities. To be consistent, the consumer must be therefore unable to decide the ranking of some commodities.

One way to formally show the incompleteness of choice is to use a result proved by da Costa and Doria [25]. Suppose choice is modeled as a formal system $\mathcal{C}$, and let P be a predicate (roughly, the formal version of a property) on the objects in $\mathcal{C}$. Let $x_1 \neq x_2$ be terms (e.g., commodities or bundles of commodities) in $\mathcal{C}$. $P(x)$ means "x has property P," and $\neg P(x)$ means "x doesn't have property P." Then da Costa and Doria show the following:

> *Given any P so that, for $x_1 \neq x_2$, $P(x_1)$ and $\neg P(x_2)$ hold, then there is a term x_3 such that $P(x_3)$ is undecidable.*[e]

That is, no algorithm can decide if either $P(x_3)$ or $\neg P(x_3)$ holds. Thus, if the consumer is able to decide that commodity x_1 has property P, for example, while commodity x_2 doesn't, then there will be a commodity x_3 for which the consumer will be logically unable to decide if it does have the property P or not.

5. Paradoxes of Transitivity

5.1. *Sorites paradox*

Many, if not most, of the properties of interest to consumers are fuzzy or *vague*. To describe a shirt, for example, as "cool" or "warm" or "comfortable," is to indicate a property without well-defined boundaries. Similarly, properties such as "ease of use," "hard work," or

[e]For a detailed discussion, and the original references to this result, see [26].

"high quality," while having clear meanings, lack sharp borders or boundaries.

The vagueness of such properties leads to a paradox called Sorites paradox, aka the paradox of the heap [40, 63]. Suppose we have a heap of sand. If we take away one grain of sand, what remains is still a heap. Taking out two grains will have the same effect. This apparently obvious and uncontroversial supposition appears to lead to the paradoxical conclusion that taking out all but one grain will still result in a heap. But if one grain is a heap, then how come taking out only one will not make any difference?

To apply this to commodities is straightforward: Two shirts with very similar colors might be considered to be equivalent, despite a negligible or insignificant difference. The negligible difference, however, adds up, so that after an ordered sequence of, say, 20 shades, we end up with a clearly different color than the one we started with. While the consumer is indifferent between any two adjacent shades, he or she might definitely prefer one shade over a distant one. Transitivity breaks down (see [53, p. 7]). This is not because the consumer is irrational. It is because the property itself has indeterminate boundaries.

Vague properties are very common that they cannot be neglected or ignored, and they explain why contracts might be incomplete. The two parties of the contract may be able to provide less than ideal performance while staying within the terms of the contract; Hart and Moore [38] refer to less than ideal performance as "shading." Since shading is non-contractible, the contract will be incomplete.

If the consumer is consistent, then he or she must be undecided concerning the range of transitivity. Thus, if $x_1 \sim x_2$, $x_2 \sim x_3, \ldots,$ $x_n \sim x_{n+1}$, then there exists k, $2 < k < n$, such that $x_1 \nsim x_k$. That is, transitivity ceases at shade k. Because of vagueness, k is uncomputable. In other words, for the consumer to be consistent, he or she must be indeterminate regarding the ranking of some commodities.

The Sorites paradox can be modeled following the classification scheme above concerning unique vs. substitutable commodities. Each shade is substitutable with its immediate neighbor, but not the "distant" one. There is no systematic way to identify which shades are substitutable and which are not. It can also be modeled following constant vs. changing variables, where the color shade, for example, might be constant in the short range but changing in the long range, but the cutoff point between the two ranges is uncomputable.

5.2. *Aggregation paradox*

Suppose the consumer is able to decide the ranking of different commodities based on a certain property p_1. But this is not the only property the consumer cares about; there are other important properties as well. Following Muñoz-Garcia [53, p. 8], suppose the consumer ranks three commodities according to the three different properties as follows:

- For property p_1,
  $x_1 \succ_{p_1} x_2 \succ_{p_1} x_3$.
- For property p_2,
  $x_2 \succ_{p_2} x_3 \succ_{p_2} x_1$.
- For property p_3,
  $x_3 \succ_{p_3} x_1 \succ_{p_3} x_2$.

Now, suppose the consumer will give properties equal weights, and therefore will decide the final ranking based on the majority rule. From the above, we have

$$x_1 \succ_{p_1,p_3} x_2 \succ_{p_1,p_2} x_3 \succ_{p_2,p_3} x_1.$$

But this is a circular ranking! This paradox is well known for majority voting, but it is equally valid for the aggregation of commodities. To be consistent, the consumer must, therefore, abstain from some rankings or use subjective criteria.

5.3. *Meta-preferences*

Suppose a consumer learns about healthy habits and would like therefore to become healthy. He, therefore, prefers to prefer healthy food to junk food (or non-smoking to smoking, etc.). This is a *meta-preference*: the agent prefers to have some preference relation among certain goods. But his direct preferences might not be consistent with his meta-preferences.

Gödel's theorem shows that, in a rich axiomatic system, meta-sentences will inevitably arise. These are sentences about sentences of the system. If the system models consumer preferences, then such sentences will involve sentences about preferences, leading inevitably to meta-preferences. It is because of such meta-sentences that a consistent formal system cannot be complete. According to philosopher Harry Frankfurt [31, p. 7], only human beings, apparently, are able to reflect on their preferences and capable of forming second-order desires [33, p. 184]. Psychologist Thomas Suddendorf [73] argues that what sets the human race apart from other species is our ability to imagine the future and construct novel scenarios. Mental time travel, into the past and into the future, is uniquely human (p. 90). Such unique ability necessarily involves self-reference and meta-statements.

An example of meta-preferences, rarely studied by economists, is the preference for freedom. Freedom is not a commodity per se; rather, it is an attribute of the choice process itself, which, according to Amartya Sen [67], should be placed in a central position of choice theory. Issues of fairness and dignity, and values in general, can also be viewed as meta-preferences guiding the act of choice [66].[f] *A statement about the process of choice is a meta-statement*, which is inevitable in an axiomatic system as rich as that of choice. Neoclassical theory, by imposing convexity of the choice set, rules out the process or act of choice from the analysis by design [1]. This is not only inconsistent with the real world, but it is also inconsistent with the mathematics of formal systems. It is particularly puzzling that, while free choice is an essential attribute of a free market, it is deliberately excluded from the analysis of mainstream economics.

[f]To the extent that axioms of choice are normative, economists are in fact stating their meta-preferences: preferences on how preferences should be.

Because of the tension between different sets of preferences, Nobel laureate Thomas Schelling suggests modeling the relationship as one between "multiple selves" [65, Chapters 3 and 4]. A person, for example, might prefer healthy habits, but he might prefer smoking. It is as if there exist two (or more) selves, each is "voting" for the final choice. Once modeled in this manner, then the paradox of voting among these selves will naturally arise. If the agent is consistent, then there will necessarily be unsolvable choice problems. Self-imposed constraints on the future choice set are one way to address such undecidabilities. A rational agent who prefers not to prefer smoking, yet prefers smoking, will self-impose restrictions to avoid inconsistent preferences [33].

Schelling argues that changing preferences explain the multiple selves model. This does not seem to be acceptable to mainstream economics. Why assume unstable preferences? But Gödel's theorem might provide a better explanation: meta-preferences that inevitably arise in such a rich and complex system such as that of consumer's choice.

5.4. *Incompleteness vs. irrationality*

Incompleteness of preferences doesn't mean that agents in the real world will fail to make a decision when facing undecidable problems.[g] Agents, in general, *will* make decisions; however, these decisions may not be determined purely by the mechanics of rational choice. Instead, social or behavioral considerations will play a role in guiding the decision-making process, particularly in undecidable situations. This is one reason why some of the fundamental assumptions behind the revealed preference approach might not be valid (see [68, Chapter A2*]).

Bewley [10] argues that the "law of inertia," or the preference for the status quo, will supplement the decision-making process when

[g]In other words, they will not be in the position of Buridan's ass. This is an ass placed at equal distances from two identical piles of hay, who is unable to decide which pile to eat and consequently dies from hunger.

preferences are incomplete. This rule will prevent agents from making intransitive choices that would expose them to a "money pump."[h] Experimental evidence shows that agents following norms of social behavior do not violate transitivity [19, pp. 31–32]. Social norms help coordinate players' actions to achieve "correlated equilibria" in situations where Nash equilibrium does not exist [19]. Arrow [4, p. 70] points out that social norms, including moral and ethical codes, work to compensate for market incompleteness.

If all choice problems were decidable, then behavioral rules will lead to sub-optimal "biases" or irrational behavior. But the ubiquitousness of undecidable or unsolvable problems means that these behavioral rules can, in principle, enhance our rationality. Gigerenzer [34] calls them: Simple Heuristics that Make Us Smart.

Many economists find it difficult to give up rationality, as Hart [36, p. 385] points out. But to be consistent with behavioral or social rules, we don't need to give up rationality. Gödel's main insight is that undecidability is the price we pay for maximum rationality.

6. Limits to Contracts

6.1. *Contracts as algorithms*

A contract is essentially a set of "if–then" statements: If, for example, the date is $t = 0$, then party A shall pay $\$x$ to B; if $t = 1$, then party B shall deliver y quantity of commodity z to party A. We tend to think that, if we could write contracts as computer programs, and if we are able to execute them just like computers do, then we will be able to eliminate all kinds of "irrationality" and inefficiency. This is the ideal world of economic theory.

In his 1985 Marshall Lecture at Cambridge University, *On the Mechanics of Economic Development*, Nobel laureate Robert Lucas

[h] A money pump occurs if a person chooses B minus a little money over A, C over B and A over C. If this cycle were repeated, the person would lose a little on each round [10, p. 85].

aims to construct a neoclassical theory of growth and development. He writes [49, p. 5]:

> I prefer to use the term "theory" in a very narrow sense, to refer to an explicit dynamic system, something that can be put on a computer and run. This is what I mean by the "mechanics" of economic development — the construction of a mechanical, artificial world, populated by the interacting robots that economics typically studies, that is capable of exhibiting behavior the gross features of which resemble those of the actual world that I have just described.

So suppose that the rational consumer is a sophisticated robot, as Lucas suggests. Does this make the consumer able to decide all choice problems? In 1936, Alonzo Church and Alan Turin, independently, strengthened Gödel's result by making it applicable to *any* computing system [9, 27]. The utmost level of sophistication in computing power is achieved by a Universal Turing Machine (UTM), which basically includes all today's computers. Yet, there are limits to the abilities of a UTM. In particular, there are infinitely many problems that the UTM (or any supercomputer with infinite resources) will be unable to compute in a finite time. This means that, even if equipped with supercomputing powers, the rational consumer robot will not be able to rank all commodities in the choice set (see [76, Chapter 3]). Another example of uncomputable problems is the best response to another rational agent or robot (a UTM). In general, there is no systematic or algorithmic procedure to decide how to respond to another robot even if the strategy of the opponent robot is fully algorithmic [72, pp. 152–153].

When a problem is undecidable, the robot will fail to complete the computation, i.e., to halt. If the robot insists to complete all computations within a defined period of time, then it must produce false results. Binmore [14, p. 208] explains as follows:

> The important point is that the insistence that machines *always* produce an output means that their predictions will necessarily sometimes be in error regardless of the sophistication of the stopping advice.

Being a robot, therefore, does not eliminate uncertainty or unpredictability. Lloyd [48, p. 36] rightly points out:

> Ironically, it is customary to assign our own unpredictable behavior and that of others to irrationality: were we to behave rationally, we reason, the world would be more predictable. In fact, it is just when we behave rationally, moving logically like a computer from step to step, that our behavior becomes provably unpredictable. Rationality combines with the capacity of self-reference to make our actions intrinsically paradoxical and uncertain.

6.1.1. *An economy of finite automata*

Suppose that the economy consists of robots with less powerful computational abilities than Turing machines. One well-known kind of such machines is finite automata (FA). For such machines, there are no undecidable problems [9, p. 46]. Although agents in this economy are less powerful than *homo economicus*, at least they face no unsolvable problems. In an economy of finite automata, every problem is algorithmically solvable, and thus, a centralized supercomputer can do the job of the market. If anything, a centralized system will be more efficient than a decentralized one. In such a world, therefore, a free market has no advantage over a centralized system. Although many economists will not agree with the conclusion, they inadvertently accept its logical premises. If a free market is ideally superior to a centralized system, then agents must possess powers beyond finite automata, and thus the market will be incomplete.[i] A complete market lacks the essential feature of a free market.

6.2. *The quantifiability problem*

A commercial contract must specify the commodity to be exchanged and its price. We have already seen the impossibility of a complete

[i] See [75].

specification of all commodities if such specification is to be consistent.

What about the price?

Gödel's incompleteness theorem applies to formal systems with sufficient arithmetic (e.g., Piano Arithmetic). In such systems, there will be many true but unprovable statements involving quantity variables. Such variables are therefore unquantifiable, i.e., their quantities cannot be systematically computed in advance. A system that requires quantifiability, like the economy, when modeled as a formal system, must, therefore, involve many instances of unquantifiable activities.

A price is obviously a quantity, and thus it is subject to Gödel's theorem. But why would it be impossible, in some cases at least, to systematically quantify the price? One reason is the nature of money in which prices are expressed. Money is primarily an accounting system for the commensurability of commodities in terms of each other [41]. So, to price a commodity means to decide the commensurability of a commodity in terms of other market commodities. As we have seen (Section 4), substitutability and commensurability of commodities are generally undecidable. Treating incommensurable goods as exchangeable leads to "money pump" and logical contradictions (see [56, pp. 182–185]). Some goods or services might be considered so unique that they are "priceless." An obvious example is human organs like kidneys. Are these "goods" commensurable so that they can be priced in money like standard market commodities? Or are they incommensurable and thus should be donated outside the market system?

This has been a controversial issue. Mainstream economic theory puts no limits on what can be offered for a market price. The argument, advanced most notably by Nobel laureate Kenneth Arrow [5], is that, if one can donate blood, for example, what is the problem of offering him an option to sell it? Expanding the set of options should be welfare-enhancing, not reducing. Critics, like Sandel [64, Chapter 3] and Bowles [18], argue that adding the commercial option causes "crowding out" of the cooperative motive. The crowding-out effect is inexplicable in neoclassical theory where options are independent and self-reference plays an insignificant role.

6.3. *Pricing vs. meta-pricing*

If some goods, like human organs, are considered mostly incommensurable, then an alternative compensation mechanism is required. Since resources are scarce, a one-way donation cannot be sustainable. A rational benevolent therefore will ask for a *reciprocal* donation to make such incommensurable objects available to the community.[j] With a little reflection, reciprocity can be viewed as a *meta-price*: instead of pricing the incommensurable good itself, it is the *act* of giving such a good that is priced instead. The meta-price of such an act is an analogous act of a similar nature, i.e., donating another incommensurable good to someone who is in need. Reciprocity, therefore, provides a sustainable mechanism for meta-pricing of incommensurable goods. Recall (Section 5.3) that a meta-statement about choice implies a statement about the process or act of choice. Hence, pricing the act of giving the incommensurable good, instead of the good itself, is in line with the Gödelian perspective.

Since reciprocity allocates incommensurable goods based on need rather than wealth, it tends to produce more value to community members than the competitive market mechanism. Moreover, the market is not likely to converge to a fair or stable value of the good, simply because, by definition, it is not describable in terms of other market commodities. Accordingly, pricing of such goods will likely be subject to unethical practices, as empirical studies show (see [64, pp. 123–124]).

The commercial option, i.e., the option to sell human organs, is inconsistent with reciprocity; the person who sells his organ breaks away from reciprocal relations. Hence, it is likely that there will be a tradeoff between reciprocal donation and commercial sale of

[j]For a successful application of reciprocity in donating human organs, see the work of Nobel laureate, Alvin Roth [62].

incommensurable goods. Introducing the option to sell blood or kidney, therefore, does not expand the choice set. Instead, it crowds out reciprocity. Thus, we end up with a kind of a Prisoner's Dilemma game. Cooperation in this game is donating the organ, while defection is commercial sale. A consumer is willing to donate if others are willing to reciprocate. But if others opt for commercial trade, the consumer will shift to the commercial option as well. The commercial choice, therefore, does crowd out social giving. The crowding-out effect has been documented in many studies (see [42, 57]).

6.4. *Expectations*

There is another reason why it would be impossible in some cases to systematically quantify the price of a good or an asset. A price of an asset or a durable commodity must involve expectations about the future not only of the asset but also of the market as well. A statement about the future inevitably involves self-reference. Thus, *a statement about the future is essentially a meta-statement.* It follows that, if the market system is consistent, there will be many instances where the future is undecidable or uncomputable. This is the main conclusion of da Costa and Doria [26] in their paper: "Computing the future." The future, in general, is uncomputable.

It is not difficult to see why statements about the future, or expectations, can become undecidable. When the price depends on the expectations of other agents' expectations, who in turn base their expectations on others' expectations, we end up with an infinite regress. This may open the door for bubbles and "sunspot equilibria" that are generally not Pareto efficient. In a competitive environment, there is no compelling reason for profit-maximizing agents to coordinate their expectations to converge to the Pareto-efficient equilibrium based on the fundamentals of the assets. Conversely, agents are motivated to deviate or defect to make additional gains. It becomes a Prisoner's Dilemma game.

Experimental evidence seems to support the above line of reasoning. The work of Nobel laureate Vernon Smith and his collaborators shows a stark difference between markets for consumption goods and markets for assets. Markets for consumption goods, where goods are traded to be consumed, not to be re-traded, are found to be in line with predictions of economic theory. In asset markets, in contrast, the asset can be resold, and thus its value depends on the expectations of others' expectations, etc. In these experiments, substantial bubbles were generated during trading sessions, even for experienced traders (see [35, Chapter 2]). Hommes [39, Chapter 8] provides experimental evidence of substantial deviations of expectations from the fundamental value of speculative assets, in contrast to non-storable commodities (see [2]).

With undecidable expectations about future prices, current prices may also become undecidable. This is another dimension of the limitations of contracting.

6.5. *Incomplete contracts*

Given the above considerations, it is clear that a contract cannot be complete. Since commodities and prices cannot be completely describable in a consistent manner, contracts will frequently fail to identify the subject matter of the contract in a systematic manner. To follow the classification scheme above, suppose we want to classify all market activities into two groups: contractable and non-contractable. So, let

- $A = \{x \mid K(x) = 1\}$, the set of all contractable market variables;
- $B = \{x \mid K(x) = 0\}$, the set of all non-contractable variables.

Since the set B is well defined, then it must be included in any contract. That is, a contract to trade members of set A must include a clause saying, for example, that "no non-contractable element is considered in this contract." In other words, the function K must be contractable, or else we would not be able to write a contract of the set A in the first place. But if K is contractable, then $B \in A$ — a contradiction. Hence, the function K is undecidable, and contracts, in principle, cannot be logically complete.

6.5.1. *Value of ownership*

Nobel laureate Oliver Hart [37] describes how did he, with Sanford Grossman, discover the relationship between incompleteness of contracts and the value of ownership. Why would a firm decide to sign a contract with another firm rather than buying it? In a complete market with complete contracts, why would we have firms in the first place? Hart reports that he and Grossman worked on the difference between firms and contracts for 10 very intense days; "these were ten days that shook my world," Hart writes (p. 372). The result is that *ownership is valuable because it fills the gap in incomplete contracts.* If a contract is incomplete, who has the right to decide about the missing parts of the contract? They call this right "the residual control or decision right" (p. 373). This basically is what ownership is. In a complete contracting world, ownership and residual rights of control are irrelevant since all decisions are specifiable in the contract (p. 381).

Ownership is essential in market systems. Yet, if the market were complete, ownership becomes irrelevant! If contract incompleteness is viewed as simply arising from frictions or inefficiencies, then in an ideal world, a market system is indistinguishable from a communist system. If they were ideally different, then there must be a fundamental reason why contracts are incomplete. Gödel's incompleteness theorems provide exactly that answer. In an ideal world, markets are inherently incomplete, and thus ownership is inherently valuable.

7. Are Markets Completable?

Arrow [4] introduced the concept of market completeness as follows: Commodities are defined based on the state of the world. For example, one gram of wheat when it rains on 20 May 2025 is a commodity, but the same wheat, if it does not rain on the same date, is another commodity. When we list all states of the world for the infinite future, and fully describe all commodities for all states, we can trade these "state-contingent commodities" in one go (see [50, p. 2]). A complete market, therefore, requires the following:

- complete preferences;
- complete descriptions of all commodities;
- complete descriptions of all states of the world, present and future;
- complete contracts.

However, if the market system is logically consistent, none of these requirements is feasible. No improvement in technology, resources, or regulations can overcome these obstacles because they are logical boundaries.

Given Gödel's incompleteness theorems, a formal axiomatic system representing the market cannot be both complete and consistent. Hence, if the market system is consistent, it must be incomplete. This incompleteness is inexhaustible, as pointed out earlier. Hence, *markets are incompletable.*

Bowles [18] argues that the incompleteness of markets cannot be, in general, remedied by taxes and subsidies. Trying to complete the market when it is inherently incompletable might make things even worse. To fill in the gaps, we need trust and norms that go beyond market mechanism.

Incomplete markets differ substantially from (hypothetically) complete ones. For example, when markets are incomplete, competitive equilibrium generally may not exist, and if it does, it may not be Pareto optimal [36]. Moreover, Hart [36] shows how in an incomplete market, permitting some trades that were previously prohibited may make everybody worse off. Institutions play a fundamentally different role in an incomplete system as compared to a complete one. Furthermore, equilibria of incomplete markets are highly sensitive to agents' beliefs, leading to "sunspot" equilibria, where any inefficient outcome might be an equilibrium (see Laffont [46, pp. 57–58]).

Economists generally agree that, in the real world, markets are incomplete. However, they do not agree on why they are so and how to deal with that. Jean-Jacques Laffont [46] reviews the main possible reasons behind market incompleteness. On the top of the list is the unpredictability of the future, or as he states: "Unforecastable Events" (p. 55). Laffont cites the argument of Karl Popper [58, pp. xi–xii]:

> The course of human history is strongly influenced by the growth of human knowledge. ... We cannot predict, by rational or scientific methods, the future growth of our scientific knowledge. We cannot therefore predict the future course of human history.[k]

Laffont notes that "Economists have little to say about this but more thinking should be devoted to this issue" (p. 55). The "issue" Laffont refers to is nothing but Gödel's theorem or, equivalently for that matter, Turing's theorem. Basically, in an interdependent system, the future is undecidable. As pointed out earlier, a statement about the future is necessarily a meta-statement, whereby self-reference is unavoidable in a rich and complex system. This interdependence is not due to frictions, transaction costs, asymmetric information, etc. It is an essential feature of the system even if we strip it down to pure numbers. Social interactions, moreover, add many layers of interdependence that make the system even less predictable. This is not to say that the economy is impossible to predict. Incompleteness implies that there will be many, many events or cases that are undecidable. This is an intrinsic feature of any rich and complex system as the market.

8. Conclusion

Is Gödel's incompleteness theorem relevant to economic theory? Does it add to our understanding of how markets work? The above analysis is an attempt to answer these questions. Briefly, in an ideal world, if markets were complete, then the following apply:

- All economic problems are algorithmically solvable, and thus there is no difference between centralized and decentralized systems.

[k]See also [59].

- All contracts are complete, and thus ownership is irrelevant. Hence, there is no difference between a private ownership system and a state-ownership system.

It follows that a complete market lacks the essential features of a free market: free choice and private ownership. This result is certainly not acceptable to many economists, but it is an inevitable conclusion of the completeness premises. If market systems are intrinsically valuable, then they must be incomplete, and Gödel's theorem shows why this would be the case.

Incompleteness doesn't involve irrationality; it is almost the opposite. It reflects the powerful abilities of economic agents. Self-imagination is a characteristic feature of mankind, which inevitably leads to self-reference and undecidabilities. The wonder of the market system is that it is able to achieve something beyond the abilities of individual agents. This can be true only with interdependent interactions. This interdependence necessarily leads to undecidabilities and incompleteness.

Incompleteness of mathematics was one of the greatest discoveries of the 20th century. The implications of such a discovery for social sciences are yet to be fully internalized. We hope that this work provides a step towards bringing economic theory to the science of the 21st century.

Bibliography

[1] S. Al-Suwailem, Behavioural Complexity, *J. Econ. Surveys* **25** (2010) 481–506.

[2] S. Al-Suwailem, Is economics a science? A Gödelian perspective, in *The Limits of Mathematical Modeling in the Social Sciences: The Significance of Gödel's Incompleteness Phenomenon*, ed. F. A. Doria (World Scientific, 2017), pp. 207–265.

[3] S. Al-Suwailem, F. A. Doria, and M. Kamel, Is risk quantifiable? in *Complex Systems Modeling and Simulation in Economics and Finance*, eds. S.-H. Chen, Y.-F. Kao, R. Venkatachalam and Y.-R. Du (Springer, 2018), pp. 275–303.

[4] K. Arrow, The organization of economic activity: Issues pertinent to the choice of market versus non-market allocation, in *Public Expenditures and*

Policy Analysis, eds. R. Haveman and J. Margolis (Markham Publishing, 1969) pp. 59–73.

[5] K. Arrow, Gifts and exchange, *Philos. Public Affairs* **1** (1972) 343–362.

[6] R. Aumann, Utility theory without completeness axiom, *Econometrica* **30** (1962) 445–462.

[7] B. Baaquie, *Quantum Finance: Path Integrals and Hamiltonians for Options and Interest Rates* (Cambridge University Press, 2004).

[8] S. Bade, Nash equilibrium in games with incomplete preferences, *Economic Theory* **26** (2005) 309–332.

[9] C. Bernhardt, *Turing's Vision: The Birth of Computer Science* (MIT Press, 2016).

[10] T. Bewley, Knightian decision theory: Part 1, Cowles Foundation Discussion Paper 807 (1986). Reprinted in *Decis. Econ. Finance* **25** (2002) 79–110.

[11] T. Bewley, Knightian decision theory: Part 2, Cowles Foundation Discussion Paper 835 (1987).

[12] T. Bewley, Knightian decision theory and econometric inference, Cowles Foundation Discussion Paper 868 (1988).

[13] T. Bewley, Market innovation and entrepreneurship: A knightian view, Cowles Foundation Discussion Paper 905 (1989).

[14] K. Binmore, Modeling rational players: part I, *Econ. Philos.* **3** (1987) 179–214.

[15] M. Blaug, The formalist revolution of the 1950s, *J. Hist. Econ. Thought* **25** (2003) 145–156.

[16] T. Bolander, Self-reference and logic, *Φ News* **1** (2002) 9–44, http://phinews.ruc.dk.

[17] G. Boolos, Gödel's second incompleteness theorem explained in words of one syllable, *Mind* **103** (1994) 1–3.

[18] S. Bowles, *The Moral Economy: Why Good Incentives are No Substitutes for Good Citizens* (Yale University Press, 2016).

[19] S. Bowles and H. Gintis, *A Cooperative Species: Human Reciprocity and Its Evolution* (Princeton University Press, 2011).

[20] W. Byers, *The Blind Spot: Science and the Crisis of Uncertainty* (Princeton University Press, 2011).

[21] G. Chaitin, Berry paradox, *Lecture given at a Physics–Computer Science Colloquium at the University of New Mexico*, Wednesday 27 October 1993 (1994), arxiv.org/abs/chao-dyn/9406002v1.

[22] G. Chaitin, Paradox of randomness, *Talk given at Monash University in Melbourne, Australia*, Monday 13 May 2002 (2002), www.cs.auckland.ac.nz/CDMTCS/chaitin.

[23] G. Chaitin, *Meta Math!: The Quest for Omega* (Vintage Books, 2005).

[24] C. Calude and M. Stay, From Heisenberg to Gödel via Chaitin, *Int. J. Theoret. Phys.* **46** (2007) 2013–2025.

[25] N. C. A. da Costa and F. A. Doria, Undecidability and incompleteness in classical mechanics, *Int. J. Theoret. Phys.* **30** (1991) 1041–1073.

[26] N. C. A. da Costa and F. A. Doria, Computing the future, in *Computability, Complexity and Constructivity in Economic Analysis*, ed. K. Velupillai (Blackwell Publishers, 2005), pp. 15–50.
[27] M. Davis, *Engines of Logic* (W.W. Norton, 2000).
[28] J. Dawson, Logical Dilemmas: The Life and Work of Kurt Gödel, A. K. Peters (Wellesley, Massachusetts, 1997).
[29] F. A. Doria, Axiomatics: the social sciences, and the Gödel phenomenon: a toolkit, in *The Limits of Mathematical Modeling in the Social Sciences*, ed. F.A. Doria (World Scientific, 2017), p. 1–90.
[30] T. Eggertsson, *Economic Behavior and Institutions* (Cambridge University Press, 1990).
[31] H. G. Frankfurt, Freedom of the will and the concept of a person, *J. Philos.* **68** (1971) 5–20.
[32] T. Galaabaatar, Essays on the representation of incomplete preferences, Ph.D. dissertation, The Johns Hopkins University (2011).
[33] D. George, Coping rationally with unpreferred preferences, *Eastern Econ. J.* **24** (1998) 181–194.
[34] G. Gigerenzer and the ABC Group, *Simple Heuristics that Make Us Smart* (Oxford University Press, 1999).
[35] S. Gjerstad and V. Smith, *Rethinking Housing Bubbles* (Cambridge University Press, 2014).
[36] O. Hart, On the optimality of equilibrium when the market structure is incomplete, *J. Econ. Theory* **11** (1975) 418–443.
[37] O. Hart, Incomplete contracts and control, *Prize Lecture* (2016), www.nobelprize.org.
[38] O. Hart and J. Moore, Contracts as reference points, *Quart. J. Econ.* **123** (2008) 1–48.
[39] C. Hommes, *Behavioral Rationality and Heterogeneous Expectations in Complex Economic Systems* (Cambridge University Press, 2013).
[40] D. Hyde and D. Raffman, Sorites paradox, *The Stanford Encyclopedia of Philosophy*, ed. E. N. Zalta (2018), plato.stanford.edu.
[41] G. Ingham, *The Nature of Money* (Polity Press, 2004).
[42] M. Janssen and E. Mendys-Kamphorst, The price of a price: on the crowding out and in of social norms, *J. Econ. Behavior Organization* **55**, (2004) 377–395.
[43] G. Jehle and P. Reny, *Advanced Microeconomic Theory*, 3rd edn. (Pearson, 2011).
[44] G. Kolata, Does Gödel's theorem matter to mathematics? *Science* **218** (1982) 779–780.
[45] D. M. Kreps, *Microeconomic Foundations I: Choice and Competitive Markets* (Princeton University Press, 2013).
[46] J.-J. Laffont, A Brief Overview of the Economics of Incomplete Markets. *Economic Record* **65** (1989) 54–65.
[47] A. Leijonhufvud, Keynes as a marshallian, in *The Cambridge Companion to Keynes*, eds. R. Backhouse and B. Bateman (Cambridge University Press, 2006).

[48] S. Lloyd, *Programming the Universe* (Vintage Books, 2006).
[49] R. Lucas, On the mechanics of economic development, *J. Monetary Econ.* **22** (1988) 3–42.
[50] M. Magill and M. Quinzii, *Theory of Incomplete Markets* (Cambridge MA: MIT Press, 2004).
[51] M. Mandler, Status quo maintenance reconsidered: Chaning or incomplete preferences? *The Economic Journal*, **114** (November) (2004) pp. F518–F535.
[52] C. Mortensen, Change and inconsistency, in *The Stanford Encyclopedia of Philosophy*, ed. E. N. Zalta (2016), plato.stanford.edu.
[53] F. Muñoz-Garcia, *Advanced Microeconomic Theory* (MIT Press, 2017).
[54] P. Nelson, Information and consumer behavior, *J. Political Econ.* **78** (1970) 311–329.
[55] E. Ok, Utility representation of an incomplete preference relation, *Journal of Economic Theory* **104** (2002) 427–449.
[56] M. Peterson, *An Introduction to Decision Theory*, 2nd edn. (Cambridge University Press, 2017).
[57] D. Pink, *Drive: The Surprising Truth about What Motivates Us* (Riverhead Books, 2009).
[58] K. Popper, *The Poverty of Historicism* (Routledge, 1957/1961).
[59] K. Popper, *The Open Universe: An Argument for Indeterminism* (Routledge, 1956/1988).
[60] G. Priest, The structure of the paradoxes of self-reference, *Mind* **103** (1994) 25–34.
[61] G. Priest, *Beyond the Limits of Thought*, 2nd edn. (Oxford University Press, 2000).
[62] A. Roth, *Who Gets What — and Why: The New Economics of Matchmaking and Market Design* (Mariner Books, 2016).
[63] R. M. Sainsbury, *Paradoxes* (Cambridge University Press, 2008).
[64] M. Sandel, *What Money Can't Buy: The Moral Limits of Markets* (Farrar, Straus and Giroux, 2012).
[65] T. C. Schelling, *Choice and Consequences* (Harvard University Press, 1984).
[66] A. Sen, Rational fools: a critique of the behavioral foundations of economic theory, *Philos. Public Affairs* **6** (1977) 317–344.
[67] A. Sen, Maximization and the act of choice, *Econometrica* **65** (1997) 745–779.
[68] A. Sen, *Collective Choice and Social Welfare*, Expanded edn. (Penguin, 2017).
[69] C. Shapiro and H. Varian, *Information Rules: A Strategic Guide to the Network Economy* (Harvard University Press, 1999).
[70] K. Sigmund, *Exact Thinking at Demented Times: The Vienna Circle and the Epic Quest for the Foundations of Science* (Basic Books, 2017).
[71] P. Smith, *An Introduction to Gödel's Theorems*, 2nd edn. (Cambridge University Press, 2013).
[72] Y. Shoham and K. Leyton-Brown, *Multiagent Systems: Algorithmic, Game-theoretic, and Logical Foundations* (Cambridge University Press, 2009).

[73] T. Suddendorf, *The Gap: What Separates Us From Other Animals* (Basic Books, 2013).

[74] K. Svozil, *Randomness & Undecidability in Physics* (World Scientific, 1993).

[75] M. Tsuji, N. C. da Costa and F. A. Doria, The incompleteness of the theory of games, *J. Philos. Logic* **27** (1998) 553–568.

[76] K. V. Velupillai, *Computable Economics* (Oxford University Press, 2000).

[77] J. von Neumann and O. Morgenstern, *Theory of Games and Economic Behavior* (Princeton University Press, 1953).

[78] H. Wang, *A Logical Journey: From Gödel to Philosophy* (Cambridge, Massachusetts: MIT Press, 1996).

[79] E. R. Weintraub, *How Economics Became a Mathematical Science* (Duke University Press, 2002).

[80] O. Williamson, *The Institutions of Capitalism* (The Free Press, 1985).

[81] N. Yanofsky, *The Outer Limits of Reason: What Science, Mathematics, and Logic Cannot Tell Us* (MIT Press, 2013).

[82] R. Zach, Hilbert's program, in *The Stanford Encyclopedia of Philosophy*, ed. E. N. Zalta (2016), plato.stanford.edu.

Gallery

"El torito," 1951, Bear Mountain, NY. Photo by the well-known NY photographer Victor Laredo, a friend of Greg's father.

Greg Chaitin's parents: Norman Chaitin and Sara Chaitin, Astor Cafe.

Family: Norman Chaitin, Gregory Chaitin and Sara Chaitin.

Gregory Chaitin and Sara Chaitin.

Passport photo: Gregory Chaitin and Sara Chaitin.

Norman Chaitin, Gregory Chaitin and Sara Chaitin.

Former Chaitin residence in NY, Greg's 60th birthday, 2007. Fisheye camera photo by Veikko Keränen.

At the 2016 Copernicus Festival in Kraków. Photo courtesy of the Copernicus Festival.

Bear Mountain, NY, 2008. Photo by Karol Jałochowski courtesy of *Politkya* magazine.

Opening the Buenos Aires Mathematics Festival with the Minister of Culture of Buenos Aires (right) and the head of the University of Buenos Aires Mathematics Department (left), 2009. Press photo courtesy of the Math Festival.

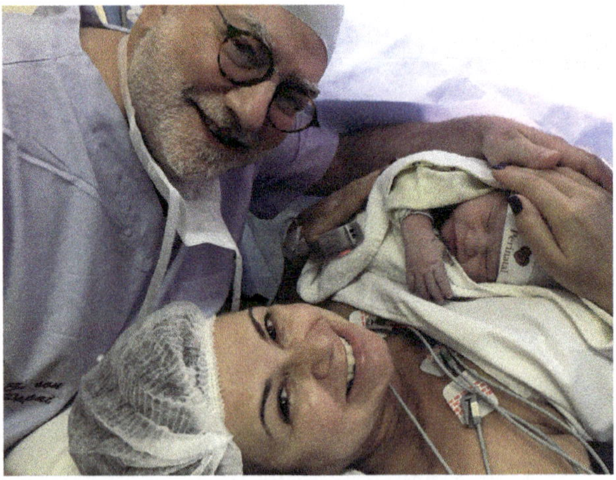

Birth of João Bernardo, Rio, 2017. Photo courtesy of the surgical staff.

João Bernardo standing up.

Index

A

A New Kind of Science, 49, 70, 80
addition calculator, 62
Alan Turing, 16, 317
Alfred Bregman, 320
algorithmic complexity, 3, 141–142, 214–218, 220, 226, 233
algorithmic information content, 21
Algorithmic Information Theory, 139, 167, 319–321, 323–324
algorithmic mutation, 321, 324
algorithmic networks, 143, 146, 152, 156
algorithmic network cycle, 145
algorithmic probability, 203, 213, 215–217, 227, 229, 231, 236
algorithmic synergistic information, 165
algorithmic synergy, 165–166, 168
algorithmic world, 219
Alighieri, D., 92
anthropic principle, 293–294
arithmetic degrees, 104
arithmetic machine, 63
ars characteristica, 53
Ars Magna, 52
asset specificity, 379
automata, 330, 332
autopoiesis, 175
average local algorithmic synergy, 169
axiom of completeness, 373–374
axiomatic system, 3
axioms of choice, 372

B

Berry's paradox, 376
BGS set, 95
Bifurcation point, 279
binary numbers, 68–69, 76
biological creativity, 338
biological evolution, 24
biology, 318, 325
Blaise Pascal, 62
Block Decomposition Method (BDM), 216–217
blurred intervals, 167
Boltzmann, Ludwig, 16
Borel–Cantelli Lemma, 85
Busy Beaver, 140
Busy Beaver function, 319, 322, 327, 334
Busy Beaver imitation game (BBIG), 136, 146, 148, 152, 157
Busy Beaver Problem, 342

C

calculus ratiocinator, 53
category theory, 186
cell-cycle, 306–312
cellular automata, 70, 211, 218, 220–221, 236, 299–300, 313
central node, 162

Chaitin, 319
Chaitin's Omega number, 86
characteristica universalis, 53, 68
Charles Babbage's, 63
Christiaan Huygens, 53
Church-Turing Thesis, 181
Claude Shannon, 317
Coding Theorem Method (CTM), 216
combinatorial mathematics, 52
combinatorics, 53, 60
commensurability, 378, 391–392
communication channels, 144
communication rounds, 147, 152, 160
competition, 158
complex systems, 175
complexity, 175
complexity of proof, 102
comprehenders, 294–296, 298, 313
compression, 201, 216, 220, 226, 233
computability of the Universe, 132
computable, 317
computable (or recursive) relation, 135
computable Life, 132
computable system, 175
computation time, 140, 142, 173
computational law, 80
computational resource, 173
computational thinking, 79, 81
computational universe, 80
concatenations, 141
conceptual complexity, 21, 339
conditional prefix algorithmic complexity, 142
consciousness, 34, 320
conserved or invariant characteristics, 381
context, 241
continuous "geometrical" mathematics, 80
continuous functions, 61
creation, 136
creativity, 8, 175, 320–324, 327, 332
Cristian Calude, 329
crowding-out effect, 391, 393
cumulative random evolution, 323, 333
Curtis Roads, 322

D

Darwin, Charles, 16
Darwinian evolution, 28, 319, 327
David Chalmer, 320
David Hilbert, 16
Descartes, 52
describability, 379
deterministic, 26
deterministic systems, 165
diffusion, 148
diffusion density, 151
digital computing, 69
digital phenomenology, 324
digital philosophy, 317–318, 320
digital physics, 317
discrete mathematics, 61, 80
discrete systems, 60
distance, 323
distributed computing, 147, 153
distributed processing, 162
DNA, 318–319, 332, 338
dynamic networks, 137, 148, 153

E

economic theory, 347
Edmund Husserl, 320
emergence, 149, 158
emergent open-endedness, 149
Emergentism, 359
Entscheidungsproblem, 378
emergent algorithmic complexity, 149–150
epistemology, 317, 322, 335
evolution, 136, 175, 320, 324
evolutionary open-endedness (or open-ended evolution), 149
evolutionary systems, 135
exist, 322
expected emergent algorithmic complexity, 150

expected (local) emergent open-endedness, 150
experience, 320–321, 324–325, 327
experience goods, 379
exponential Diophantine, 92

F

field equations, 275
final output, 145
finite automata, 390
first incompleteness theorem, 371
fitness, 136, 154, 158
Formal axiomatic systems, 369
formal languages, 139, 155
formal notation, 80
Formalist Revolution, 370
foundations of economics, 370, 372
Frances Crick, 318
freedom, 386
functionality, 135

G

Gödel's incompleteness theorems, 318
Gödel, Kurt, 2, 318
Gödel numbering, 61
Galileo, 52
game theory, 158
genotype, 319
Georg Cantor's, 318
George I of Hanover, 75
Gestalt psychology, 320
Giuseppe Peano, 54
global information-sharing (or communication) protocol, 154
god, 326–327
god/intelligent designer, 333
Gottfried Leibniz, 49, 52, 205, 317
Gottlob Frege, 54
graphs, 137
Greg Chaitin, 76

H

habitability window, 290–291
halting computation, 172
halting function, 94
halting probability, 3, 16, 151, 323, 327
halting problem, 133, 157, 160
Hao Wang, 332
Heraclitus, 104
high-order networks, 137
hypercomputable system, 147
hypercomputation, 136, 175, 196
hypercomputers, 133

I

IBM Watson Research Center, 12
I Ching, 68
imitation-of-the-fittest protocol, 147
incomplete contracts, 394
incomplete preferences, 374
incompleteness, 134, 318, 320, 322, 332
incompleteness of contracts, 379
Incompleteness Theorems, 370
infinite trials, 85
infinitesimal calculus, 61
information, 290, 294, 296–299, 305–306, 308–313
Information Theory, 319
information-sharing protocol, 145, 155
intelligence, 289–297, 306, 312–314
intelligent design, 323, 332–333
intentionality, 320, 334
intuition, 320
irrationality, 387, 398
irreducible information, 165

J

Jacob Bernoulli, 74
James Tenney, 320, 322, 333
James Watson, 318
Jarkko Kari, 332
John von Neumann, 317
joint prefix algorithmic complexity, 142
jurisprudence, 107, 118

K

Kurt Gödel, 70
Kurt Koffka, 320

L

Lagrangian, 272
Larry Polansky, 322, 333
law of inertia, 387
law of large numbers, 159
laws of motion, 381
legislation, 107
Leibniz, 205, 320, 326
Leibniz's calculator, 67
Leibniz, Gottfried, 202
Leibniz, Gottfried Wilhelm, 6
Leonhard Euler, 322
imitation-of-the-fittest protocol, 147

M

mapping, 144
Martin Heidegger, 320
Mathematica, 49, 54, 56, 81
mathematical logic, 134
mathematique universelle, 53
Maurice Merleau-Ponty, 320
mechanical calculator, 61–63, 75
meta-preferences, 385
Meta+phenomenology, 325, 327, 332
metabiology, 7, 133, 136, 319–321, 323–327, 331–333
minimal program, 323, 325
monads, 70
Monadology, 37
money pump, 388
multi-agent systems, 158
MultiAspect Graph, 137, 143, 156
multiple selves, 387
multivariate information theory, 165, 174
multiverse, 293–294, 298
mutation algorithm, 325, 333
mutations, 29, 149, 323, 326, 329
mutual algorithmic information, 142, 167
mutual information, 324–325, 327

N

np-complete, 331
Nash equilibrium, 388
negative probabilities, 262
neighbors, 146
network, 137, 306–312
network input, 145
network science, 162
Newton, 52, 75, 80
node, 138, 144
node centralities, 162
node cycle, 145
normal numbers, 88
Novelty, 326

O

o-nodes, 173
Omega, 16, 329, 332
Omega number, 203
ontological, 317
open-ended evolution, 133
open-endedness, 149
optimization, 158
Oracle, 323, 332
oracle Turing machine, 143, 147
oracle-sensitiveness, 135, 157
Oracles, 327, 333
output, 145

P

pairing function, 141
panpsychism, 35
parallel universes, 330–331
partial function, 140
partial information diagrams, 165
partial output, 145
partial recursive function, 140
partial relation, 134
partial structure, 134
payoff, 154, 158
perception, 320–321
phase transition, 152
phenomenology, 318, 320–322, 334
phenotype, 319
Pierre–Simon Laplace, 15
Poincaré recurrence time, 302–304

polymones, 70
population, 144, 146, 153, 156
Principia, 80
prefix algorithmic complexity of information, 142
Principle of Computational Equivalence, 81
Prisoner's Dilemma, 393
program size complexity, 319
program-size, 329–330
program-size probability distribution, 159
programming language, 141–142, 172

Q

quantifiability, 390
quantifiers, 59
quantum gravity, 36
quantum mechanics, 239, 331
quasi-truth, 134

R

Ramon Llull, 52
random walk, 7
randomly generated self-delimiting programs, 147, 153
randomly generated solution, 158
randomness, 4, 92, 380
reciprocity, 392
recursive function, 158
relative complexity, 323
resource-bounded Turing machines, 174
revealed preference, 387
Rice's theorem, 93
Richardson, D., 94
risk, 380
Robert Berger, 332
Robert Hooke, 63
run-time, 329–330
Russell's paradox, 375–376

S

Sangalli, A., 91
scientia generalis, 53
second incompleteness theorem, 371
self-delimiting (or prefix-free) language, 141
self-organization, 175
selfish genes, 340
SETI, 292
shading, 384
Shannon, 319
Shannon entropy, 215–216
small-diameter phenomenon, 151–152
Solomonoff–Kolmogorov–Chaitin complexity, 142
sorites paradox, 383
Spinoza, 73
Srinivasa Ramanujan, 322
state-dependent dynamics, 299, 306, 313
state-dependent laws, 301, 312
Stephen Wolfram, 49, 202, 204, 211, 317
statistical, 26
stochastic dynamical systems, 165
stochastic processes, 165, 174
subcomputation, 136
subjectivity, 320, 325–326
summa, 58
sunspot equilibria, 393
Suppes, P., 99
Susceptible–Infected–Susceptible contagion, 137
Sydney Brenner, 318
synchronous, 147, 153
synchronous algorithmic networks, 154
synergistic information, 165
synergistic imitation-of-the-fittest protocol, 153–154
synergy, 175
system, 134
systematic search, 323
systemics, 134

T

temporal diffusion diameter, 139, 153, 160
theoretical machines, 143

Theta function, 97
Thorp, N., 91
Tibor Rado's, 322
time hierarchies, 174
time-reachability centrality, 163
time-varying graphs, 138, 153
topological complexity, 115
topos, 187
transcendence, 327
transfer entropy, 309
transitivity, 372, 383–384
trust, 379
Turing machine, 60, 140, 319, 332
Turing's, 318

U

Uncertainty Principle, 381
uncomputable Nature, 132–133
undecidable, 94
Unforecastable Events, 397
universal, 331
universal calculator, 62
universal computation, 60–61, 80
universal elements, 61, 79
universal Turing machine, 141–142, 389
unknowability, 16

V

vagueness, 384
value of ownership, 395
vertex, 138, 144
von Neumann, 318–319, 322

W

Wang-tilings, 332
Wolfram Language, 54
Wolfram, Stephen, 202
Wolfram|Alpha, 81

Z

Zeno's Arrow paradox, 381

Lightning Source UK Ltd.
Milton Keynes UK
UKHW020758190220
358970UK00001B/61

9 789811 200069